Advances in Intelligent, Flexible, and Lean Management and Engineering

Carolina Machado
University of Minho, Portugal

J. Paulo Davim
University of Aveiro, Portugal

A volume in the Advances in Logistics, Operations, and Management Science (ALOMS) Book Series

IGI Global
PUBLISHER of TIMELY KNOWLEDGE

Published in the United States of America by
IGI Global
Business Science Reference (an imprint of IGI Global)
701 E. Chocolate Avenue
Hershey PA, USA 17033
Tel: 717-533-8845
Fax: 717-533-8661
E-mail: cust@igi-global.com
Web site: http://www.igi-global.com

Library of Congress Cataloging-in-Publication Data

Names: Machado, Carolina, 1965- editor. | Davim, J. Paulo editor.
Title: Advances in intelligent, flexible, and lean management and engineering / Carolina Machado and J. Paulo Davim, editors.
Description: Hershey, PA : Engineering Science Reference, [2021] | Includes bibliographical references and index. | Summary: "This book offers a variety of contribution for the exchange of experiences and thoughts about the state of the research in the intelligent, flexible and lean management and engineering areas, as well as the future direction of these fields of research"-- Provided by publisher.
Identifiers: LCCN 2021019404 (print) | LCCN 2021019405 (ebook) | ISBN 9781799857686 (h/c) | ISBN 9781799868552 (s/c) | ISBN 9781799857792 (eISBN)
Subjects: LCSH: Flexible manufacturing systems.
Classification: LCC TS155.65 .A43 2021 (print) | LCC TS155.65 (ebook) | DDC 670.42/7--dc23
LC record available at https://lccn.loc.gov/2021019404
LC ebook record available at https://lccn.loc.gov/2021019405

This book is published in the IGI Global book series Advances in Logistics, Operations, and Management Science (ALOMS) (ISSN: 2327-350X; eISSN: 2327-3518)

British Cataloguing in Publication Data
A Cataloguing in Publication record for this book is available from the British Library.

Advances in Logistics, Operations, and Management Science (ALOMS) Book Series

ISSN:2327-350X
EISSN:2327-3518

Editor-in-Chief: John Wang Montclair State University, USA

MISSION

Operations research and management science continue to influence business processes, administration, and management information systems, particularly in covering the application methods for decision-making processes. New case studies and applications on management science, operations management, social sciences, and other behavioral sciences have been incorporated into business and organizations real-world objectives.

The **Advances in Logistics, Operations, and Management Science** (ALOMS) Book Series provides a collection of reference publications on the current trends, applications, theories, and practices in the management science field. Providing relevant and current research, this series and its individual publications would be useful for academics, researchers, scholars, and practitioners interested in improving decision making models and business functions.

COVERAGE

- Political Science
- Risk Management
- Information Management
- Organizational Behavior
- Decision analysis and decision support
- Networks
- Computing and information technologies
- Operations Management
- Production Management
- Services management

The Advances in Logistics, Operations, and Management Science (ALOMS) Book Series (ISSN 2327-350X) is published by IGI Global, 701 E. Chocolate Avenue, Hershey, PA 17033-1240, USA, www.igi-global.com. This series is composed of titles available for purchase individually; each title is edited to be contextually exclusive from any other title within the series. For pricing and ordering information please visit http://www.igi-global.com/book-series/advances-logistics-operations-management-science/37170. Postmaster: Send all address changes to above address.

Titles in this Series

For a list of additional titles in this series, please visit:
http://www.igi-global.com/book-series/advances-logistics-operations-management-science/37170

Global Perspectives on Change Management and Leadership in the Post-COVID-19 Era
Ebtihaj Al-Aali (University of Bahrain, Bahrain) and Meryem Masmoudi (University of Bahrain, Bahrain)
Business Science Reference • © 2021 • 374pp • H/C (ISBN: 9781799869481) • US $215.00

Handbook of Research on Innate Leadership Characteristics and Examinations of Successful First-Time Leaders
Matthew Waritay Guah (South Carolina State University, USA)
Business Science Reference • © 2021 • 500pp • H/C (ISBN: 9781799875925) • US $285.00

Handbook of Research on Recent Perspectives on Management, International Trade, and Logistics
Güner Koç Aytekin (Ufuk University, Turkey) and Çağlar Doğru (Ufuk University, Turkey)
Business Science Reference • © 2021 • 490pp • H/C (ISBN: 9781799858867) • US $345.00

Handbook of Research on Multidisciplinary Perspectives on Managerial and Leadership Psychology
Rick D. Johnson (Johnson Professionals, Inc., USA)
Business Science Reference • © 2021 • 742pp • H/C (ISBN: 9781799838111) • US $285.00

Transforming Corporate Governance and Developing Models for Board Effectiveness
Qaiser Rafique Yasser (Centre for Rural Economy, Pakistan) and Abdullah Al-Mamun (University of Wollongong, Australia)
Business Science Reference • © 2021 • 345pp • H/C (ISBN: 9781799866695) • US $195.00

Encyclopedia of Organizational Knowledge, Administration, and Technology
Mehdi Khosrow-Pour D.B.A. (Information Resources Management Association, USA)
Business Science Reference • © 2021 • 2734pp • H/C (ISBN: 9781799834731) • US $3,950.00

For an entire list of titles in this series, please visit:
http://www.igi-global.com/book-series/advances-logistics-operations-management-science/37170

IGI Global
PUBLISHER of TIMELY KNOWLEDGE

701 East Chocolate Avenue, Hershey, PA 17033, USA
Tel: 717-533-8845 x100 • Fax: 717-533-8661
E-Mail: cust@igi-global.com • www.igi-global.com

Editorial Advisory Board

List of Reviewers

Table of Contents

Preface xiii

Acknowledgment xv

Introduction xvi

Chapter 1
Like a Cog in a Machine: The Effectiveness of AI-Powered Human Resourcing . 1
Ali B. Mahmoud, The Peter J. Tobin College of Business, St. John's University, USA & University of Wales Trinity Saint David, UK

Chapter 2
Green Transformational Leadership as a Redefinition of the Organizational Psychological Contract: Psychological Capital in Green Corporate Performance 21
Diana Fernandes, University of Minho, Portugal
Carolina Feliciana Machado, Interdisciplinary Centre of Social Sciences (CICS.NOVA.UMinho), University of Minho, Portugal

Chapter 3
Towards Sustainable Factories 51
Gema Calleja, Universitat Politècnica de Catalunya, Spain
Jordi Olivella Nadal, Universitat Politècnica de Catalunya, Spain

Chapter 4
Pull Production Systems: Link Between Lean Manufacturing and PPC 80
Eduardo Guilherme Satolo, São Paulo State University (UNESP), Brazil
Milena Estanislau Diniz Mansur dos Reis, Federal University of Rio de Janeiro (UFRJ), Brazil
Robisom Damasceno Calado, Fluminense Federal University (UFF), Brazil

Chapter 5
Seru Production System: A Conceptual Model Proposal for the Clothing Industry 113
Emre Bilgin Sarı, Faculty of Economics and Administrative Sciences, Dokuz Eylul University, Turkey
Sabri Erdem, Dokuz Eylul University, Turkey

Chapter 6
Human Resources Management in the Portuguese Education System: A Strategic Contribution to an Intelligent and Flexible Management 139
Diana Fernandes, University of Minho, Portugal
Carolina Feliciana Machado, Interdisciplinary Centre of Social Sciences (CICS.NOVA.UMinho), University of Minho, Portugal

Chapter 7
Assessing the Soft Skills of IT Professionals: A Proposal for the Information and Communication Technology Sector 185
Sílvia Diana Moreira de Araújo, Polytechnic Institute of Cávado and Ave, Portugal
Pedro Novo Melo, Polytechnic Institute of Cávado and Ave, Portugal & CICS.NOVA, Univeristy of Minho, Portugal

Chapter 8
Leadership: A Fundamental Tool for Dynamic and Intelligent Management – An Application 209
Maria Heliodora Matos, University of Minho, Portugal
Carolina Feliciana Machado, Interdisciplinary Centre of Social Sciences (CICS.NOVA.UMinho), University of Minho, Portugal

Compilation of References 257

About the Contributors 289

Index 294

Detailed Table of Contents

Preface xiii

Acknowledgment xv

Introduction xvi

Chapter 1
Like a Cog in a Machine: The Effectiveness of AI-Powered Human Resourcing . 1
Ali B. Mahmoud, The Peter J. Tobin College of Business, St. John's University, USA & University of Wales Trinity Saint David, UK

Similar to its sisters, the Fourth Industrial Revolution (Industry 4.0) has sparked varying sentiments and views regarding the ethicality and effectiveness of employing artificial intelligence (AI) tools in human resource management (HRM) in a way that triggers the need for a synthesis of current published work on the different views in that respect. This chapter presents an attempt to engage the different cogs with each other so the millstone will go around, and an updated understanding of AI-powered HRM from different angles is provided. This work reviews the main concepts revolving around AI and Industry 4.0. Also, it offers an up-to-date investigation of AI uses in HRM (e.g., People Analytics) and what risks or ethical concerns are being argued in contemporary discourse.

Chapter 2
Green Transformational Leadership as a Redefinition of the Organizational Psychological Contract: Psychological Capital in Green Corporate Performance 21
Diana Fernandes, University of Minho, Portugal
Carolina Feliciana Machado, Interdisciplinary Centre of Social Sciences (CICS.NOVA.UMinho), University of Minho, Portugal

This study grounds on the contemporary concern towards sustainability, reviewing literature to ascertain the personality traits of current leaders, framing them under the most effective leadership style in order to leverage organizational green consciousness and performance. It provides insights on the multilevel social dynamics outlining individual citizenship behaviors at work, as it advances that organizations shall rely on green transformational leadership to enhance the workforce green cognitions and behaviors, providing workplace opportunities to engage in environmental management related activities, thus leveraging green consciousness and performance. Hence, this study maps the current leader's personality traits so that their leadership paradigm may be best captured in terms of the value congruence between leaders/followers, solidifying a green psychological climate at the organization, impacting on its psychological capital and ownership, reconfiguring the organizational psychological contract by defining it as a collaborative learning process.

Chapter 3
Towards Sustainable Factories..51
Gema Calleja, Universitat Politècnica de Catalunya, Spain
Jordi Olivella Nadal, Universitat Politècnica de Catalunya, Spain

Sustainability is becoming a major objective of factories worldwide in an effort to eliminate or reduce the use of toxic materials and greenhouse emissions, extend the life of products, reuse waste, and conserve energy, not only for meeting the needs of consumers, but also the multi-stakeholders from both within and beyond the supply chain. The increasing global demand for scarce natural resources is posing a great challenge for producing companies. In recent years, the advent of new technologies and the changes in demography and working conditions as well as the desire for individualized products has added greater complexity to manufacturing. In this context, this chapter provides an overview of the main issues affecting the industrial sector on the way ahead towards sustainability, including contemporary trends triggering the requirements for factories of the future as well as the main research and innovation lines necessary to answer such requirements.

Chapter 4
Pull Production Systems: Link Between Lean Manufacturing and PPC.............80
Eduardo Guilherme Satolo, São Paulo State University (UNESP), Brazil
Milena Estanislau Diniz Mansur dos Reis, Federal University of Rio de Janeiro (UFRJ), Brazil
Robisom Damasceno Calado, Fluminense Federal University (UFF), Brazil

This chapter aims to organize knowledge about pull production systems by presenting the underlying concepts of lean manufacturing as for its origin, principles, and relations with PPC. Pull production is one the fundamental principles of lean manufacturing, and its implementation can bring positive impacts. For such a purpose, sequential and mixed supermarket pull systems stand out in which the integration between pull production systems and PPC and its various levels is a main subject of discussion. The JIT model or Kanban method and hybrid systems, such as conwip and lung-drum-string theory, are mechanisms for managing pull production systems. Finally, a pull production system implementation is presented for illustration purposes. At the end of this chapter, it is expected that skills are developed by readers, which are going to assist them in using the tools presented to model production systems and aid decision-making processes.

Chapter 5
Seru Production System: A Conceptual Model Proposal for the Clothing Industry 113
Emre Bilgin Sarı, Faculty of Economics and Administrative Sciences, Dokuz Eylul University, Turkey
Sabri Erdem, Dokuz Eylul University, Turkey

Seru production system is a flexible, cost-effective, workforce competence-oriented manufacturing management system that provides the opportunity to respond quickly to customer demand. As in parallel to technology and physical improvements, customer demands are also effective for development of production systems. The impact of change in demand has been seen on changeover from job shop to mass production, flexible, and lean manufacturing systems. Seru production system is more appropriate for targeting work both cost-effectively like mass production and maximum diversification like job shop production. This chapter clarifies the Seru production system and explain its use and benefits in the clothing industry. In the application, a shirt production is illustrated according to the principles of mass production, lean production, and Seru production. Thus, different types of production systems have been benchmarked. There will be potential study areas for proving the efficiency of Seru soon.

Chapter 6
Human Resources Management in the Portuguese Education System: A Strategic Contribution to an Intelligent and Flexible Management...................139
Diana Fernandes, University of Minho, Portugal
Carolina Feliciana Machado, Interdisciplinary Centre of Social Sciences (CICS.NOVA.UMinho), University of Minho, Portugal

Acknowledging the revival of the debate around school management models, following the criticism neoliberalism has brought into discussion, this work focusses on the human resources management paradigm in the Portuguese compulsory public education system. Analysing secondary quantitative and qualitative data, it describes the observed performance comparing it to the expected one (strategic human resources management). The system's main coordination mechanism is the standardization of competencies, centrally delegated from the Ministry of Education and Science, as so, "mobilization" is the vector in which the system is more constrained towards the expected performance, harming its potential for flexibilization and adaptation. This work concludes there is no strategic human resources management in the Portuguese compulsory public education, despite some positive efforts already being taken to overcome such situation. These insights can be used as an analytical guide to the implementation of changes towards the expected performance in the Portuguese education system.

Chapter 7
Assessing the Soft Skills of IT Professionals: A Proposal for the Information and Communication Technology Sector 185
Sílvia Diana Moreira de Araújo, Polytechnic Institute of Cávado and Ave, Portugal
Pedro Novo Melo, Polytechnic Institute of Cávado and Ave, Portugal & CICS.NOVA, Univeristy of Minho, Portugal

This research aims to assess the soft skills of information technology professionals in the information and communication technologies sector. The research method used was a combination of qualitative techniques to survey data about the most valued skills. As major results of this research, the following skills stand out: teamwork, problem solving, organization, leadership, communication skills, dynamism, initiative, autonomy, motivation, creativity, time management, attitude, responsibility, and proactivity. Based on collected data, a cross-cutting skills framework was constructed for five functional areas: trainee, junior consultant, consultant, senior consultant, and business development consultant. It was found that skills are transversal to all functions, and only the indicators vary in relation to the categories.

Chapter 8
Leadership: A Fundamental Tool for Dynamic and Intelligent Management – An Application 209
Maria Heliodora Matos, University of Minho, Portugal
Carolina Feliciana Machado, Interdisciplinary Centre of Social Sciences (CICS.NOVA.UMinho), University of Minho, Portugal

The main aim of this chapter is to examine the role played by leaders as determinant agents of a special type of banking institution such as the mutual agricultural credit banks (or Caixas de Crédito Agrícola Mútuo [CCAM]). In other words, it looks to study and understand to what extent employees with leadership roles use their position to maximize CCAM performance through their direct intervention on the teams. Considering that both leaders and subordinates have their own aims (organizational/ professional and personal, respectively), and in order to find the necessary equilibrium between them, it also seems to be relevant to better understand the way leaders can mobilize and support organizational subordinates. In sum, issues like leadership, motivation, satisfaction, and personal development are studied in this chapter.

Compilation of References 257

About the Contributors 289

Index 294

Preface

Looking to provide a channel of communication to disseminate the knowledge of how to manage in a green environment, between academics/researchers and managers the subjects developed in this book assume a particular emphasis in nowadays organizations. Words like sustainability, not only in an environmental perspective but also in an organizational perspective, recycling, re-use, low costs, waste reduction, high productivity, are only some, among many others, about which organizations and researchers are concerned, requiring a break in management traditional paradigms. Indeed, present organizations need to be managed in different models where flexibility, innovation and change are key words as they drive the organization to the success. At this level intelligent and lean management appears as a new way of manage, think, and understand organizations, making them more strategic and competitive in the markets where they are and look to be present in the near future.

Taking into account these concerns, this book, entitled *Advances in Intelligent, Flexible, and Lean Management and Engineering* looks to give a contribution to the exchange of experiences and thoughts about the state of the research in the intelligent, flexible and lean management and engineering areas, as well as the future direction of these fields of research. It also aims to provide a support to academics and researchers, as well as all those that working in the engineering and management field need to deal with practices, policies and strategies related to management and engineering work issues.

In sum, attentive to the market challenges and changes, this book is essential to all those that are working in the management and engineering field because:

- It introduces new lines of research in management and engineering areas;
- The models, theories and tools presented and discussed by the different contributions allow management to take a more strategic role in organizations.
- In the business world there is a growing importance of sophisticated analysis for managers to support decision making, to use strategic information and

> innovative tools in order to guide thinking and behavior as well as to manage more strategically to adapt environments and achieve the organization's aims.

Looking to promote research related to these new trends and developments in the field of management and engineering, the present book, divided in eight chapters covers, in Chapter 1 "Like a Cog in a Machine: The Effectiveness of AI-Powered Human Resourcing," while Chapter 2 discusses "Green Transformational Leadership as a Redefinition of the Organizational Psychological Contract: Psychological Capital in Green Corporate Performance." Chapter 3 focuses "Towards Sustainable Factories" at the same time that Chapter 4 deals with "Pull Production Systems: Link Between Lean Manufacturing and PPC." Chapter 5 contains information about "Seru Production System: A Conceptual Model Proposal for Clothing Industry" while Chapter 6 highlights "Human Resources Management in the Portuguese Education System: A Strategic Contribution to an Intelligent and Flexible Management." Finally, Chapter 7 speaks about "Assessing Soft Skills of IT Professionals: A Proposal for the Information and Communication Technology Sector," and Chapter 8 ends the book by developing a reflection about "Leadership: A Fundamental Tool for Dynamic and Intelligent Management – An Application."

Able to be used by academics, researchers, human resources managers, managers, engineers, and other professionals in related matters with management and engineering, this book can also be used as a research book for final undergraduate engineering and management courses (for example, management, industrial engineering, international business, production engineering, etc.) or as a subject on management and engineering at the postgraduate level.

The Editors acknowledge their gratitude to IGI Global for this opportunity and for their professional support. Finally, we would like to thank to all chapter authors for their interest and availability to work on this project.

Carolina Machado
University of Minho, Portugal

J. Paulo Davim
University of Aveiro, Portugal

Acknowledgment

The Editors acknowledge their gratitude to IGI Global for this opportunity and for their professional support. Finally, we would like to thank to all chapter authors for their interest and availability to work on this project.

Carolina Machado
University of Minho, Portugal

J. Paulo Davim
University of Aveiro, Portugal

Introduction

The growing challenges we are facing require that organizations develop intelligent and flexible management models, capable of more quickly and proactively positioning themselves in the markets in which they are inserted.

Competition is getting stronger, the different stakeholders are progressively more demanding, customers are more and more informed and with higher levels of education, society demands more responsible organizations. In all of this context, we must keep in mind that resources are scarce, which requires organizations to be able to use these resources efficiently and effectively. Sustainable organizations are needed. In fact, today's organizations are required to be able to use these resources efficiently in order to meet current needs without compromising, however, the ability of future generations to meet their own needs.

It follows that only with dynamic, flexible and intelligent models and management practices will organizations be able to make better use of employees' time and increase the productivity of their teams. Only in this way will they be able to achieve levels of performance and profitability capable of making them more competitive compared to their more direct competitors.

Aware of this reality and the continuous demands that organizations are facing, we developed this book, which, entitled "Advances in Intelligent, Flexible, and Lean Management and Engineering" seeks to contribute to a better understanding of these problems, acting as a critical tool for achieving more efficient and effective management. To this end, we resort to the contributions of different authors, who, based on the studies developed by them, seek to share with all potential readers of this book some of the main theories, models, strategies, practices and policies that dynamic and competitive managers and engineers look to develop in the different types of organization with a view to achieving increasingly higher levels of success.

In other words, this book, organized in eight chapters looks:

- To focuses on the latest and more recent research findings that are occurring in this field in different countries;

- To show in what ways different types of organizations, in different countries, are facing today's challenges and developments in intelligent, flexible and lean management and engineering;
- To share knowledge and insights on an international scale;
- at the same time that it looks to help researchers and practitioners to select among the different options and strategies the more relevant priorities to manage competitive organizations.

Finally, it is important to highlight that this subject, with a specific emphasis in the Management and Engineering areas, will be critical and of great usefulness for all those - managers, engineers, researchers, human resource managers, academics, as well as professionals from different areas - that in their daily, academic and professional routines need to lead with intelligent, flexible and lean management and engineering issues.

Carolina Machado
University of Minho, Portugal

J. Paulo Davim
University of Aveiro, Portugal

Chapter 1
Like a Cog in a Machine:
The Effectiveness of AI-Powered Human Resourcing

Ali B. Mahmoud
https://orcid.org/0000-0002-3790-1107
The Peter J. Tobin College of Business, St. John's University, USA & University of Wales Trinity Saint David, UK

ABSTRACT

Similar to its sisters, the Fourth Industrial Revolution (Industry 4.0) has sparked varying sentiments and views regarding the ethicality and effectiveness of employing artificial intelligence (AI) tools in human resource management (HRM) in a way that triggers the need for a synthesis of current published work on the different views in that respect. This chapter presents an attempt to engage the different cogs with each other so the millstone will go around, and an updated understanding of AI-powered HRM from different angles is provided. This work reviews the main concepts revolving around AI and Industry 4.0. Also, it offers an up-to-date investigation of AI uses in HRM (e.g., People Analytics) and what risks or ethical concerns are being argued in contemporary discourse.

INTRODUCTION

Upon the birth of the second decade of the twenty-first century, the exponential increase of AI employability has reached new records regarding almost every function and department at modern businesses. Nevertheless, the human resource, i.e., people, remains the *most valuable asset* to any firms, especially within the services sector.

DOI: 10.4018/978-1-7998-5768-6.ch001

Furthermore, the human factor, according to the cogent argument of many AI theorists, is still not replaceable by the machines, at least, on the foreseen future. In this regard, Human resources (HR) professionals are promoting the growing use of software-enhanced modules, to potential job applicants, for precision, cost cut, and time-saving efficiency. However, other scholars (e.g., Goodman, 2017) regard the features of such algorithms as somewhat elusive. Also, when compared to traditional evaluation and recruiting methods, work in this area is somewhat lagging in terms of data protection and the use of proper testing tools, as well as validity checking (Goodman, 2017). A recent study, by Van Esch et al. (2019), shows that adopting AI recruitment will not lower the likelihood of people applying for jobs that way as long as their attitudes are positive towards that technology. Thus, this chapter serves as a scholarly attempt to shed light on AI human resourcing literature, mainly by synthesising how companies are using artificial intelligence in managing human resources. Besides, what opportunities and threats to individual and organisations could the adoption of AI-enabled human resource management bring in terms of effectiveness, efficiency and ethics.

Background

There is already a significant and structural evolution impacting economies across the nations, and much more disruption and turmoil will arise in the foreseeable future. Artificial intelligence (AI) has been widely noted to impose a significant growing change onto our way of life and that it has become an integral part of the future (Mahmoud, Tehseen, et al., 2020). Samuel (1959) defines machine learning as the area of science that empowers the machines, i.e., the computers, to learn without being excessively programmed. AI refers primarily to computational technology inspired by ways in which humans use their brains' neurons and the central nervous system to reason and make judgments and decisions, even though they usually work much differently (Mahmoud, Tehseen, et al., 2020). Again, AI is the employment of machines, i.e., computers, to take part in human-like thinking processes, such as learning, reasoning, and self-correction (Armstrong & Taylor, 2020). Basically, AI allows us to do stuff efficiently and effortlessly. AI systems operate without actually being overtly programmed (Mahmoud, Tehseen, et al., 2020). They get on the data they gather and the assessment of how the systems are managed (ICO, 2017; Plageras et al., 2018). AI can analyse data ('data mining') and process the queries by searching for an answer in a data set (Armstrong & Taylor, 2020).

However, Machine Intelligence has been bungled in popular culture to describe almost any sort of digital or programmed processing. To deter confusion, AI researchers and specialists prefer to use the term Artificial General Intelligence (AGI) to refer to human-level or higher intelligence systems, adequately adept at deriving concepts from little experience and sharing knowledge between branches of science (Mahmoud, Tehseen, et al., 2020). AGI is also described as the "Strong AI" to differentiate from "Weak AI" or "Narrow AI," which is associated with machines or software designed for Addressing particular tasks that cannot be funnelled by far to other systems (Yao et al., 2018). Put it another way, the differences between human-like general intelligence and the limited or narrow artificial intelligence are that we can draw useful conclusions from one activity and link them to another (Mahmoud, Tehseen, et al., 2020). Also, dissimilar to AGI, narrow AI lacks choice and does what we *order* it to, whilst, a general AI will know how to reflect on its targets and determine whether to change them or not (Chace, 2015). The artificial intelligence exponential rise is believed to bring, considerably, more tremendous changes to the human race (Mahmoud, Tehseen, et al., 2020). That is, AI will be the next stage of human evolution that the tech expert Peter H. Diamandis notes by saying: "We are going from Darwinism evolution by natural selection to evolution by human direction." (Ashley, 2019). That is not far away from OpenAI's announcement in 2018, a US-based AI-research lab, about creating a robotic arm that had trained itself how to play with and crack the Rubik's cube tapping a 'reinforcement-learning' algorithm. This method articulates how animals learn (Hao, 2019).

Following the Internet revolution that led to e-commerce in the early 2000s, the digitalisation empowered by the Internet of Things (IoT) and other technologies has in recent times steered the business sector towards the smart era (Bissola & Imperatori, 2020) that characterises what so-called the Fourth Industrial Revolution or Industry 4.0 is. Whilst the First Industrial Revolution utilised steam power to industrialise production; the Second employed electricity into elevating production to mass scale. Following that, the use of electronics and information technology for production automation identified the Third Industrial Revolution which, since the middle of the twentieth century, has paved the way to the Fourth Industrial Revolution; a new chapter in the humankind history. For more than two and a half centuries now, the first three industrial revolutions have remodelled the way humans create value and have transformed the world. The most crucial stimulus had always come in the form of technological advances which led to the development of new machines that improved industrial manufacturing (Harteis, 2018). In each, technologies, political systems and social institutions all co-evolved, reshaping not just the business world, but also people's attitudinal and behavioural patterns (Schwab & Davis, 2018). Ultimately, Industry 4.0 is a model that represents adopting digitalisation, cloud computing, internet of things and big data technologies and processes by industry to achieve competitive advantages in markets at the national and international levels (Reinhardt et al., 2020). Habraken and Bondarouk (2017, p. 240) describe the fourth generation of industries as those that "have a high degree of flexibility in production, in terms of product needs (specifications, quality, design), volume (what is needed), timing (when it is needed), resource efficiency and cost (what is required), being able to (fine)tune to customer needs and make use of the entire supply chain for value creation. It is enabled by a network-centric approach, making use of the value of information, driven by ICT and the latest available proven manufacturing techniques". In other words, Industry 4.0 is the application of IoT in the business processes. IoT is inseparable from AI. Whilst IoT devices generate big data in different forms and structures, AI-powered techniques will be needed to analyse such enormous sets of user-generated data.

As a result, many businesses have become in the search for the correct adaptation to the Fourth Industrial Revolution. In this regard, the conversion to Industry 4.0 is, according to Salkin et al. (2018, p. 5), based on eight fundamental advances in technology: adaptive robotics, data analytics and artificial intelligence (big data analytics), simulation, embedded systems, communication and networking such as the Industrial Internet, cloud systems, additive manufacturing and virtualisation technologies. These tools need the support of both essential technologies such as internet security, sensors and actuators, Radio-frequency identification (RFID), Real-time Locating System (RTLS) and portable technologies, as well as seven design principles, including such real-time data management, interoperability, virtualisation, decentralisation, agility, service orientation and embedded business operations (Salkin et al., 2018; Wang et al., 2016). Such design considerations and technologies allow practitioners to identify the levels of adaptation to Industry 4.0. However, a structured implementation blueprint for the transformation to Industry 4.0 remains debatable (Salkin et al., 2018). In this regard, it is worth mentioning that this chapter uses 'digitalisation' to refer to any improvements in the business processes as a result of transforming data into digital format, i.e., digitisation. According to Bissola and Imperatori (2020), digitisation is a disruptive innovation that offers new social and business opportunities, whilst at the same time challenging the conventional way of inventing 'work' in a way that would require both individuals and organisations to adjust.

ARTIFICIAL INTELLIGENCE AND HUMAN RESOURCE MANAGEMENT

Throughout many organisations, a multitude of HR-related activities, such as filling in performance reports (Mahmoud, Sitladin, et al., 2020), interviewing job applicants, making suggestions on pay raises, and reporting job-related news to raise challenges to HR given the dynamics reshaping the business environment. Therefore, HR managers need to be able to behave in a circumstance of incomplete and asymmetric information whilst maintaining the ability to choose from a variety of options and choices (Bernazzani et al., 2019; López-Cotarelo, 2009). HRM line managers play an essential role in organisations. This is because they are called upon to handle situations where making sophisticated and non-routine decisions in unique and non-frequent circumstances (such as creative decision making and problem-solving) is the norm (Bernazzani et al., 2019; Fish & Hardy, 2015; Marcolin et al., 2016). Managerial decisions made in the HR department have an impact on the organisation's whole value-creation process (Bernazzani et al., 2019; Sondhi, 2018). Undeniably, the roles of the HR department are directly relevant to both the development and implementation of organisational strategies (Bernazzani et al., 2019; Kuipers & Giurge, 2017). Human Resource Management (HRM) is everything about decision-making that have direct impact organisations' competitive performance (Bernazzani et al., 2019; Shih & Chiang, 2005; Walger et al., 2016). The strategic significance of HRM is evident in the essential alignment between HR and organisational strategy, as well as by the fact that HR strategies, policies, and practices are always needed (Holbeche, 2009; Kuipers & Giurge, 2017; Shih & Chiang, 2005; Wright & Snell, 1998). Doing more with less alongside other talent-special challenges are confronting HR leaders all over the world every day (Eubanks, 2018). The foreseeable future of HR represents both digital technologies as well as humans as HR leaders concentrate on boosting the mix of human resources and automation (Tambe et al., 2019). This creates new precedence for HRM (Meister, 2019) which obliges leaders and teams to build proficiency in artificial intelligence (Khan & Millner, 2020) at the same time as they re-envision HR to be more personal, human and intuitive (Meister, 2019).

On the whole, HRM is a highly administrative function and possibly, AI can significantly reduce the number of tasks pertaining to this kind of work (Armstrong & Taylor, 2020; Tambe et al., 2019). On a more positive note, it is possible to create data that informs the decision-making process on people management matters (Khan & Millner, 2020). It is anticipated that AI will continue to change work profoundly, and consequently, the field of human resource management (Habraken & Bondarouk, 2017). And that this chapter echoes that "the world of work is not going to be disrupted it is disrupted" (Ross Peter et al., 2017, p. xvii)

AI Uses in Human Resource Management

Tata (2018) argues that AI is able to offer a unique cognitive view in order to raise challenges as much as inform decisions; provide a different perspective; serve as a devil's advocate to handle the stakeholders' involvement and process their views. In this respect, Armstrong and Taylor (2020) summarise the specific applications of artificial intelligence in human resource management as below.

HR Analytics

Also known as People analytics and workforce analytics, is the use of people-data in analytical processes to find solutions to business challenges (CIPD, 2020b). HR analytics utilises people-data, collected through HR systems (such as absence management, payroll) as well as business information (for example, operations performance data)– (CIPD, 2020b). In simple words, HR analytics is about collecting and analysing data and information on people and HRM practices to inform the decision-making process (Armstrong & Taylor, 2020). Therefore, people analytics can substantially contribute to developing evidence-based decision making (CIPD, 2020b). Without effective HR analytics practices, processes and systems, and without continued development of the HR analytics skills and confidence of HR professionals, the HR profession risks losing out on the performance, productivity and job quality gains that can come from improvements in evidence-based decision making (CIPD, 2020a). There have been scholarly calls for people analytics to be taken out of HR and turned into part of active end-to-end business analytics, where HR is just one component in the value chains examined as this will result in more business pertinent conclusions and high-impact interventions (Rasmussen & Ulrich, 2015).

Recruitment and Selection

Typically, HR teams compete for the top talent, and managers do not want to lose the candidate (Armstrong & Taylor, 2020). Considering human capital planning as an input equivalent to the organisation's *blood*, HRM assists in providing *nutrition* to the organisation, solving organisational recruitment and staff pairing matters (Jia et al., 2018). Matching candidates situated in a database to job/person specifications; utilising natural language processing (NLP) to anticipate which phrases in job advertisements will make the most significant impact on applicants (Armstrong & Taylor, 2020). NLP can save valuable time for parsing incoming resumes and screening applicants. As a supplement to saving HR time, NLP can assist eliminate subjective bias and boost diversity in candidates. Most businesses adopting AI-applications in recruitment and selection tend to be larger, tech-focussed, and/or innovative firms (Albert Edward, 2019). A recent article in the Financial Post (Deveau, 2017) indicates that using AI in recruitment can lead to 71% cut in staffing costs and a threefold increase in recruitment efficiency alongside no unconscious bias. In spite of the exponential growth rate of AI-adoption, firms have yet to reach an inflexion point as they presently show an unwillingness to invest in that technology for recruitment and selection process (Albert Edward, 2019).

Talent Management

AI can be utilised in identifying suitability for inclusion in a talent development programme (a talent pipeline) by offering information on performance records. Human resource development represents a paramount concern of many organisations (Yabanci, 2020). This particular area of the organisation's activities seeks to develop people's knowledge, expertise, productivity, and satisfaction concerning their professions (Werner & DeSimone, 2012). For example, IBM Watson offers professionals as well as novices several solutions to such a matter all through its products, such as talent frameworks, career coaching, and talent evaluation (Yabanci, 2020). IBM Watson's talent framework is a comprehensive ready-to-use collection specifically designed to help both managers and employees from different sectors through general corporate function extension on operational and developmental matters, such as job descriptions, primary responsibilities, and required skills (IBM, 2019; Yabanci, 2020). IBM Watson also helps employees by offering interview questions, coaching instructions, objectives for development, and rigid learning suggestions for each skill and competency level (IBM, 2019). Further, the career coach aims to impact the internal mobility of personnel by advising on their career planning (IBM, 2019). It monitors the employees' interests and preferences and recommends them for existing job opportunities, and how to steer their prospective career moves by relying on the records (IBM, 2019; Yabanci, 2020).

Learning and Development

Building and operating performance support systems that give on-the-job access to learning and training material in the shape of information about techniques, procedures and skills and knowledge requirements with examples and advice on how they should be developed and implemented (Armstrong & Taylor, 2020). For example, Sidekick, an AI-operated coaching platform via remote-ready managers, uses a combination of data points to supervise employees to detect micro-behaviour changes which need interventions such as one-on-one coaching meetings (Armstrong & Taylor, 2020; Sidekick, 2020). AI-based training systems are intelligent, smart and specialised with handling queries. Previous research shows that these systems can curate content, grade, evaluate, and give feedback to trainees, thereby making learning both context-sensitive and flexible (Upadhyay Ashwani & Khandelwal, 2019).

Retention Management

Machine learning and natural language processing can help in anticipating the future (Mahmoud, Tehseen, et al., 2020). AI can help with employee retention by eliminating repetitive tasks, understanding employee engagement factors, thus enhancing the overall performance of businesses (Sekhri & Cheema, 2019). Utilising neural networks to help predict employee turnover, AI can help organisations identify potential leavers so that measures can be taken as necessary in order to retain those with turnover intentions (Armstrong & Taylor, 2020). AI can predict the exit of employees based on various types of employee-generated data. AI also helps in finding out the patterns and reasons for employee stress by using chatbots and virtual assistants (Clutterbuck, 2020). Several AI software packages are available to track the activities of the employee, such as emails and browsing history, which serve as an information supplier for management to help identify those who are leaving the organisation (Sekhri & Cheema, 2019). Moodometers are AI-built platforms established to assess job attitudes amongst employees over a period, and that can further assist the HR managers in investigating possible reasons for turnover and hence developing strategies for employee retention (Bhardwaj, 2019; Sekhri & Cheema, 2019).

Employee Sentiment Analysis

Actual employee engagement is a puzzle to nearly all organisations. One of the primary reasons for that is the use of standard engagement-measuring instruments —employee questionnaires and manually checked feedback channels are insufficient to get a complete understanding of how employees actually feel (Maurer, 2019). In particular, when it comes to human resources, AI has been tackling one of the most critical problems facing managers for many years-improving employee engagement and retention levels (Jani, 2019). Powered by natural language processing and machine learning sentiment analysis technology interprets huge amounts of feedback to uncover organisational strengths and weaknesses and detect positive and negative feelings toward a new policy, a change in benefits, management practices or workplace culture (Maurer, 2019). AI enables organisations to learn about employee attitudes, moods or sentiments through analysing the large amounts of text, for example, e-mails, HR can then make well-informed changes and decisions to respond properly to employees' concerns, and that will ultimately lead to higher engagement encourages and more open communication. (Armstrong & Taylor, 2020).

Workforce Scheduling

Human capital is typically one of the most valuable resources of a company. Shift planning thus plays a crucial role in operations efficiency, and an adequate staffing level must be ensured while legal and practical constraints are recognised for each individual schedule (Becker, 2020). One of the most widespread conditions laid down in shift scheduling is to work several consecutive days, as this typically implies several consecutive days off (Becker, 2020). AI can be used in scheduling employees onto shifts and showing where it is best to allocate workers with particular skills (Armstrong & Taylor, 2020). Previous research shows that if some relief breaks are never assigned or if relief-timing restrictions are loosened so that more reliefs may be allocated in real-time, there will be a frustrated and less productive labour force and probably breaches of contractual commitments (Thompson & Pullman, 2007).

Payroll

Payroll refers to payments made to employees for their work alongside the company's record of payments previously made to employees, including salaries and wages, bonuses and deferred taxes, or to the company's department which calculates and pays out those amounts (Bragg, 2003). People that have worked with payroll realise that the actual process itself is ready for disruption by AI; a highly specialised, repetitive process that has to be run on a regular timetable with data that is substantially the same across multiple periods (Eubanks, 2018). AI algorithms can be used to verify initial payroll entries and calculations (Armstrong & Taylor, 2020).

Employee Queries and Self-Service Matters

HR can consider installing chatbots (virtual assistants) in order to answer inquiries from employees on issues such as holiday and paternity leave entitlement or to offer assistance with issues, e.g., with computer operating systems (Armstrong & Taylor, 2020). It is expected that chatbots will allow employee users to feel comfortable and when confident about how to use this cost-efficient service irrespective of the employee's computer knowledge owing to the natural language processing technology employed in the messaging software (Tadvi et al., 2020).

Ethical Considerations and AI Perils

While the benefits of the extensive application of Artificial Intelligence in business are undeniable the rise of AI in HRM can be looked at as a double-edged sword (Mahmoud, Tehseen, et al., 2020). 'Well-designed' procedures may be corrupted in operation as users manipulate systems to suit their needs, particularly if they find that it does not provide the responses it wants. Therefore, without being a 'Luddite' and destroying machinery, people can overlook or undermine it (Reilly, 2018). This might occur where managers disregard the recommendation of the AI staffing system and instead use their intuition, in spite of the evidence that AI is a better predictor of applicant success than humans (Agrawal et al., 2018). In a related matter, users might enter incorrect information on the knowledge that it will deliver the desired answer (Reilly, 2018). According to Morgan McKinley (2018), the most prominent risk faced when using artificial intelligence kits for recruitment reasons is the distinct paucity of transparency which surrounds the processes - strategies need to be set up from the very beginning. Below are the main negative repercussions that might be driven by AI-impowered human resourcing.

Personal Attributes and Candidate Privacy

AI and big data will continue to be able to identify *proxy* variables for personal or private attributes with increased accuracy (Prince & Schwarcz, 2019). Facebook pages or groups liked by social media users can be used to extrapolate race and sexual orientation with substantial precision (Dattner et al., 2019; Fox, 2018). Discrimination can happen in any socio-technical system in which someone decides to use an algorithmic process to inform decision-making (Williams et al., 2018). Employers may become susceptible to legal exposure if it makes adverse hiring decisions based on protected categories such as place of birth, race, or native language (Dattner et al., 2019). It is illegal to take an unfavourable action based upon certain protected or personal characteristics, no matter how these were learned or implied (Dattner et al., 2019). The courts will handle situations where companies have relied upon means using these proxy variables, and it remains unclear how the law will deal with it (Dattner et al., 2019). For example, the Employee Polygraph Protection Act of 1988 (EPPA) in the U.S. (U.S. Department of Labor, 2020) forbids employers typically from using lie detector tests (polygraphs) as a pre-employment assessment instrument (Dattner et al., 2019).

Sexist Hiring Algorithms

In 2015, Amazon realised that its new system was not scoring applicants for software developer jobs and other technical positions in a gender-neutral manner. That is because Amazon's computational models were trained to evaluate applicants by observing patterns in résumés submitted to Amazon throughout ten years. An overwhelming majority of résumés were submitted by men, a manifestation of male dominance throughout the technology industry (Dastin, 2018).

Data Protection Concerns

Law is not out of play here from discrimination-resistant regulation. Though the use and misuse of user-generated data by social media companies have dominated the media headlines for years, privacy activists have started raising concerns in response to the growing gathering and processing of employee-generated data in the workplace (Wall, 2019). In recent years, many employers have moved beyond the conventional forms of monitoring in the interest of trade secret protection and policy enforcement. They have started to take into account how they can use advanced data analytics for a variety of organisational priorities and goals (Wall, 2019). By embarking on data protection regulation like the General Data Protection Regulations (GDPR), and the heavy penalties that may result from errors in processing personal data and this is a particular worry when robotics is introduced to employment (Morgan McKinley, 2018). For example, GDPR permit candidates to find out whether computerised decision making has been applied and challenge an outcome if the process did not include human involvement (Armstrong & Taylor, 2020).

Threatening Humans Employment and Employability

An overarching theme has, on many occurrences, been noted over the humankind history. That theme has revolved around inventing new machines that take over jobs that once used to be performed by humans. Robots pose an overwhelming threat embodied in replacing human labour involved in such repetitive tasks and duties (Achieng et al., 2019). In the next few decades, the concept of work will hinge on predictability, whether a human or a virtual being does it. As they were set to start a few years ago, machines will handle the routine whilst humans take on unpredictable tasks (Grasso, 2015).

Nonetheless, recent technology advances, e.g., quantum computing, could accelerate the evolution of Artificial General Intelligence (AGI) and represent a new-entrant threat to human resource in the job market (Mahmoud, Tehseen, et al., 2020). A recent report by Oxford Economics shows that robots could replace up to twenty million manufacturing jobs around the globe by 2030 (BBC, 2019). Further such drastic changes may be unfair on older generations in the workplace who might tend to be less adaptable to the 'new normal' imposed by the machines (Mahmoud, Fuxman, et al., 2020; Mahmoud, Grigoriou, et al., 2020; Mahmoud et al., 2021; Mahmoud, Reisel, et al., 2020). Besides, the drive to replace humans with robots is accelerating as businesses struggle to prevent workplace infections of COVID-19 and to keep operational costs low making at least 42% of the jobs lost gone forever (Semuels, 2020). Such concerns are in line with Elon Musk's claims upon unwrapping his Neuralink's new brain chip. He uncovered his large-coin-sized microchips late in August this year, promoting it as a brain implant. More importantly, he shared his thoughts on the need for developing "superhuman cognition," boosting the human brain in part in the *battle* against the threat, that Elon Musk sees from artificial intelligence (Cellan-Jones, 2020).

CONCLUSION

This essay has offered a scholarly of AI human resourcing literature. In this investigation, it has been synthesised how businesses are employing artificial intelligence in managing human resources. Returning to the objectives introduced at the beginning of this script it is now possible to determine what opportunities and threats to individual and organisations that the adoption of AI-enabled human resource management establishes against the standards of effectiveness, efficiency and ethics.

In conclusion, AI-based products are revolutionising how human resource management functions. AI-powered HRM Innovations are aimed for improving the efficiency of HRM operations whilst improving or maintaining the quality of attendance to the repetitive tasks and duties as if performed by humans. Organisations have no choice but to embrace the emerging innovations in order to boost the efficiency of its HRM operations and hence defend their competitiveness. All of this being done with an announced notion that AI is not assumed to replace humans.

Nevertheless, the most significant risk faced when using artificial intelligence kits for recruitment reasons is the distinct paucity of transparency. Employees with a presence on social media networks can be susceptible to having their personal attribute to their employers and hence affecting their employment/employability. Also, discrimination can happen in any socio-technical system in which algorithmic tools are used to inform decision-making, resulting in adverse outcomes like *sexist hiring* Algorithms. The ambitious initiatives to develop *superhuman cognition* to help the human brain reaches the competency of an evolved AGI raises grave concerns about what the future hides for humans. Besides, the latest advances in technology, e.g., quantum computing, could accelerate the evolution of AGI.

Therefore, this chapter endorses that governments and policymakers observe such developments that *agile and smart* legislation is instituted to govern the growth of the machine's general intelligence and protect humanity from any potential threats that might be imposed by evolving mutations in robots. Additionally, employers should adhere to norms and values that would commit them to be, genuinely, socially responsible businesses whose investments into AI-powered innovation and implementation are not driven only by their interests but also by the common good.

REFERENCES

Achieng, S., Majuto, C. M., Aseka, P., & Astiaya, E. (2019). Replacing Humans with Machines: Threats and Opportunities. *East African Journal of Business and Economics*, *1*(2), 54–68.

Agrawal, A., Gans, J., & Goldfarb, A. (2018). The economics of artificial intelligence. *The McKinsey Quarterly*.

Albert Edward, T. (2019). AI in talent acquisition: A review of AI-applications used in recruitment and selection. *Strategic HR Review*, *18*(5), 215–221. doi:10.1108/SHR-04-2019-0024

Armstrong, M., & Taylor, S. (2020). *Armstrong's handbook of human resource management practice*. Kogan Page Publishers.

Ashley, M. (2019). Forget Darwinian Evolution. Humanity May Soon Evolve Itself Through A.I. *Forbes*. Retrieved 28 July from https://www.forbes.com/sites/cognitiveworld/2019/01/15/forget-darwinian-evolution-humanity-may-soon-evolve-itself-through-a-i/#3790240c466d

BBC. (2019). *Robots 'to replace up to 20 million factory jobs' by 2030*. BBC. Retrieved 04 September from https://www.bbc.co.uk/news/business-48760799

Becker, T. (2020). 2020/06/30). A decomposition heuristic for rotational workforce scheduling. *Journal of Scheduling*, *23*(5), 539–554. Advance online publication. doi:10.100710951-020-00659-2

Bernazzani, R., Cantoni, F., & Piva, M. (2019). The Future Role of Machine Learning in HR Development. In F. Cantoni & G. Mangia (Eds.), *Human Resource Management and Digitalization*. Routledge.

Bhardwaj, R. (2019). How AI is revolutionizing the human resource functions. *Entrepreneur Asia Pacific, 4*.

Bissola, R., & Imperatori, B. (2020). HRM 4.0: The digital transformation of the HR Department. In F. Cantoni & G. Mangia (Eds.), *Human Resource Management and Digitalization*. Routledge.

Bragg, S. M. (2003). *Essentials of Payroll: Management and Accounting*. John Wiley & Sons.

Cellan-Jones, R. (2020). *Is Elon Musk over-hyping his brain-hacking Neuralink tech?* BBC. Retrieved 02 September from https://www.bbc.co.uk/news/technology-53987919

Chace, C. (2015). *Surviving AI: The promise and peril of artificial intelligence*. Three Cs.

CIPD. (2020a). *People analytics: Explore the CIPD's point of view on people analytics, including actions for Government and recommendations for employers*. Chartered Institute of Personnel and Development. Retrieved 19 August from https://www.cipd.co.uk/news-views/viewpoint/people-analytics

CIPD. (2020b). *People analytics: Understand what people analytics is, different types of HR data, and the importance of people analytics strategy and process*. Chartered Institute of Personnel and Development. Retrieved 19 August from https://www.cipd.co.uk/knowledge/strategy/analytics/factsheet

Clutterbuck, D. (2020). The challenges of coaching and mentoring in a digitally connected world. *Coaching im digitalen Wandel*.

Dastin, J. (2018). *Amazon scraps secret AI recruiting tool that showed bias against women*. Reuters. Retrieved 03 September from https://www.reuters.com/article/us-amazon-com-jobs-automation-insight/amazon-scraps-secret-ai-recruiting-tool-that-showed-bias-against-women-idUSKCN1MK08G

Dattner, B., Chamorro-Premuzic, T., Buchband, R., & Schettler, L. (2019). The legal and ethical implications of using AI in hiring. *Harvard Business Review*, 25.

Deveau, D. (2017). Machine learning helps large companies hire better, potentially cutting turnover. *Financial Post*. Retrieved 23 August from https://financialpost.com/entrepreneur/0508-biz-dd-hr

Eubanks, B. (2018). *Artificial intelligence for HR: use AI to support and develop a successful workforce*. Kogan Page Publishers.

Fish, S., & Hardy, M. (2015). Complex issues, complex solutions: Applying complexity theory in social work practice. *Nordic Social Work Research, 5*(sup1), 98-114.

Fox, A. A. (2018). *Discrimination concerns in job ads: Important social media changes you need to know*. Recruitics. Retrieved 03 September from https://info.recruitics.com/blog/discrimination-concerns-in-job-ads-important-social-media-changes-you-need-to-know

GDPR. (2016). *What is GDPR, the EU's new data protection law?* Retrieved 03 September from https://gdpr.eu/what-is-gdpr/

Goodman, E. (2017). Biometrics won't solve our data-security crisis. *Harvard Business Review*. Retrieved 1 March from https://hbr.org/2017/12/biometrics-wontsolve-our-data-security-crisis

Grasso, A. (2015). Will machines replace us or work with us? *Wired*. Retrieved 03 September from https://www.wired.com/insights/2015/01/will-machines-replace-us-or-work-with-us/

Habraken, M., & Bondarouk, T. (2017). Smart industry research in the field of HRM: Resetting job design as an example of upcoming challenges. In T. Bondarouk, H. Ruël, & E. Parry (Eds.), *Electronic HRM in the smart era* (pp. 221–259). Emerald Publishing Limited. doi:10.1108/978-1-78714-315-920161009

Hao, K. (2019). A robot hand taught itself to solve a Rubik's Cube after creating its own training regime. *MIT Technology Review*. Retrieved 28 July from https://www.technologyreview.com/2019/10/15/75292/a-robot-hand-taught-itself-to-solve-a-rubiks-cube-after-creating-its-own-training-regime/

Harteis, C. (2018). Machines, Change and Work: An Educational View on the Digitalization of Work. In C. Harteis (Ed.), *The Impact of Digitalization in the Workplace* (pp. 1–10). Springer International Publishing. doi:10.1007/978-3-319-63257-5_1

Holbeche, L. (2009). *Aligning human resources and business strategy*. Routledge. doi:10.4324/9780080941424

IBM. (2019). *Prepare for re-entry with a resilient talent management strategy*. Retrieved 02 September from https://www.ibm.com/talent-management

ICO. (2017). *Big data, artificial intelligence, machine learning and data protection*. Information Commissioner's Office. Retrieved 20 August from https://ico.org.uk/media/for-organisations/documents/2013559/big-data-ai-ml-and-data-protection.pdf

Jani, N. (2019). Sentiment Analysis and Employee Engagement: How Companies can Leverage AI? *Medium*. Retrieved 1 September from https://becominghuman.ai/sentiment-analysis-and-employee-engagement-how-companies-can-leverage-ai-3aa7d2e0dd4b

Jia, Q., Guo, Y., Li, R., Li, Y., & Chen, Y. (2018). A conceptual artificial intelligence application framework in human resource management. *Proceedings of the International Conference on Electronic Business*.

Khan, N., & Millner, D. (2020). *Introduction to People Analytics: A Practical Guide to Data-driven HR*. Kogan Page Publishers.

Kuipers, B. S., & Giurge, L. M. (2017). Does alignment matter? The performance implications of HR roles connected to organizational strategy. *International Journal of Human Resource Management, 28*(22), 3179–3201. doi:10.1080/09585192.2016.1155162

López-Cotarelo, J. (2009). *HR discretion: understanding line managers' role in Human Resource Management*. Industrial Relations Unit, Warwick Business.

Mahmoud, A. B., Fuxman, L., Mohr, I., Reisel, W. D., & Grigoriou, N. (2020). "We aren't your reincarnation!" workplace motivation across X, Y and Z generations. *International Journal of Manpower, 42*(1), 193–209. doi:10.1108/IJM-09-2019-0448

Mahmoud, A. B., Grigoriou, N., Fuxman, L., Reisel, W. D., Hack-Polay, D., & Mohr, I. (2020). A generational study of employees' customer orientation: A motivational viewpoint in pandemic time. *Journal of Strategic Marketing*, •••, 1–18. doi:10.1080/0965254x.2020.1844785

Mahmoud, A. B., Reisel, W. D., Fuxman, L., & Mohr, I. (2021). A motivational standpoint of job insecurity effects on organizational citizenship behaviors: A generational study. *Scandinavian Journal of Psychology, 62*(2), 267–275. doi:10.1111jop.12689

Mahmoud, A. B., Reisel, W. D., Grigoriou, N., Fuxman, L., & Mohr, I. (2020). The reincarnation of work motivation: Millennials vs older generations. *International Sociology, 35*(4), 393–414. doi:10.1177/0268580920912970

Mahmoud, A. B., Sitladin, M., Tehseen, S., Fuxman, L., Reisel, W. D., Purkayastha, H., & Grigoriou, N. (2020). *Job satisfaction as a mediator between performance appraisal effectiveness and organisational outcomes in the Caribbean. British Academy of Management.* Alliance Manchester Business School, University of Manchester.

Mahmoud, A. B., Tehseen, S., & Fuxman, L. (2020). The Dark Side of Artificial Intelligence in Retail Innovation. In E. Pantano (Ed.), *Retail Futures* (1st ed., pp. 165–180). Emerald Publishing Limited. doi:10.1108/978-1-83867-663-620201019

Marcolin, L., Miroudot, S., & Squicciarini, M. (2016). *Routine jobs, employment and technological innovation in global value chains*. OECD Publishing. Retrieved 20 August from https://www.oecd.org/sti/ind/GVC-Jobs-Routine-Content-Occupations.pdf

Maurer, R. (2019). *Employee Sentiment Analysis Shows HR All the Feels*. SHRM. Retrieved 1 September from https://www.shrm.org/resourcesandtools/hr-topics/technology/pages/employee-sentiment-analysis-shows-hr-all-the-feels.aspx

Meister, J. (2019). Ten HR Trends In The Age Of Artificial Intelligence. *Forbes*. Retrieved 20 August from https://www.forbes.com/sites/jeannemeister/2019/01/08/ten-hr-trends-in-the-age-of-artificial-intelligence/

Morgan McKinley. (2018). *The risks of Artificial Intelligence recruitment tools*. Retrieved 03 September from https://www.morganmckinley.com/uk/article/risks-artificial-intelligence-recruitment-tools

Plageras, A. P., Psannis, K. E., Stergiou, C., Wang, H., & Gupta, B. B. (2018). Efficient IoT-based sensor BIG Data collection–processing and analysis in smart buildings. *Future Generation Computer Systems*, *82*, 349–357. doi:10.1016/j.future.2017.09.082

Prince, A. E. R., & Schwarcz, D. (2019). Proxy discrimination in the age of artificial intelligence and big data. *Iowa Law Review*, *105*, 1257.

Rasmussen, T., & Ulrich, D. (2015). Learning from practice: How HR analytics avoids being a management fad. *Organizational Dynamics*, *44*(3), 236–242. doi:10.1016/j.orgdyn.2015.05.008

Reilly, P. (2018). *The impact of artificial intelligence on the HR function*. Institute for Employment Studies. Retrieved 04 September from https://www.employment-studies.co.uk/system/files/resources/files/mp142_The_impact_of_Artificial_Intelligence_on_the_HR_function-Peter_Reilly.pdf

Reinhardt, I. C., Oliveira, J. C., & Ring, D. T. (2020). Current perspectives on the development of Industry 4.0 in the pharmaceutical sector. *Journal of Industrial Information Integration*, *18*, 100131. doi:10.1016/j.jii.2020.100131

Ross Peter, K., Ressia, S., & Sander Elizabeth, J. (2017). Prelims. In *Work in the 21st Century* (pp. i–xxvi). Emerald Publishing Limited., doi:10.1108/978-1-78714-577-120171009

Salkin, C., Oner, M., Ustundag, A., & Cevikcan, E. (2018). A conceptual framework for Industry 4.0. In Industry 4.0: Managing the Digital Transformation (pp. 3-23). Springer.

Samuel, A. L. (1959). Some Studies in Machine Learning Using the Game of Checkers. *IBM Journal of Research and Development*, *3*(3), 210–229. doi:10.1147/rd.33.0210

Schwab, K., & Davis, N. (2018). *Shaping the future of the fourth industrial revolution*. Currency.

Sekhri, A., & Cheema, D. J. (2019). The new era of HRM: AI reinventing HRM functions. *International Journal of Scientific Research and Review*, *7*(3).

Semuels, A. (2020). Millions of Americans Have Lost Jobs in the Pandemic—And Robots and AI Are Replacing Them Faster Than Ever. *Time*. Retrieved 04 September from https://time.com/5876604/machines-jobs-coronavirus/

Shih, H. A., & Chiang, Y. H. (2005). Strategy alignment between HRM, KM, and corporate development. *International Journal of Manpower*, *26*(6), 582–603. doi:10.1108/01437720510625476

Sidekick. (2020). *Organizations thrive with remote-ready managers*. Retrieved 2 September from https://sidekick-hq.com/

Sondhi, P. (2018). *5 Important Guidelines to Increase HR Values In Your Organization*. Entrepreneur. Retrieved 02 August from https://www.entrepreneur.com/article/312696

Tadvi, S., Rangari, S., & Rohe, A. (2020). *HR Based Interactive Chat bot (PowerBot)*. In *2020 International Conference on Computer Science, Engineering and Applications (ICCSEA)*, Gunupur, India. 10.1109/ICCSEA49143.2020.9132917

Tambe, P., Cappelli, P., & Yakubovich, V. (2019). Artificial intelligence in human resources management: Challenges and a path forward. *California Management Review*, *61*(4), 15–42. doi:10.1177/0008125619867910

Tata. (2018). *Cognitive Diversity: AI & The Future of Work*. Tata Consultancy Services. Retrieved 20 August from https://www.tatacommunications.com/wp-content/uploads/2018/09/Report_Cognitive-Diversity_AI-and-The-Future-of-Work.pdf

Thompson, G. M., & Pullman, M. E. (2007). Scheduling workforce relief breaks in advance versus in real-time. *European Journal of Operational Research*, *181*(1), 139–155. doi:10.1016/j.ejor.2006.05.018

Upadhyay Ashwani, K., & Khandelwal, K. (2019). Artificial intelligence-based training learning from application. *Development and Learning in Organizations*, *33*(2), 20–23. doi:10.1108/DLO-05-2018-0058

U.S. Department of Labor. (2020). *Employee Polygraph Protection Act*. U.S. Department of Labor. Retrieved 03 September from https://www.dol.gov/agencies/whd/polygraph

Van Esch, P., Black, J. S., & Ferolie, J. (2019). Marketing AI recruitment: The next phase in job application and selection. *Computers in Human Behavior*, *90*, 215–222. doi:10.1016/j.chb.2018.09.009

Walger, C., Roglio, K. D. D., & Abib, G. (2016). HR managers' decision-making processes: A "reflective practice" analysis. *Management Research Review*, *39*(6), 655–671. doi:10.1108/MRR-11-2014-0250

Wall, F. M. (2019). 2019 Law Journal: Navigating the risks of processing employee-generated data. *Business North Carolina Magazine*. Retrieved 04 September from https://businessnc.com/2019-law-journal-navigating-the-risks-of-processing-employee-generated-data/

Wang, S., Wan, J., Zhang, D., Li, D., & Zhang, C. (2016). Towards smart factory for industry 4.0: A self-organized multi-agent system with big data based feedback and coordination. *Computer Networks*, *101*, 158–168. doi:10.1016/j.comnet.2015.12.017

Werner, J. M., & DeSimone, R. L. (2012). *Human resource development*. Cengage Learning.

Williams, B. A., Brooks, C. F., & Shmargad, Y. (2018). How Algorithms Discriminate Based on Data They Lack: Challenges, Solutions, and Policy Implications. *Journal of Information Policy*, *8*, 78–115. doi:10.5325/jinfopoli.8.2018.0078

Wright, P. M., & Snell, S. A. (1998). Toward a unifying framework for exploring fit and flexibility in strategic human resource management. *Academy of Management Review*, *23*(4), 756–772. doi:10.5465/amr.1998.1255637

Yabanci, O. (2020, 2020/04/20). From human resource management to intelligent human resource management: a conceptual perspective. *Human-Intelligent Systems Integration*. doi:10.100742454-020-00007-x

Yao, M., Zhou, A., & Jia, M. (2018). *Applied Artificial Intelligence: A Handbook for Business Leaders*. Topbots Inc.

ADDITIONAL READING

Broad, E. (2018). *Made by Humans*. Melbourne University Publishing.

Carmona, D. (2020). *The AI Organization*. O'Reilly Media.

KEY TERMS AND DEFINITIONS

AI-Powered Human Resource Management: Deploying AI tools to handle repetitive or routine tasks and duties that used to be performed by humans.

Digitalisation: It refers to improvements in the business processes as a result of digitisation.

Digitisation: It is a disruptive innovation that offers new social and business opportunities, whilst at the same time challenging the conventional way of inventing 'work' in a way that would require both individuals and organisations to adjust.

Employee Sentiment Analysis: The use of natural language processing and machine learning sentiment to learn about employee attitudes, moods, or sentiments through analysing the massive amounts of text (e.g., e-mails), and thus inform the decisions made by HR.

HR Analytics: The use of people-data in analytical processes to find solutions to business challenges.

People: The most valuable asset to any firms.

Superhuman Cognition: Using AI-powered chips to boost human intelligence to perform at the level of or outperform the evolving artificial general intelligence.

Chapter 2
Green Transformational Leadership as a Redefinition of the Organizational Psychological Contract:
Psychological Capital in Green Corporate Performance

Diana Fernandes
University of Minho, Portugal

Carolina Feliciana Machado
https://orcid.org/0000-0002-9685-1576
Interdisciplinary Centre of Social Sciences (CICS.NOVA.UMinho), University of Minho, Portugal

ABSTRACT

This study grounds on the contemporary concern towards sustainability, reviewing literature to ascertain the personality traits of current leaders, framing them under the most effective leadership style in order to leverage organizational green consciousness and performance. It provides insights on the multilevel social dynamics outlining individual citizenship behaviors at work, as it advances that organizations shall rely on green transformational leadership to enhance the workforce green cognitions and behaviors, providing workplace opportunities to engage in environmental management related activities, thus leveraging green consciousness and performance. Hence, this study maps the current leader's personality traits so that their leadership paradigm may be best captured in terms of the value congruence between leaders/followers, solidifying a green psychological climate at the organization, impacting on its

DOI: 10.4018/978-1-7998-5768-6.ch002

psychological capital and ownership, reconfiguring the organizational psychological contract by defining it as a collaborative learning process.

INTRODUCTION

Across literature on sustainable management, reference is made to the vital role leadership represents as a backbone on the transition towards a sustainable economy, even though, scholars have only recently begun to study the antecedents of green personality traits and behaviors in the workplace (Boiral, 2009; Boiral et al., 2009; Guest, 2011; Kim et al., 2014): this work aims to fill in that gap. It grounds on the contemporary concern towards sustainability, reviewing literature to ascertain the personality traits on current leaders, framing them under the most effective leadership style to leverage organizational green consciousness and performance. It provides insights on the multilevel social dynamics that outline individual citizenship behaviors at work, as it advances that organizations shall rely on green transformational leadership to implement initiatives enhancing the workforce green cognitions and behaviors to provide them with workplace opportunities in order to advance engagement in environmental management related activities, hence leveraging green consciousness and performance. Thus, reviewing literature articulating different scientific fields (Sociology, Psychology, Economics and Management), this study maps the current leader's personality traits so that their leadership paradigm may be best captured and explained in terms of the value congruence between leaders and followers, solidifying a green psychological climate at the organization, which will impact on its psychological capital and ownership, reconceptualizing the organizational psychological contract by (re)defining it as a collaborative learning process.

Theoretically, the argumentation grounds on the overarching framework of the Upper-echelons, which is built on the assumption that complex strategic decisions consist largely on the results of behavioral factors, thus, it emphasizes personality traits as main predictors of organizational outcomes. So, the subject will be theoretically framed also making use of the insights provided by the Ability, Motivation and Opportunities as well as by the Leader-Member Exchange theories, and it will also be framed within the postulates of the resource-based-view of the firm.

Taking into account the organizations' multilevel nature, the chapter argues that workplace green behavior of individual employees is amplified through the interpersonal dynamics that occur within work groups, in the sense that leaders' workplace green behavior and work group green advocacy, conveyed by Green Transformational Leadership, can be viewed as contextual forces in a top-down cross-level model. The reminder of the chapter is the following: after an initial contextualization of the topic, literature will be reviewed in order to theoretically frame the subject. Afterwards, the distinctive personality traits of current leaders will be presented, which will then be framed into a leadership style: green transformational leadership will in here be emphasized and further examined in order to argument that it assumes as an efficient means by which such leaders' personality traits can best be highlighted and enhanced in order to boost follower's green consciousness, thus improving corporate green performance. Solutions and main recommendations will then be exposed, as well as future research directions.

Background

The Ability-Motivation-Opportunity Theory (AMO) states that personal ability and motivation rest on the basis for action, which will then happen if proper opportunities are provided by the organization, so, it postulates that corporate performance will be a function of employees' ability, motivation and opportunity to participate in organizational initiatives, by an integrative perspective illustrating why and how leaders and strategic Human Resources practices promote organizational performance (Appelbaum et al., 2000; Bos-Nehles et al., 2013; Boxall & Purcell, 2003; Colbert, 2004; Gerhart, 2015).

It can be argued that green corporate consciousness and performance may be boosted by the cooperative influence of organizational contexts (e.g. leadership) and individual vectors (e.g. employee's seldom environmental awareness), because it is the symbiosis of both groups of factors that affects workforce ability, motivation and opportunities for green performance (Hülsheger et al., 2009).

In this sense, this theory is indeed widely applied by scholars in the green management literature supporting the association between the human capital and social, ethical, and ecological performance. It entails a multi-dimensional approach aiming at fostering sustainable organizational outcomes based on a combination of four vectors: the proactive engagement in green activities, the willingness to foster an eco-friendly atmosphere inside and outside the workplace, the awareness that increasing motivation for societal activities is a communal responsibility of the employees and the organization, and the opportunity to provide employees with an appropriate work environment and a supportive organizational culture that promotes a green attitude and nurtures involvement in volunteering activities (Renwick et al., 2013).

Hence, expanding AMO theory allows to advocate that leadership plays a crucial role in enhancing workforce green consciousness: for example, research has proven that transformational leadership significantly contributes to this aim (Gumusluoglu & Ilsev, 2009; Mittal & Dhar, 2016), but efforts will be in vain if employees lack the enthusiasm to engage (Zhu et al., 2018).

An alternative approach to understand leaders' influence on individual followers is driving attention towards the dyadic relationships between leaders and each of the subordinates. Initially coined Vertical Dyad Linkage (Dansereau et al., 1975), Leader-Member Exchange Theory (LMX) draws from social exchange theory (Blau, 1964) and differs from other leadership theories by its explicit emphasis on the dyadic connection and the unique relationships leaders develop with each follower. Social exchange theory posits a seeming obligation of subordinates to return high-quality relationships established with their leaders (Blau, 1964; Gerstner & Day, 1997; Gouldner, 1960), so, dyadic relationships and work roles are (re)constructed over time via a multiplicity of exchanges (Dienesch & Liden, 1986).

Empirical research undeniably has demonstrated that LMX entails substantial impacts on outcomes such as task performance, satisfaction, turnover, and organizational commitment. Thus, the central premise of this construct theorizes that a leader and a follower develop reciprocal obligation, respect, influence, and trust in their relationship: the dyad is defined not only by the quality of exchange, but as well by itself the exchange of valued resources (Graen & Uhl-Bien, 1995; Wayne & Green, 1993).

Research on the antecedents of LMX for increased LMX quality is wide-ranging, even though, the role that a given leadership style represents in foreseeing the quality of LMX has remained understudied.

High-quality leader-member relationships entail high levels of interaction, support, trust, and the exchange of material and nonmaterial rewards beyond what is specified in the formal job description (Dienesch & Liden, 1986; Liden & Graen, 1980). Thus, to reciprocate high LMX relationships subordinates will go beyond the required in-role behavior to further engage in citizenship attitudes in order to maintain an equitable social exchange (Law et al., 2010; Liden & Graen, 1980; Settoon et al., 1996; Wayne et al., 2002). Therefore, the opportunity to experience task challenges allows followers to contribute to the meaningfulness of work in the work context, because high-quality LMX would impel employees to take on duties beyond formal job requirements, leading them to the possibility of exhibiting a higher level of extra-role performance. Hackett et al. (2018) also hypothesized that high-quality LMX should increase "organizational citizenship behaviors" on the part of subordinates, term previously coined by Smith et al. (1983). In its essence, the concept postulates that subordinates in higher quality LMX relationships return value to their leaders by engaging in citizenship behaviours that benefit not only the leader, but as well other actors in a broader scale (Liden et al., 1997; Settoon et al., 1996).

The Resource-Based View of the Firm (RBV) shifts the managerial emphasis from external factors (e.g. industry position) to internal resources (e.g. leadership, knowledge, dynamic capability, organizational social relationships) in justifying sustainable competitive advantage. Therefore, this approach has assumed instrumental in the process of Sustainable Management's solidification as a theoretical and managerial perspective (Peteraf, 1993; Wernerfelt, 1984). The value created when human capital aligns with corporate strategy and fully engages in turning the organization efficient and effective has empirically reported significant positive impact on performance outcomes (Harter et al., 2002). Thus, the key precept of this framework focus on internal organizational resources that are rare, valuable, inimitable, and non-substitutable as sources of superior organizational performance and sustained competitive advantage (Amit & Schoemaker, 1993; Barney, 1991).

Later refinements of the RBV resulted in a theoretical framework encompassing the full range of resources upon which business depend, involving natural as well as human resources (Hart & Dowell, 2011), so that this framework now postulates that the development of human competencies and the preservation of natural resources are recognized essential factors for generating an organizational sustainable competitive advantage (Arulrajah & Opatha, 2016; Bowman & Ambrosini, 2000; Lockett et al., 2009).

Therefore, it is plausible to consider leadership and employees as critical organizational resources, working by an ongoing symbiosis, idea rooting on green management postulates once its policies, procedures and practices aim at developing, motivating, and providing opportunities aimed at exhibiting superior cognitions and job behaviors for firm's sustained competitive advantage and superior performance (Arulrajah & Opatha, 2016; Boxall & Steeneveld, 1999; Florea et al., 2012; Nejati et al., 2017; Takeuchi et al., 2007).

If in the old economic assumptions, performance could be allied to a certain level of output or production, nonetheless, in the current era measurement turned more challenging: the time has now come to go beyond those types of assets and embrace "positive psychological capital" (Luthans et al., 2004).

Leaders have the power and the authority to directly impact the organization (Barker & Mueller, 2002; Haleblian & Finkelstein, 1993; Herrmann & Datta, 2005; Piaskowska & Trojanowski, 2014). Henceforth, the Upper-Echelons perspective grounds on the assumption that complex strategic decisions basically consist of the outcome of behavioural factors, emphasizing personality traits as main predictors of organizational outcomes, building on the assumption that leaders are rationally bounded and susceptible to cognitive biases (Hambrick, 2007). Consequently, organizational strategic choices will mirror leaders' personal values and perceptions, thus, observable managerial characteristics can be used as proxies associated with organizational outcomes (Carpenter et al., 2004; Papadakis & Barwise, 2002).

Notwithstanding, the dynamics of how leaders' personal characteristics could shape other outcomes such as corporate social responsibility and organizational green performance has received less research attention (Waldman & Siegel, 2008).

Green Transformational Leadership as a Means to Enhance Green Management

Personality Traits of Contemporary Leaders

Personality traits have been acknowledged as a protuberant feature of individual difference that influences human behavior, as so, postulating that personality traits outline individual behaviors through proximal motivational variables (Barrick et al., 2002), with more recent work focusing on the psychological mechanisms that account for the interconnections among personality traits, other individual characteristics, and behavior (Funder, 2006). Reviewing literature, it is possible to summarize the leadership personality traits current organizations search for, as follows:

- **Integrity**: The authors postulate that integrity is a valuable personality trait in current leaders once, by it, moral reflectiveness can be comprehended as an inner process explaining the individual struggle towards the accomplishment of prosocial performance, extrapolated from the individual level to as well cover the organizational action, as such ideas and behaviors depend on the individual's moral reflectiveness which arises from individual conscientiousness (Kim et al., 2014). Hence, empirical studies have found that people who attribute considerable relevance to morality tend to be concerned with collective well-being (Reynolds, 2008), thus engaging in more prosocial behavior at work as an external expression of individual's prosocial value, once they understand it as a means to achieve a more equitable and sustainable world (Aquino & Reed, 2002). So, by this personality trait an individual's apprehensions relating to social and environmental topics have been attributed to moral purposes, echoing a basic respect for human dignity (Aguilera et al., 2007).
- **Confidence**: Stajkovic and Luthans (2003) define confidence (or self-efficacy) as the individual's conviction about the self-abilities, which then allows such agent to mobilize the motivation, cognitive resources, and courses of action intrinsic to successfully execute a specific task/carry along a certain responsibility within a given context. Based on it and alluding to Bandura (1997), the authors postulate that confidence, assuming as a positive psychological capital capacity, positively relates to work-related performance because, through this aspect, leaders and followers are impelled at integrity, evidencing not only technical competence, but (with growingly importance) self-knowledge, kindness, and maturity. In a remarkable contribute, Arena et al. (2018) signify the relevance of confidence in current leadership to advance hubris as a projecting personality trait of organizational leaders in a position

of authority and power that manifests itself via an inflated self-evaluation, self-confidence, and self-worthiness, which drives them to the overestimation of their real ability, level of control, performance, or possibility of success (Hayward & Hambrick, 1997; Hayward et al., 2006; Hribar & Yang, 2016). The analysis can be related to the early contributes by Shamir et al. (1994) on charisma as a vector for the leaders' pro-environmental cognitions and behaviours. Scholars stand divided on the benefits of hubris on organizational outcomes, reporting both value-destroying (Hayward & Hambrick, 1997; Malmendier & Tate, 2005) and value-enhancing effects (Hirshleifer et al., 2012; Tang et al., 2015).

- **Empathy**: current leaders are required to excel at establishing an emotional connection with a plethora of actors, thus, they are challenged to recognize the assets that people can bring (as well as acknowledging their needs) – this is their core responsibility. Therefore, they are required to have passion for what they are doing, clearly being enthusiastic about it, and furthering so, to recognize people's needs, concerns, fears, expectations and aims and, moreover, offer active support towards attaining them. Hence, they are fostered to make and inspire an emotional connection with innovation to connect with people's hopes and passions, working with people in order to build and solidify their confidence to try things out and take personal responsibility, and empower them to become civic leaders themselves. Subsequently, todays' leaders are encouraged to vividly advocate the importance of values, investing as much energy as possible into relationships and behaviours as into delivering tasks. At its very essence, it requires such leaders to acknowledge their personal enjoyment in the process of working towards goals, and not solely focussing on the final attainment. Synthetizing, leadership largely origins in the resilience to keep edifying strong local relationships via mundane activities in order to involve the workforce's motivation and tacit knowledge Grint (2008).
- **Critical thinking**: basilar to this personality trait is the ability to conceive situations in an holistic perspective, acting in a world framed on an intercomplex systems' logics, which requires navigating through ambiguity and an adaptive learning beyond traditional boundaries. Hence, these leaders' cognitions evolve from mere reactive problem-solving to a level of increased commitment, imagination, patience and perseverance creating the desired future, thus, to proactive problem-solving. So, this skill will support leaders in addressing the challenge of supporting commitment to both their social mission and business plan, reducing the risk the organization may lose its dual focus (Smith et al., 2012; Sotarauta, 2016).
- Today's leaders are impelled towards the acceptance of uncertainty, adopting an open and constantly enquiring mindset. They shall be dialogue driven,

accepting the importance of relationships over predefined structures, implementing reflection before reaction, which will impel them to encourage constructive dissent over destructive consent. This personality trait impels such leaders to be prone at deconstructing hierarchies: they will be required to embody a vibrant guiding vision, both professionally and personally. Hence, they shall identify and support boundary-spanners, ensuring leadership and decision-making are distributed throughout all levels and functions, fostering collective intelligence over individual genius.

- **Embracing diversity**: as mentioned above, such leaders are eager at recognizing, accepting and integrating heterogeneity, being positive and open-minded about change, identifying creative synergies between contradictory elements as a means to go beyond competing and even paradoxical demands, dealing with them as an invitation to creative action instead of as an obstacle, so that conflict becomes productive rather than intractable. Working with conflict as an ally to produce new insight and create deeper meaning and shared understanding provides organizations added value which would not be generated if resolving conflict through compromise – narrative, engagement and connectivity, hence, assume as fundamental leader's skills (Smith et al., 2012). Detecting integrative options entails complex thinking (Suedfeld, 1992), driven by two specific capabilities: interpersonal skills that build trust, openness, and cultural sensitivity; and decision-making skills that empower leaders to pursue synergies rather than rapid solutions (Smith et al., 2012). Such leaders are then challenged to create settings that bring top-down and bottom-up place-based leaders together to fit the several pieces of the organizational puzzle together, each part being respected and included in its diversity, contributing and being valued by their own perspective and knowledge.
- **Resilience**: according to Coutu (2002) and Luthans (2002), the common profile of resilient individuals entails core characteristics, namely a faithful acceptance of reality, a profound belief that life is meaningful, and a refined ability to adapt to continuous and substantial change via an enlarging process of constant learning. Nevertheless, except for its application to stress resistance, resilience has received scant attention in organizational behavior, even though those studies stress that this leadership trait can contribute to positive psychological capital with a return of desired performance outcomes (Luthans et al., 2004). Based on Seligman (2002) and Snyder et al. (2003), this personality trait can find its roots in others, namely on hope and optimism, by which it can manifest itself as a positive psychological state rooted on a sense of fruitful focus-oriented energy to achieve increasingly demanding goals. Hence, such leaders can act as catalysts in the sense that they actively and

positively impel organizations to capitalize on change, based on the premise that the agency (or willpower) and pathways (or waypower) components will propel a positively effect on the leaders' and followers' self-motivation, autonomy, and contingency. However, only a few studies have examined the impact of this this personality trait on workplace performance, nonetheless evidencing positive results (Peterson & Luthans, 2003; Youssef & Luthans, 2005).

Reconfiguring the Organizational Psychological Contract by Green Transformational Leadership: Greening Corporate Performance

Literature has described the influence of different leadership styles on employee's extra role attitudes and behaviors towards the environment, but empirical research is still recent and scarce.

The concept of green transformational leadership concerns to the behaviors of leaders who motivate followers to achieve environmental goals, inspiring them to perform beyond expected levels of environmental performance (Chen & Chang, 2013). Green transformational leadership, then, emphasizes green vision and shapes employees' harmonious passion for pro-environmental behaviors that positively affect employees' engagement in pro-environmental behaviors (Robertson & Barling, 2013). So, in here grounds the premise that, if a transformational leader embodies green values, via an intellectual stimulation, individualized consideration, charisma, and inspirational motivation (Bass, 1985) such leader will be able to energize a higher level of motivation, cohesion, commitment, trust and performance among followers, inspiring them to shift attention to the achievement of the aims favorable to the long-term development of the organization (Kark et al., 2003; Robertson & Barling, 2013). When this organizational actor makes pro-environmental decisions and conducts pro-environmental behavior in the daily operations of the organization, he/she assumes an eco-role model for subordinates.

So, core in this leadership style is the concept of ***Green Identity***. Green transformational leaders highpoint the relevance of their green values, which then spill-over to environmental related issues associated with company's long-term objectives, evidenced via their high-level expectations when communicating with subordinates. So, transformational leadership can positively impact on employees' pro-environmental cognitions and actions because such leaders delegate tasks and responsibilities to subordinates regarding the achievement of sustainable objectives, also monitoring their green performance and reward them for pro-environmental behaviors (Robertson & Barling, 2013).

Communication is vital in this expression of environmental beliefs as a means for the subordinates' value internalization (Tajfel, 1979): if followers display higher levels of identification with sustainability, they will be more sensitive to the environmental information conveyed by the leader and, subsequently, more prone to internalize the leaders' green cognitions. Therefore, in here the authors suggest to the interconnection between the concepts of ***Value Congruence*** and ***Green Identity***.

Values, which represent a chief role in transformational leadership theory (House, 1996), consist of abstract notions mirroring what individuals conceive as important in their lives. Literature defends that green transformational leaders devote an increased significance to the organization's green vision, making subordinates as well consider work as a reflection of their own values. Hence, the authors postulate that employees' green self-identity refers to the extent which people regard themselves as a pro-environmental agent (Van der Werff et al., 2014), so, such employees reveal a higher propensity to interpret the vision and values conveyed by their green transformational leaders as sustainable and ecological, undergoing a profounder congruence between the perceived value of the leader and their own (Hoffman et al., 2011). Summing up, employees' green self-identity will affect the profundity of their value congruence to their green transformational leader (Graves et al., 2019; Jung & Avolio, 2000; Robertson & Barling, 2013).

Aiming at establishing robust and durable relationships with their leaders (Aguilera et al., 2007), followers may endeavor to express analogous values, and in this sense, the authors postulate that leader-follower value congruence sponsors higher quality of LMX relationships (Robertson & Barling, 2013).

In an archetypal workplace, the leader is likely to be a salient referent right away due to his/her higher status and power at the organizational hierarchy (Bass, 1985), which also characterizes them as essential sources of information and appropriate behavior in the group. As such, the behavior of a leader is highly likely to influence the motivation of followers, including their discretionary pro-social behaviors (Ilies et al., 2009). Nonetheless, in this case the authors postulate that the ***Internal Motivation*** is a core vector to trigger individual and organizational green performance, advocating that the followers' internal motivation is implicated in the achievement of the organization's green objectives rather than external motivation by itself only, because only grounding the core on internal motivation is it possible to assist to the full embodiment of their values and correspondent behaviors in a complete consistence with their self-concepts.

Following this reasoning, Chams and García-Blandón (2019) hypothesize that leadership style and personality traits are imbricated within the establishment of a ***Green Organizational Culture***, grounding on the assumption that leaders are responsible for proactively enhancing a collective and ethical atmosphere at the organizational level. Indeed, it is crucial to highpoint that, in addition to individual motivational processes, social contexts also motivate behavior at work (Barrick et al., 2013; Chen & Kanfer, 2006). In here, the authors propose the intricate relationship between the already mentioned concepts and the notion of ***Green Corporate Consciousness***.

Relatable, the authors then suggest that a ***Green Psychological Climate***, which entails employees' interpretations of their organization's policies, procedures, and practices concerning environmental sustainability (Norton et al., 2015), mediates the link between green transformational leadership, green self-identity and green corporate performance. Notably, a ***Green Psychological Climate*** originates from the fact that the employees' perception of an organizational green climate will positively impact their pro-environmental cognitions and behaviors: hereby, this concept entails the workforce joint awareness that the organization's environmental policies and procedures enhance green values and environmental sustainability (Moore et al., 2019; Norton et al., 2015; Ramus & Steger, 2000), an opposite effect observed when applying a vice-versa logics (Tuan, 2019; Zientara & Zamojska, 2018).

The relational motive is, undeniably, a primary psychological function capable of motivating employees to reply to direct and indirect social signals indicating that green cognitions and behaviors are desirable and valued by their peers, furthering the value congruence argument: exposed to the interaction with a green transformational leader, an individual evidencing a green identity may be predictable to assume as more prone to identify with green information regarding the leader's cognitions and actions, which will positively boost followers' propensity to follow him/her and show respect, appreciation, and loyalty (Kidwell et al., 1997; Kim et al., 2014; Kulik & Ambrose, 1992; Shamir et al., 1994). Thus, and grounding the social identity theory and alluding to the above already explained, the authors suggest that the green self-identity may act as a moderator on the relationship between value congruence, transformational leadership and employee green behavior.

So, expanding the above argumentation it is then possible to relate the already mentioned concepts with the other one of ***Psychological Capital***. With the intensifying acknowledgement of human resources as a competitive advantage in today's economy, human capital is being advertised as a key organizational asset towards the attainment of a sustainable competitive advantage (Fernandes & Machado, 2020), even though, to date psychological capital has been nearly ignored (Luthans et al., 2004). Indeed, the contemporary challenge is to find pioneering means of capitalizing on and developing human, social, and especially the psychological capacities of human resources in order to boost organizational sustainable competitive advantage.

Framed on positive psychology's postulates, this concept articulates that, by eschewing a genuine concern with personal dysfunctions and concentrating on personal qualities, current leaders can develop confidence, optimism, hope, and resilience, capacities which are measurable, open to development, and that can be managed for more effective work performance, according to Luthans (2002). Hence, transformational leaders will be able to improve both individual and organizational performance. As a comprehensive definition, this means that Psychological Capital roots beyond human and social capital to basically consist of who people are, conceiving those agents in an holistic perspective open to constant development (Luthans et al., 2007; Stajkovic & Luthans, 2003).

The authors stress the relevance of this concept and its interconnectedness with the other concepts above exposed also by focusing on the notion of ***Persuasive Mapping***. It describes the ability to construct and make use of mental models to encourage lateral thinking on the surrounding agents, thus boosting a cognitive spill-over effect (Barbuto & Wheeler, 2006). Henceforth, the authors theorize that green transformational leaders who excel in persuasive mapping are skilled at articulating issues and deconstructing problems, conceptualizing creative options by sharing their path of thought, continually deepening it in order to have the most rigorous and wide-ranging possible understanding of the case. Excelling on this skill, such leaders possess the required knowledge to effectively assist and support their followers towards the achievement not only of personal goals, but by as well imprinting valuable contributes towards the enhancement of organizational performance, making use of persuasive methods rather than legitimization, positively enhancing stronger relationships with followers (Druskat & Pescosolido, 2002).

The authors as well suggest that the concept of ***Psychological Empowerment*** assumes relevance as a mediator in the relationship between the above mentioned concepts. Psychological Empowerment theory upholds that when employees feel empowered, they are expected to take proactive initiatives towards their work and, subsequently, deliver extra-role performance, acting beyond their mandate (Spreitzer, 2008). Alluding to Spreitzer (1995), this concept refers to a motivational variable encompassing four dimensions: meaning (which provides employees a notion their jobs' value), competence (which refers to employees' confidence in their knowledge and skills, essential to accomplish their responsibilities in the workplace), self-determination (which regards to the feeling employees experience when they can choose in the decision-making process relevant to their jobs), and impact (concerning the level to which employees trust their jobs can influence in the accomplishment of organizational goals).

Following this line, Vandewalle et al. (1995) reported that ***Psychological Ownership*** positively relates to organizational citizenship behaviors, boosting overall organizational performance, which roots on the fact, alluding to Van Dyne and Pierce (2004), that psychological ownership in the organization was positively related with employee increased levels of job satisfaction, enhancing organization-based self-esteem and organizational commitment, culminating in positive effects on work behavior. Indeed, by concentrating on the satisfaction of their needs, offering autonomy in their domains, equitable treatment, accountability and transparency in the connections, as well as personal development opportunities (Newman et al.,

2017; Stone et al., 2004), such leaders nurture their subordinates' psychological empowerment, positively impacting followers' outcomes. Even if empirical evidence has so far been scarce to verify this argument (Newman et al., 2017), psychological empowerment is expected to positively link extra-role initiatives when employees conceive such as valuable for environmental sustainability (Lamm et al., 2015).

Grounding on the assumption that Psychological Capital offers a dynamic resource potential that can grow and be sustainable over time, all the above exposed allows to advocate that a new psychological contract has emerged from the turbulent corporate landscape marked by the constant pressures in downsizing personal identity. In this new psychological contract, employees are assigned the strategic management of their own careers, architecting a dynamics of self-management based on proactive ownership. Employer's responsibility towards employees assumes as mainly of a service provider, which limits to the duration and capacity of which and where costs can be justified.

Postulating this new organizational contract requires that managing today's organizations entails a vital shift towards a refined paradigm, where sustainable competitive advantage no longer roots on traditional (physical, financial, or technological) resources: it lies on the constant identification, development, leverage, and management of psychological capital. This new capital approach grounds on the generally accepted fact that most organizations today are not grasping the full potential of their human resources (Gardner et al., 2005): the short term-oriented crisis-management model which traditionally marked organizational management led to a negative perspective, hence, in the contemporary milieu such sustainable competitive advantage can best be accomplished via context-specific, renewable, cumulative, and thus hard-to-imitate factors, not disregarding the continued presence of traditional material resources.

This new configuration of the ***Organizational Psychological Contract*** will foster ***Positive Organizational Behavior***, which consists of the examination and subsequent application of positively oriented human resources' psychological capacities capable of being measured, managed, and enhanced in order to lift organizational performance (Ahn et al., 2018; Khan et al., 2019; Luthans, 2002; Usman et al., 2018; Wright, 2003). Hence, green transformational leadership will be expected to positively impact om ***Green Corporate Performance***, because this leadership style acts as a catalyst in promoting employees' prosocial behaviors, as these leaders are likely to represent a significant role on supporting positive green corporate initiatives as a means by which it instigates and stimulates followers to achieve sustainable organizational goals, thus, may incite a positive social atmosphere boosting green advocacy among followers (Berrone & Gomez-Mejia, 2009; Chen & Chang, 2013; Eiadat et al., 2008; Lim & Ployhart, 2004).

SOLUTIONS AND RECOMMENDATIONS

This study stresses that leaders in contemporary organizations are embodying the concept of sustainability as a source of competitive advantage and as a method to incessantly improve overall organizational performance. The argumentation thus proposed advances that, via green transformational leadership, organizational leaders can improve green corporate consciousness and performance by acting as messengers or representatives in the process of leading green consciousness and performance. The argumentation integrates psycho-social elements shedding light on vectors such as dependence-independence-interdependence dynamics, holism, duty, ethics, cooperation, conditioning, and power, so that the current leader's personality traits can be best mapped and understood, as well, so that their leadership paradigm may be understood by the values defining both the leaders and the followers, via a consequent spill-over effect. This allows to advance green corporate consciousness and performance by converging value between organizational actors, impacting their psychological capital and ownership, reconfiguring the organizational psychological contract redefining it as a collaborative learning process.

Such perceptions, vital antecedents of pro-environmental behaviors, then solidify a green psychological climate at the organization and can help this structures to overcome business-related threats to societal sustainability and environmental conservation. In order to capitalize on such capital, organizations, in the figure of their leaders, shall display an acute emphasis on the employee's autonomous motivation towards the environment and societal welfare, continuously taking initiatives that can enhance this motivation, which this study claims to be best conveyed by leaders who excel at skills such as integrity, confidence, empathy, critical thinking, resilience, and embracing diversity, articulating such cognitions and behaviors to their followers via green transformational leadership. Hence, this work suggests that organizations shall rely on green transformational leadership to install and implement policies and practices to enhance employee's green motivations and abilities, providing them with workplace opportunities to engage in environmental management related activities, which, by their end, will positively affect the organization's potential towards green performance.

Acknowledging the value-based essence of transformational leadership, this study extends knowledge on green transformational leadership and employee organizational citizenship behaviors by openly exploring value-based mechanisms that connect green transformational leadership to employees' green cognitions and behaviors, stressing that the main mechanism consists on the value channel, subsequently focusing on the leader's personality traits as a means to conceptualize green identity as a boundary condition for green transformation leadership's effects, aspect which had remained largely underexamined in the extant literature. Doing so, this study stresses that this leadership style roots on the belief in creating value for the community, via regular interaction with subordinates, power delegation, and followers' proactive empowerment towards self-development: hence, it impels a vivid moral assumption towards ethics, having the symbiosis between sustainability and common welfare as a core vector in organizational cognitions and actions.

FUTURE RESEARCH DIRECTIONS

Although literature on green leadership has shed light on the relevance, motivations and impacts of managers' involvement in organizational greening, its nature and scope are still lacking further study: first, the analysis has so far mainly been conducted from a theoretical perspective; second, green leadership and its organizational manifestations are yet scarcely explored, foremost focusing on an instrumental approach. So, answering these limitations would as well require different approaches than those predominating in the extant literature, applying quantitative analysis and replacing cross-sectional designs with experience-sampling or diary designs.

Research as well indicates that environmental knowledge and attitudes differ around the world, as so presenting a frame of societal norms condensed in the cognitive-sociocultural pressure at the institutional level, resulting from the inherent differences in the regulatory, cultural, social and physical environments that differentially impact organizations in various countries (Aoyagi-Usui et al., 2003). It would be relevant to further investigate this because, by understanding such variances, organizations could expand their ability to leverage the natural tendencies of their most conscientious employees, which presumes a profounder examination of the leaders' background and a further focus on the meaning systems of individuals. Indeed, such differences may enhance or constrain the interpersonal influences. Expanding this idea, it would as well be pertinent to consider the inverse causal path, exploring if followers' green advocacy may impact organizational leaders' green cognitions and behaviors (Aguinis & Glavas, 2012).

CONCLUSION

The main aim and contribution of this study grounds on the fact that it uncovers a central psychological dimension of green leadership that has remained largely underexplored. Hence, it allows fruitful inputs towards a basis for developing appropriate leadership practices that sustain green corporate culture.

It predicts that the leaders of the next generation would display certain personality traits in order to efficiently face the challenges brought by the Globalization:

broad knowledge and experience, boundless curiosity and enthusiasm, contagious optimism, belief in people and teamwork, willingness to take risks, focus on long-term growth rather than short-term profit, commitment to excellence, adaptive capacity, authenticity, integrity, and confidence.

Hence, it posits that transformational leaders can inspire employees' green consciousness and performance through concentrating efforts on the correspondence of values between both actors. As well, it advocates that value congruence itself is mediated by green transformational leadership's effect on employees' green cognitions and behaviors. The significance of employees' psychological empowerment for their autonomous motivation towards the engagement in voluntary green behavior in the workplace is, then, highlighted: so far, research has reported that organizations commonly enhanced workforce green behaviors through external rewards, but this study proposes that, when subjected to a green transformational leadership style, subordinates are more prone to accept and internalize the leader's green values, subsequently enhancing their own cognitions around environmental conservation and societal sustainability, solidifying a green corporate consciousness by greening psychological capital, which culminates with improving green corporate performance via an increased green psychological ownership. Thus, the authors defend that green transformational leadership reconfigures the organizational psychological contract by redefining it as a collaborative learning process.

By doing so, the traditional monolithic perspective of environmental leaders is critically scrutinized.

This study as well contributes to advance the RBV and AMO theories to understand and explain what causes organizational green innovation and environmental performance. It also stresses the acuteness of LMX postulates to better explain the dynamics of green transformational leadership in enhancing follower's green consciousness and performance. In what concerns to the RBV, it posits that green transformational leadership consists of a strategic resource that organizations should leverage to design and to implement policies, processes and practices capable of influencing green awareness and performance. Expanding this idea, this study suggests that leaders, leadership dynamics and employees are critical organizational resources which should be valued in a manner that it becomes difficult for the competing organizations to imitate them. The advances on the AMO theory provided by this study ground on the fact that it suggests that organizations should design and implement policies, procedures and practices in order to attract, train, motivate, and retain green conscious employees, which can be achieved via the incessant monitoring role of green transformational leadership, as a moto to enhance overall organizational green awareness and performance.

REFERENCES

Aguilera, R. V., Rupp, D. E., Williams, C. A., & Ganapathi, J. (2007). Putting the S back in corporate social responsibility: A multilevel theory of social change in organizations. *Academy of Management Review, 32*(3), 836–863. doi:10.5465/amr.2007.25275678

Aguinis, H., & Glavas, A. (2012). What we know and don't know about corporate social responsibility: A review and research agenda. *Journal of Management, 38*(4), 932–968. doi:10.1177/0149206311436079

Ahn, J., Lee, S., & Yun, S. (2018). Leaders' core self-evaluation, ethical leadership, and employees' job performance: The moderating role of employees' exchange ideology. *Journal of Business Ethics, 148*(2), 457–470. doi:10.100710551-016-3030-0

Amit, R., & Schoemaker, P. J. (1993). Strategic assets and organizational rent. *Strategic Management Journal, 14*(1), 33–46. doi:10.1002mj.4250140105

Aoyagi-Usui, M., Vinken, H., & Kuribayashi, A. (2003). Pro-environmental attitudes and behaviors: An international comparison. *Human Ecology Review, 10*(1), 23–31.

Appelbaum, E., Bailey, T., Berg, P., Kalleberg, A. L., & Bailey, T. A. (2000). *Manufacturing advantage: Why high-performance work systems pay off*. Cornell University Press.

Aquino, K. F., & Reed, A. (2002). The self-importance of moral identity. *Journal of Personality and Social Psychology, 83*(6), 1423–1440. doi:10.1037/0022-3514.83.6.1423 PMID:12500822

Arena, C., Michelon, G., & Trojanowski, G. (2018). Big egos can be green: A study of CEO hubris and environmental innovation. *British Journal of Management, 29*(2), 316–336. doi:10.1111/1467-8551.12250

Arulrajah, A. A., & Opatha, H. H. (2016). Analytical and theoretical perspectives on green human resource management: A simplified underpinning. *International Business Research, 9*(12), 153–164. doi:10.5539/ibr.v9n12p153

Bandura, A. (1977). Self-efficacy: Toward a unifying theory of behavioral change. *Psychological Review, 84*(2), 191–215. doi:10.1037/0033-295X.84.2.191 PMID:847061

Barbuto, J. E. Jr, & Wheeler, D. W. (2006). Scale development and construct clarification of servant leadership. *Group & Organization Management, 31*(3), 300–326. doi:10.1177/1059601106287091

Barker, V. L. III, & Mueller, G. C. (2002). CEO characteristics and firm R&D spending. *Management Science*, *48*(6), 782–801. doi:10.1287/mnsc.48.6.782.187

Barney, J. B. (1991). Firm resources and sustained competitive advantage. *Journal of Management*, *17*(1), 99–120. doi:10.1177/014920639101700108

Barrick, M. R., Mount, M. K., & Li, N. (2013). The theory of purposeful work behavior: The role of personality, higher-order goals, and job characteristics. *Academy of Management Review*, *38*(1), 132–153. doi:10.5465/amr.2010.0479

Barrick, M. R., Stewart, G. L., & Piotrowski, M. (2002). Personality and job performance: Test of the mediating effects of motivation among sales representatives. *The Journal of Applied Psychology*, *87*(1), 43–51. doi:10.1037/0021-9010.87.1.43 PMID:11916215

Bass, B. M. (1985). *Leadership and performance beyond expectations*. Free Press.

Berrone, P., & Gomez-Mejia, L. R. (2009). The pros and cons of rewarding social responsibility at the top. *Human Resource Management*, *48*(6), 959–971. doi:10.1002/hrm.20324

Blau, P. M. (1964). *Exchange and power in social life*. John Wiley and Sons.

Boiral, O. (2009). Greening the corporation through organizational citizenship behaviors. *Journal of Business Ethics*, *87*(2), 221–236. doi:10.100710551-008-9881-2

Boiral, O., Cayer, M., & Baron, C. M. (2009). The action logics of environmental leadership: A developmental perspective. *Journal of Business Ethics*, *85*(4), 479–499. doi:10.100710551-008-9784-2

Bos-Nehles, A. C., Van Riemsdijk, M. J., & Kees Looise, J. (2013). Employee perceptions of line management performance: Applying the AMO theory to explain the effectiveness of line managers' HRM implementation. *Human Resource Management*, *52*(6), 861–877. doi:10.1002/hrm.21578

Bowman, C., & Ambrosini, V. (2000). Value creation versus value capture: Towards a coherent definition of value in strategy. *British Journal of Management*, *11*(1), 1–15. doi:10.1111/1467-8551.00147

Boxall, P., & Purcell, J. (2003). *Strategy and human resource management*. Blackwell.

Boxall, P., & Steeneveld, M. (1999). Human resource strategy and competitive advantage: A longitudinal study of engineering consultancies. *Journal of Management Studies*, *36*(4), 443–463. doi:10.1111/1467-6486.00144

Carpenter, M. A., Geletkanycz, M. A., & Sanders, W. G. (2004). Upper echelons research revisited: Antecedents, elements, and consequences of top management team composition. *Journal of Management, 30*(6), 749–778. doi:10.1016/j.jm.2004.06.001

Chams, N., & García-Blandón, J. (2019). On the importance of sustainable human resource management for the adoption of sustainable development goals. *Resources, Conservation and Recycling, 141*, 109–122. doi:10.1016/j.resconrec.2018.10.006

Chen, G., & Kanfer, R. (2006). Toward a systems theory of motivated behavior in work teams. *Research in Organizational Behavior, 27*, 223–267. doi:10.1016/S0191-3085(06)27006-0

Chen, Y. S., & Chang, C. H. (2013). The determinants of green product development performance: Green dynamic capabilities, green transformational leadership, and green creativity. *Journal of Business Ethics, 116*(1), 107–119. doi:10.100710551-012-1452-x

Colbert, B. A. (2004). The complex resource-based view: Implications for theory and practice in strategic human resource management. *Academy of Management Review, 29*(3), 341–358. doi:10.5465/amr.2004.13670987

Coutu, D. L. (2002). How resilience works. *Harvard Business Review, 80*(5), 46–56. PMID:12024758

Dansereau, F. Jr, Graen, G. B., & Haga, W. (1975). A vertical dyad linkage approach to leadership in formal organizations. *Organizational Behavior and Human Performance, 13*(1), 46–78. doi:10.1016/0030-5073(75)90005-7

Dienesch, R. M., & Liden, R. C. (1986). Leader-member exchange model of leadership: A critique and further development. *Academy of Management Review, 11*(3), 618–634. doi:10.5465/amr.1986.4306242

Druskat, V. U., & Pescosolido, A. T. (2002). The content of effective teamwork mental models in self-managing teams: Ownership, learning, and heedful interrelating. *Human Relations, 55*(3), 283–314. doi:10.1177/0018726702553001

Eiadat, Y., Kelly, A., Roche, F., & Eyadat, H. (2008). Green and competitive? An empirical test of the mediating role of environmental innovation strategy. *Journal of World Business, 43*(2), 131–145. doi:10.1016/j.jwb.2007.11.012

Fernandes, D. S., & Machado, C. F. (2020). How to Look at Organizations and Human Resource Management in the Economy of the Future? In Circular Economy and Engineering A New Ecologically Efficient Model (pp. 89-117). Springer.

Florea, L., Cheung, Y. H., & Herndon, N. C. (2013). For all good reasons: Role of values in organizational sustainability. *Journal of Business Ethics*, *114*(3), 393–408. doi:10.100710551-012-1355-x

Funder, D. C. (2006). Towards a resolution of the personality triad: Persons, situations, and behaviors. *Journal of Research in Personality*, *40*(1), 21–34. doi:10.1016/j.jrp.2005.08.003

Gardner, W. L., Avolio, B. J., Luthans, F., May, D. R., & Walumbwa, F. (2005). "Can you see the real me?" A self-based model of authentic leader and follower development. *The Leadership Quarterly*, *16*(3), 343–372. doi:10.1016/j.leaqua.2005.03.003

Gerhart, B. (2005). Human resources and business performance: Findings, unanswered questions, and an alternative approach. *Management Review*, *16*, 174–185.

Gerstner, C. R., & Day, D. V. (1997). Meta-Analytic review of leader–member exchange theory: Correlates and construct issues. *The Journal of Applied Psychology*, *82*(6), 827–844. doi:10.1037/0021-9010.82.6.827

Gouldner, A. W. (1960). The norm of reciprocity: A preliminary statement. *American Sociological Review*, *25*(2), 161–178. doi:10.2307/2092623

Graen, G. B., & Uhl-Bien, M. (1995). Development of leader-member exchange (LMX) theory of leadership over 25 years: Applying a multi-level multi-domain perspective. *The Leadership Quarterly*, *6*(2), 219–247. doi:10.1016/1048-9843(95)90036-5

Graves, L. M., Sarkis, J., & Gold, N. (2019). Employee proenvironmental behavior in Russia: The roles of top management commitment, managerial leadership, and employee motives. *Resources, Conservation and Recycling*, *140*, 54–64. doi:10.1016/j.resconrec.2018.09.007

Grint, K. (2008). Wicked problems and clumsy solutions: The role of leadership. *Clinical Leader*, *1*(2), 11–25.

Guest, D. E. (2011). Human resource management and performance: Still searching for some answers. *Human Resource Management Journal*, *21*(1), 3–13. doi:10.1111/j.1748-8583.2010.00164.x

Gumusluoglu, L., & Ilsev, A. (2009). Transformational leadership, creativity, and organizational innovation. *Journal of Business Research*, *62*(4), 461–473. doi:10.1016/j.jbusres.2007.07.032

Hackett, R. D., Wang, A. C., Chen, Z., Cheng, B. S., & Farh, J. L. (2018). Transformational Leadership and Organisational Citizenship Behaviour: A Moderated Mediation Model of Leader-Member-Exchange and Subordinates' Gender. *Applied Psychology*, *67*(4), 617–644. doi:10.1111/apps.12146

Haleblian, J., & Finkelstein, S. (1993). Top management team size, CEO dominance, and firm performance: The moderating roles of environmental turbulence and discretion. *Academy of Management Journal*, *36*(4), 844–863.

Hambrick, D. C. (2007). Upper echelons theory: An update. *Academy of Management Review*, *32*(2), 334–343. doi:10.5465/amr.2007.24345254

Hart, S. L., & Dowell, G. (2011). Invited editorial: a natural-resource-based view of the firm: fifteen years after. *Journal of Management*, *37*(5), 1464–1479. doi:10.1177/0149206310390219

Harter, J. K., Schmidt, F. L., & Hayes, T. L. (2002). Business-unit-level relationship between employee satisfaction, employee engagement, and business outcomes: A meta-analysis. *The Journal of Applied Psychology*, *87*(2), 268–279. doi:10.1037/0021-9010.87.2.268 PMID:12002955

Hayward, M. L., & Hambrick, D. C. (1997). Explaining the premiums paid for large acquisitions: Evidence of CEO hubris. *Administrative Science Quarterly*, *42*(1), 103–127. doi:10.2307/2393810

Hayward, M. L., Shepherd, D. A., & Griffin, D. (2006). A hubris theory of entrepreneurship. *Management Science*, *52*(2), 160–172. doi:10.1287/mnsc.1050.0483

Herrmann, P., & Datta, D. K. (2005). Relationships between top management team characteristics and international diversification: An empirical investigation. *British Journal of Management*, *16*(1), 69–78. doi:10.1111/j.1467-8551.2005.00429.x

Hirshleifer, D., Low, A., & Teoh, S. H. (2012). Are overconfident CEOs better innovators? *The Journal of Finance*, *67*(4), 1457–1498. doi:10.1111/j.1540-6261.2012.01753.x

Hoffman, B. J., Bynum, B. H., Piccolo, R. F., & Sutton, A. W. (2011). Person-organization value congruence: How transformational leaders influence work group effectiveness. *Academy of Management Journal*, *54*(4), 779–796. doi:10.5465/amj.2011.64870139

House, R. J. (1996). Path-goal theory of leadership: Lessons, legacy, and a reformulated theory. *The Leadership Quarterly*, *7*(3), 323–352. doi:10.1016/S1048-9843(96)90024-7

Hribar, P., & Yang, H. (2016). CEO overconfidence and management forecasting. *Contemporary Accounting Research*, *33*(1), 204–227. doi:10.1111/1911-3846.12144

Hülsheger, U. R., Anderson, N., & Salgado, J. F. (2009). Team-level predictors of innovation at work: A comprehensive meta-analysis spanning three decades of research. *The Journal of Applied Psychology*, *94*(5), 1128–1145. doi:10.1037/a0015978 PMID:19702361

Ilies, R., Fulmer, I. S., Spitzmuller, M., & Johnson, M. D. (2009). Personality and citizenship behavior: The mediating role of job satisfaction. *The Journal of Applied Psychology*, *94*(4), 945–959. doi:10.1037/a0013329 PMID:19594236

Jung, D. I., & Avolio, B. J. (2000). Opening the black box: An experimental investigation of the mediating effects of trust and value congruence on transformational and transactional leadership. *Journal of Organizational Behavior*, *21*(8), 949–964. doi:10.1002/1099-1379(200012)21:8<949::AID-JOB64>3.0.CO;2-F

Kark, R., Shamir, B., & Chen, G. (2003). The two faces of transformational leadership: Empowerment and dependency. *The Journal of Applied Psychology*, *88*(2), 246–255. doi:10.1037/0021-9010.88.2.246 PMID:12731708

Khan, M. A., Jianguo, D., Ali, M., Saleem, S., & Usman, M. (2019). Interrelations Between Ethical Leadership, Green Psychological Climate, and Organizational Environmental Citizenship Behavior: A Moderated Mediation Model. *Frontiers in Psychology*, *10*, 1–12. doi:10.3389/fpsyg.2019.01977 PMID:31555173

Kidwell, R. E. Jr, Mossholder, K. W., & Bennett, N. (1997). Cohesiveness and organizational citizenship behavior: A multilevel analysis using work groups and individuals. *Journal of Management*, *23*(6), 775–793. doi:10.1177/014920639702300605

Kim, A., Kim, Y., Han, K., Jackson, S. E., & Ployhart, R. E. (2017). Multilevel influences on voluntary workplace green behavior: Individual differences, leader behavior, and coworker advocacy. *Journal of Management*, *43*(5), 1335–1358. doi:10.1177/0149206314547386

Kulik, C. T., & Ambrose, M. L. (1992). Personal and situational determinants of referent choice. *Academy of Management Review*, *17*(2), 212–237. doi:10.5465/amr.1992.4279534

Lamm, E., Tosti-Kharas, J., & King, C. E. (2015). Empowering employee sustainability: Perceived organizational support toward the environment. *Journal of Business Ethics, 128*(1), 207–220. doi:10.100710551-014-2093-z

Law, K. S., Wang, H., & Hui, C. (2010). Currencies of exchange and global LMX: How they affect employee task performance and extra-role performance. *Asia Pacific Journal of Management, 27*(4), 625–646. doi:10.100710490-009-9141-8

Liden, R. C., & Graen, G. (1980). Generalizability of the vertical dyad linkage model of leadership. *Academy of Management Journal, 23*(3), 451–465.

Liden, R. C., Sparrowe, R. T., & Wayne, S. J. (1997). Leader-member exchange theory: The past and potential for the future. In J. Ferris (Ed.), *Research in personnel and human resource management* (pp. 47–119). JAI Press.

Lim, B. C., & Ployhart, R. E. (2004). Transformational leadership: Relations to the five-factor model and team performance in typical and maximum contexts. *The Journal of Applied Psychology, 89*(4), 610–621. doi:10.1037/0021-9010.89.4.610 PMID:15327348

Lockett, A., Thompson, S., & Morgenstern, U. (2009). The development of the resource-based view of the firm: A critical appraisal. *International Journal of Management Reviews, 11*(1), 9–28. doi:10.1111/j.1468-2370.2008.00252.x

Luthans, F. (2002). The need for and meaning of positive organizational behavior. Journal of Organizational Behavior: *The International Journal of Industrial. Occupational and Organizational Psychology and Behavior, 23*(6), 695–706. doi:10.1002/job.165

Luthans, F., Luthans, K. W., & Luthans, B. C. (2004). Positive psychological capital: Beyond human and social capital. *Business Horizons, 47*(1), 45–50. doi:10.1016/j.bushor.2003.11.007

Luthans, F., Youssef, C. M., & Avolio, B. J. (2007). Psychological capital: Investing and developing positive organizational behavior. *Positive Organizational Behavior, 1*(2), 9–24. doi:10.4135/9781446212752.n2

Malmendier, U., & Tate, G. (2005). Does overconfidence affect corporate investment? CEO overconfidence measures revisited. *European Financial Management, 11*(5), 649–659. doi:10.1111/j.1354-7798.2005.00302.x

Mittal, S., & Dhar, R. L. (2016). Effect of green transformational leadership on green creativity: A study of tourist hotels. *Tourism Management, 57*, 118–127. doi:10.1016/j.tourman.2016.05.007

Moore, C., Mayer, D. M., Chiang, F. F., Crossley, C., Karlesky, M. J., & Birtch, T. A. (2019). Leaders matter morally: The role of ethical leadership in shaping employee moral cognition and misconduct. *The Journal of Applied Psychology, 104*(1), 123–145. doi:10.1037/apl0000341 PMID:30221953

Nejati, M., Rabiei, S., & Jabbour, C. J. (2017). Envisioning the invisible: Understanding the synergy between green human resource management and green supply chain management in manufacturing firms in Iran in light of the moderating effect of employees resistance to change. *Journal of Cleaner Production, 168*, 163–172. doi:10.1016/j.jclepro.2017.08.213

Newman, A., Schwarz, G., Cooper, B., & Sendjaya, S. (2017). How servant leadership influences organizational citizenship behavior: The roles of LMX, empowerment, and proactive personality. *Journal of Business Ethics, 145*(1), 49–62. doi:10.100710551-015-2827-6

Norton, T. A., Parker, S. L., Zacher, H., & Ashkanasy, N. M. (2015). Employee green behavior: A theoretical framework, multilevel review, and future research agenda. *Organization & Environment, 28*(1), 103–125. doi:10.1177/1086026615575773

Papadakis, V. M., & Barwise, P. (2002). How much do CEOs and top managers matter in strategic decision-making? *British Journal of Management, 13*(1), 83–95. doi:10.1111/1467-8551.00224

Peteraf, M. A. (1993). The cornerstones of competitive advantage: A resource-based view. *Strategic Management Journal, 14*(3), 179–191. doi:10.1002mj.4250140303

Peterson, S. J., & Luthans, F. (2003). The positive impact and development of hopeful leaders. *Leadership and Organization Development Journal, 24*(1), 26–31. doi:10.1108/01437730310457302

Piaskowska, D., & Trojanowski, G. (2014). Twice as smart? The importance of managers' formative-years' international experience for their international orientation and foreign acquisition decisions. *British Journal of Management, 25*(1), 40–57. doi:10.1111/j.1467-8551.2012.00831.x

Ramus, C. A., & Steger, U. (2000). The roles of supervisory support behaviors and environmental policy in employee "Ecoinitiatives" at leading-edge European companies. *Academy of Management Journal, 43*(4), 605–626.

Renwick, D. W., Redman, T., & Maguire, S. (2013). Green human resource management: A review and research agenda. *International Journal of Management Reviews, 15*(1), 1–14. doi:10.1111/j.1468-2370.2011.00328.x

Reynolds, S. J. (2008). Moral attentiveness: Who pays attention to the moral aspects of life? *The Journal of Applied Psychology, 93*(5), 1027–1041. doi:10.1037/0021-9010.93.5.1027 PMID:18808223

Robertson, J. L., & Barling, J. (2013). Greening organizations through leaders' influence on employees' pro-environmental behaviors. *Journal of Organizational Behavior, 34*(2), 176–194. doi:10.1002/job.1820

Seligman, M. (2002). *Authentic happiness: Using the new positive psychology to realize your potential for lasting fulfillment*. Free Press.

Settoon, R. P., Bennett, N., & Liden, R. C. (1996). Social exchange in organizations: Perceived organizational support, leader–member exchange, and employee reciprocity. *The Journal of Applied Psychology, 81*(3), 219–227. doi:10.1037/0021-9010.81.3.219

Shamir, B., Arthur, M. B., & House, R. J. (1994). The rhetoric of charismatic leadership: A theoretical extension, a case study, and implications for research. *The Leadership Quarterly, 5*(1), 25–42. doi:10.1016/1048-9843(94)90004-3

Smith, C., Organ, D. W., & Near, J. P. (1983). Organizational citizenship behavior: Its nature and antecedents. *The Journal of Applied Psychology, 68*(4), 653–663. doi:10.1037/0021-9010.68.4.653

Smith, W. K., Besharov, M. L., Wessels, A. K., & Chertok, M. (2012). A paradoxical leadership model for social entrepreneurs: Challenges, leadership skills, and pedagogical tools for managing social and commercial demands. *Academy of Management Learning & Education, 11*(3), 463–478. doi:10.5465/amle.2011.0021

Snyder, C. R., Lopez, S. J., Shorey, H. S., Rand, K. L., & Feldman, D. B. (2003). Hope theory, measurements, and applications to school psychology. *School Psychology Quarterly, 18*(2), 122–139. doi:10.1521cpq.18.2.122.21854

Sotarauta, M. (2016). Place leadership, governance and power. *Administration, 64*(3-4), 45–58. doi:10.1515/admin-2016-0024

Spreitzer, G. M. (1995). Psychological empowerment in the workplace: Dimensions, measurement, and validation. *Academy of Management Journal, 38*(5), 1442–1465.

Spreitzer, G. M. (2008). Taking stock: A review of more than twenty years of research on empowerment at work. *Handbook on Organizational Behaviour, 1*, 54–72. doi:10.4135/9781849200448.n4

Stajkovic, A. D., & Luthans, F. (2003). Behavioral management and task performance in organizations: Conceptual background, meta-analysis, and test of alternative models. *Personnel Psychology*, *56*(1), 155–194. doi:10.1111/j.1744-6570.2003.tb00147.x

Stone, A. G., Russell, R. F., & Patterson, K. (2004). Transformational versus servant leadership: A difference in leader focus. *Leadership and Organization Development Journal*, *25*(4), 349–361. doi:10.1108/01437730410538671

Suedfeld, P. (1992). Cognitive managers and their critics. *Political Psychology*, *13*(3), 435–453. doi:10.2307/3791607

Tajfel, H. (1979). Individuals and groups in social psychology. *The British Journal of Social and Clinical Psychology*, *18*(2), 183–190. doi:10.1111/j.2044-8260.1979.tb00324.x

Takeuchi, R., Lepak, D. P., Wang, H., & Takeuchi, K. (2007). An empirical examination of the mechanisms mediating between high-performance work systems and the performance of Japanese organizations. *The Journal of Applied Psychology*, *92*(4), 1069–1083. doi:10.1037/0021-9010.92.4.1069 PMID:17638466

Tang, G., Cai, Z., Liu, Z., Zhu, H., Yang, X., & Li, J. (2015). The importance of ethical leadership in employees' value congruence and turnover. *Cornell Hospitality Quarterly*, *56*(4), 397–410. doi:10.1177/1938965514563159

Tuan, L. T. (2019). Catalyzing Employee OCBE in Tour Companies: Charismatic Leadership, Organizational Justice, and Pro-Environmental Behaviors. *Journal of Hospitality & Tourism Research (Washington, D.C.)*, *43*(5), 682–711. doi:10.1177/1096348018817582

Usman, M., Hameed, A. A., & Manzoor, S. (2018). Exploring the links between Ethical Leadership and Organizational Unlearning: A Case Study of a European Multinational Company. *Business and Economic Review*, *10*(2), 29–54. doi:10.22547/BER/10.2.2

Van der Werff, E., Steg, L., & Keizer, K. (2014). I am what I am, by looking past the present: The influence of biospheric values and past behavior on environmental self-identity. *Environment and Behavior*, *46*(5), 626–657. doi:10.1177/0013916512475209

Van Dyne, L., & Pierce, J. L. (2004). Psychological ownership and feelings of possession: Three field studies predicting employee attitudes and organizational citizenship behavior. *Journal of Organizational Behavior*, *25*(4), 439–459. doi:10.1002/job.249

Vandewalle, D., Van Dyne, L., & Kostova, T. (1995). Psychological ownership: An empirical examination of its consequences. *Group & Organization Management, 20*(2), 210–226. doi:10.1177/1059601195202008

Waldman, D. A., & Siegel, D. (2008). Defining the socially responsible leader. *The Leadership Quarterly, 19*(1), 117–131. doi:10.1016/j.leaqua.2007.12.008

Wayne, S. J., & Green, S. A. (1993). The effects of leader-member exchange on employee citizenship and impression management behavior. *Human Relations, 46*(12), 1431–1440. doi:10.1177/001872679304601204

Wayne, S. J., Shore, L. M., Bommer, W. H., & Tetrick, L. E. (2002). The role of fair treatment and rewards in perceptions of organizational support and leader-member exchange. *The Journal of Applied Psychology, 87*(3), 590–598. doi:10.1037/0021-9010.87.3.590 PMID:12090617

Wernerfelt, B. (1984). A resource-based view of the firm. *Strategic Management Journal, 5*(2), 171–180. doi:10.1002mj.4250050207

Wright, P. M., Gardner, T. M., & Moynihan, L. M. (2003). The impact of HR practices on the performance of business units. *Human Resource Management Journal, 13*(3), 21–36. doi:10.1111/j.1748-8583.2003.tb00096.x

Youssef, C. M., & Luthans, F. (2005). Resiliency development of organizations, leaders and employees: Multi-level theory building for sustained performance. *Authentic leadership theory and practice: Origins, effects and development, 3*(1), 303-343.

Zhu, J., Liao, Z., Yam, K. C., & Johnson, R. E. (2018). Shared leadership: A state-of-the-art review and future research agenda. *Journal of Organizational Behavior, 39*(7), 834–852. doi:10.1002/job.2296

Zientara, P., & Zamojska, A. (2018). Green organizational climates and employee pro-environmental behaviour in the hotel industry. *Journal of Sustainable Tourism, 26*(7), 1142–1159. doi:10.1080/09669582.2016.1206554

ADDITIONAL READING

Afsar, B., Cheema, S., & Javed, F. (2018). Activating employee's pro-environmental behaviors: The role of CSR, organizational identification, and environmentally specific servant leadership. *Corporate Social Responsibility and Environmental Management, 25*(5), 904–911. doi:10.1002/csr.1506

Derue, D. S., Nahrgang, J. D., Wellman, N. E., & Humphrey, S. E. (2011). Trait and behavioral theories of leadership: An integration and meta-analytic test of their relative validity. *Personnel Psychology*, *64*(1), 7–52. doi:10.1111/j.1744-6570.2010.01201.x

Feinberg, M., & Willer, R. (2013). The moral roots of environmental attitudes. *Psychological Science*, *24*(1), 56–62. doi:10.1177/0956797612449177 PMID:23228937

Frieder, R. E., Wang, G., & Oh, I. S. (2018). Linking job-relevant personality traits, transformational leadership, and job performance via perceived meaningfulness at work: A moderated mediation model. *The Journal of Applied Psychology*, *103*(3), 324–333. doi:10.1037/apl0000274 PMID:29016164

Green, D. D., & McCann, J. (2011). Benchmarking a leadership model for the green economy. *Benchmarking*, *18*(3), 445–465. doi:10.1108/14635771111137804

Judge, T. A., & Bono, J. E. (2000). Five-factor model of personality and transformational leadership. *The Journal of Applied Psychology*, *85*(5), 751–765. doi:10.1037/0021-9010.85.5.751 PMID:11055147

Priyankara, H. P., Luo, F., Saeed, A., Nubuor, S. A., & Jayasuriya, M. P. (2018). How does leader's support for environment promote organizational citizenship behaviour for environment? A multi-theory perspective. *Sustainability*, *10*(1), 2071–1050. doi:10.3390u10010271

Saeed, B. B., Afsar, B., Hafeez, S., Khan, I., Tahir, M., & Afridi, M. A. (2019). Promoting employee's proenvironmental behavior through green human resource management practices. *Corporate Social Responsibility and Environmental Management*, *26*(2), 424–438. doi:10.1002/csr.1694

Zhao, H., & Zhou, Q. (2019). Exploring the Impact of Responsible Leadership on Organizational Citizenship Behavior for the Environment: A Leadership Identity Perspective. *Sustainability*, *11*(4), 2071–1050. doi:10.3390u11040944

Zhu, W., Chew, I. K., & Spangler, W. D. (2005). CEO transformational leadership and organizational outcomes: The mediating role of human–capital-enhancing human resource management. *The Leadership Quarterly*, *16*(1), 39–52. doi:10.1016/j.leaqua.2004.06.001

KEY TERMS AND DEFINITIONS

Corporate Consciousness: Aggregate concept formed by the individuals that constitute an organization and their collective consciousness, also entailing in the equation the world with which such vectors interact, resulting in a relationship represented in multiple tangible and intangible ways, whose symbiosis assumes as the daily organizational culture.

Green Transformational Leadership: Behaviors of leaders who are guided by the sustainability moto, emphasizing green vision, and shaping employees' passion for pro-environmental behaviors, acting as role models to motivate followers to achieve environmental goals, inspiring them to perform beyond expected levels of environmental performance.

Organizational Citizenship Behaviors: Individual and discretionary social behaviors imprinted of an ethical character, directed at societal sustainability and well-being, contributing to improve the effectiveness of organizational sustainable management, even though not explicitly recognized by the formal organizational reward system.

Organizational Psychological Contract: Refers to the individual beliefs that are molded by the organization, concerning the terms of an exchange agreement between individuals and their organization, being at a large extent informal and implicitly rather than explicitly understood because they root on expectations.

Personality Traits: Traits reflecting people's distinctive patterns of thoughts, feelings, and behaviors, thus implying consistency and stability, reflecting basic features on which people differ. These traits reflect continuous distributions rather than distinct personality types; hence, people score relatively low or high along a continuous distribution of a given personality trait.

Positive Organizational Behavior: Examination and subsequent application of positively oriented human resources' psychological capabilities able of being measured, managed, and enhanced in order to leverage organizational performance.

Psychological Capital: Entails an individual's positive psychological state of development, defined by displaying self-assurance to prosper at challenging tasks, making a sustained and positive ascription about succeeding both in the present and in the future, and, when beset by adversity, readdressing paths, and bouncing back or even beyond in order to reach success. It is via the synergy of the individual psychological capital capacities (confidence, hope and resilience) that each adds unique variance and becomes additive to overall psychological capital.

Psychological Ownership: The state of consciousness in which individuals assume a certain target of ownership as inherently theirs, proactively assigning to themselves the accountability for a particular goal, task, or responsibility, hence experiencing feelings of concern for those.

Value Congruence: At an organizational context, refers to the consistency between an individual's and his/her organization's values, and the extent that individuals feel they have to compromise their personal values to meet organizational demands. It postulates that minimal conflict between both values results in positive outcomes for both agents, which first requires the individual to be aware of his/her own values and those of the organization.

Chapter 3
Towards Sustainable Factories

Gema Calleja
Universitat Politècnica de Catalunya, Spain

Jordi Olivella Nadal
Universitat Politècnica de Catalunya, Spain

ABSTRACT

Sustainability is becoming a major objective of factories worldwide in an effort to eliminate or reduce the use of toxic materials and greenhouse emissions, extend the life of products, reuse waste, and conserve energy, not only for meeting the needs of consumers, but also the multi-stakeholders from both within and beyond the supply chain. The increasing global demand for scarce natural resources is posing a great challenge for producing companies. In recent years, the advent of new technologies and the changes in demography and working conditions as well as the desire for individualized products has added greater complexity to manufacturing. In this context, this chapter provides an overview of the main issues affecting the industrial sector on the way ahead towards sustainability, including contemporary trends triggering the requirements for factories of the future as well as the main research and innovation lines necessary to answer such requirements.

INTRODUCTION

Sustainability has become an important issue in factories and manufacturing companies throughout the world, as well as a main driver for a positive development and prosperity of countries. Traditionally, the focus of improvement activities for industrial companies has been on generating valuable goods in defined quantities, at the highest possible quality, lowest cost and shortest time to market. In an

DOI: 10.4018/978-1-7998-5768-6.ch003

effort to achieve these requirements, industries have mostly focused on economic benefits. It has resulted in intense consumption of natural resources along with the simultaneous generation of waste and pollution (Bhanot, Rao, & Deshmukh, 2016). At the same time, in order to maintain the ecosystem strict governmental regulations are restricting industries from emitting harmful emissions to the environment (Albulescu et al., 2019). In this contradictory context, it is crucial for factories to adopt manufacturing paradigms that can achieve economic benefits while protecting the environment (Chan et al., 2016). Otherwise, governmental regulations can threat their survival in the market (Esfahbodi et al., 2016). Sustainable manufacturing is one of such paradigms, which balances and integrates environmental objectives into operational and economic aspects of the business. Sarkis (2001) emphasized that the environmental initiatives and the manufacturing function are inextricably linked. Sustainable manufacturing not only protects the environment but also adopts the processes that are beneficial for customers, suppliers, employees as well as local community and society.

Nowadays, producing companies are increasingly shifting their focus to the sustainable perspective of their activities and adding new goals of improvement. To extend the adoption of sustainability throughout the business functions, it should be strategically applied to products, processes and services of the company (Bilge et al., 2016). In the current scenario, none of the manufacturing industries can survive without incorporating sustainability in their strategic decisions of business (Bogue, 2014). This modern view of sustainability in manufacturing includes three closely connected dimensions – environment, economy and society, as shown in Figure 1 and described next.

Figure 1. The three dimensions of sustainability

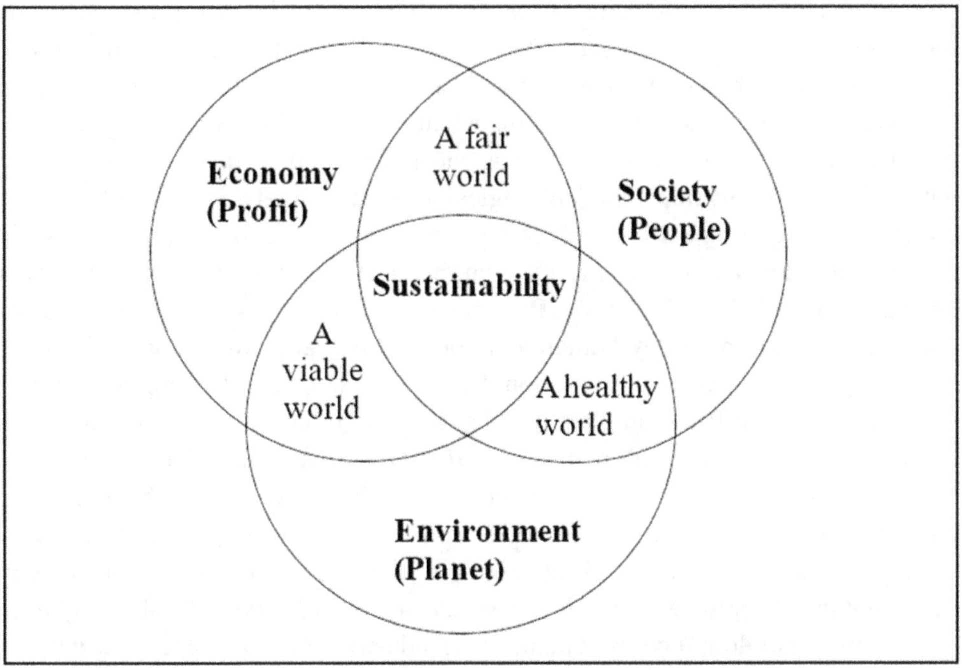

Environmental Sustainability

The environmental dimension of sustainability aims at reducing negative environmental impacts from the organization's activities, products or services. Song & Moon (2016) highlighted the positive relationship between sustainability and the manufacturing industry. The majority of research articles after 2008 define sustainable manufacturing according to the U.S. Department of Commerce as "the creation of manufactured products that use processes that are non-polluting, conserve energy and natural resources, and are economically sound and safe for employees, communities, and consumers" (Moldaska & Welo, 2017). The above definition reveals the aim of sustainable manufacturing, but to achieve it many hurdles need to be removed during the implementation stage (Xia et al., 2015). Manufacturing strongly contributes to environmental burdens such as depletion of non-renewable resources, global climate change, generation of waste materials, as well as air, land and water pollution, which question the competence and adaptableness in adopting sustainability aspects (Ocampo & Clark, 2015). These negative effects are aggravated with the challenge of coping with a limited world – an earth with a finite "carrying capacity" (Arrow et al., 1995), and expecting "limits to growth" (Meadows et al.,

1972), which will further expand along with the unprecedented growth in global population and worldwide demand for industrial goods. The disruption caused by humans on the environment has reached critical levels. On the one hand, the limited capacity of the earth's atmosphere to absorb safely the industrial emissions generated by our carbon-based economies jeopardizes natural equilibria. Global chemical flows produced by industrial activities, such as phosphorus and nitrogen, already exceed natural flows, thereby threatening biological balance (Peñuelas et al., 2013). Hence, the risk of dissipating earth's resources faster than they can be replenished, while emitting pollutants and wastes faster than the earth can absorb them is an urgent challenge to solve (D.H. Meadows, Randers, & D.L. Meadows, 2004). On the other hand, the stress imposed by human activities on natural equilibria intensifies the increasing inequality of living conditions between developed and emerging countries (Intergovernmental Panel on Climate Change, 2014). Indeed, research has shown that countries least vulnerable to negative effects of climate change are generally the highest greenhouse gases emitters and conversely those most vulnerable to climate change are the least responsible for its genesis (Althor, Watson, & Fuller, 2016).

The manufacturing sector is also responsible for a third of global energy consumption (Kepplinger, Templ, & Uppadhyaya, 2013) and 40% of greenhouse gases emissions. Global energy demand in the industrial sector is expected to increase by about 30% until 2040, mainly due to economic activity in developing countries (Davis, 2012). At the same time, although numerous initiatives have been carried out in favor of energy efficiency and renewable energy, the need for action is now greater than ever before. This is highlighted by the fact that, despite international measures to address climate change, the energy system is carbonizing due to a coal renaissance in a vast array of fast-growing countries around the world, particularly in Southeast Asia (Steckel, Edenhofer, & Jakob, 2015). As further increases in energy consumption are anticipated, the need for innovative decarbonization solutions is becoming more urgent and necessary.

In addition to its direct environmental impacts, the manufacturing industry also plays a role in sustainability since industry influences the resources consumption of its products during their life cycle (Duflou et al., 2012). This is particularly underscored by the phenomenon faced by households in developed countries called the "rise of machines". Today, the average household in developed countries may accumulate thousands of products and appliances, and, as a result, managing such a large volume of items becomes a stress factor (Arnold, Graesch, Ragazzini, & Ochs, 2012).

Economic Sustainability

The economic dimension of sustainability refers to practices that support long-term economic growth without negatively affecting social, environmental, and cultural aspects of the community. In this sense, manufacturing is a crucial sector for the development of economies. According to Central Intelligence Agency (2017), manufacturing generates 31% of the world gross domestic product (GDP). Further to the quantitative contribution of manufacturing to the GDP, whose translation into actual wealth is questionable, it has been observed that manufacturing sectors deploy specific and sequential patterns in the economic development of poor countries. Therefore, manufacturing appears to be an important enabler for the development of modern economies (Jovane et al., 2008). With today's levels of consumption and increasing resource use and environmental impacts, it seems increasingly apparent that business as usual is not viable for a sustainable future (Randers, 2012). A holistic approach is needed to address the challenges of long-term sustainability: responses to environmental changes will necessarily be developed in parallel with economic and social change (Bocken et al., 2014). According to Jackson (2009), features for a sustainable economy might be: minimize material and energy consumption, maximize societal and environmental benefit, create value from waste, deliver functionality rather than ownership, and encourage collaboration and sharing rather than aggressive competition.

These changes for a sustainable economy require a fundamental shift in the purpose and almost every aspect of a business. With careful business model redesign it is possible for enterprises and start-ups to more readily integrate sustainability into their business, as suggested by Stubbs & Cocklin (2012). In this sense, business model innovation offers the opportunity to deliver the required change through re-conceptualizing the purpose and re-thinking the value creation logic of the business (Bocken et al., 2014). Business models and business model innovation is receiving growing attention from researchers and practitioners (Spieth, Schneckenberg, & Ricart, 2014) and it is increasingly highlighted that business model innovation is a key to business success (Zott, Amit, & Massa, 2011).

Amidst the rising global sustainability pressures, new types of collaboration between firms and other stakeholders are needed (Hu et al., 2019). Looking at value creation by firms, the traditional approach based on firms acting autonomously is beginning to shift to a new approach where firms act together with parties external to the firm through informal arrangements or formal alliances. The business model may be viewed as a conceptual tool for business analysis, which takes into account these collaborative ties (Beattie & Smith, 2013).

Social Sustainability

Finally, the social dimension of sustainability focuses on the promotion of wellbeing within an organization's members today as well as into the future by fostering healthy communities. The social dimension of sustainability has often been associated with equity among people and equal opportunities (Colantonio, 2007). In a broad sense, social sustainability reflects the philosophy that all people in a community should have a decent quality of life, such as equity, security, health and access to basic needs (Littig & Griessler, 2005). Marafa (2002) highlighted the social impacts of sustainability as the social and cultural consequences to society from both short-term and long-term perspectives.

Clearly, manufacturing plays an important role in society across the globe. As evidence of its societal value, in 2017 industry was responsible for 30% of the $127.8 trillion global GDP (Central Intelligence Agency, 2017), and much of this economic performance is due to manufacturing. As a result, manufacturers produced all sort of consumer products intended to meet the needs of and improve the quality of life for society. In addition, the manufacturing sector also contributed directly to employ 23.5% of the 3.43 billion people in the world workforce (Central Intelligence Agency, 2017).

Despite the steady improvement of working conditions in the manufacturing sector over the last decades, the latest figures indicate an overall increase in the number of deaths attributed to poor working conditions: from 2.3 million deaths in 2014 to 2.8 million deaths in 2017 (Hämäläinen, Takala, & Tan, 2017). In addition, recent estimates from the World Health Organization (2018) show that the fraction of the global disease burden in the general population due to occupational risks amounts to 2.7%. Globally, industries account for 11.9% of child labor, representing 18 million people (International Labour Organization, 2017). Poor working conditions in manufacturing are a serious problem that is difficult to address. New approaches accompanied focusing on enhancing social sustainability are needed.

The last years have witnessed a significant increase of the importance of the social dimension of sustainability (Mani, Gunasekaran, & Delgado, 2018). As a result, there is a growing level of interest from both public and private companies in corporate social responsibility (Ortiz-Avram et al., 2018).

Contribution and Objective of the Chapter

Envisioning sustainable manufacturing three-dimensionally requires the exploration of many distinctive features of sustainability for a better understanding of the research field (Garetti & Taish, 2012). According to the literature review by Malek & Desai (2020), the research on sustainable manufacturing is mature but suffers from a paucity of holistic investigations. The past review studies have neglected to articulate the holistic concept of sustainable manufacturing (Pantelic, Sakal, & Zehetner, 2016) rather focusing on some specific areas including sustainability indicators (Ahmad, Wong, & Rajoo, 2019), sustainability definitions (Moldavska & Welo, 2017), sustainability assessment methods (Thies et al., 2019), and sustainable product development techniques (Gbededo, Liyanage, & Garza-Reyes, 2018).

Against this background, the objective of this chapter is to present an overview of previous research on how sustainability affects the manufacturing industry, what are the main strategies and initiatives adopted by factories in the field and which are the most promising lines of research and innovation towards sustainable manufacturing future. For this purpose, a search of academic and grey literature was conducted for relevant publications on sustainable manufacturing. The main search used Web of Science and Scopus databases. A supplementary grey literature search was conducted using the general Internet to identify governmental reports and documents not published commercially. Based on the available literature, this work identifies the main new paradigms for the adaptation of factories to sustainable manufacturing, and emphasizes the need for adopting a holistic approach to sustainability.

Structure of the Chapter

The chapter starts from the general concept and relevancy of sustainability, with a focus on its three main dimensions (environment, economy and society). Subsequently, the gaps in the literature are defined to highlight the need for an integrated approach of sustainable manufacturing. Section 2 (Definitions) provides the definition of manufacturing, sustainability and sustainable manufacturing adopted in this work. Section 3 (Sustainability strategies and initiatives) discusses the main strategies and initiatives that companies are adopting for addressing the challenges and trends arising in manufacturing and building new solutions towards a sustainable future. Drawing upon these strategies, Section 4 (The new manufacturing paradigms) describes the four manufacturing paradigms that enable the holistic makeover of factories towards sustainability, and Section 5 (Future research and innovation lines) offers suggestions for the research and innovation lines that are arising from the academic and industrial communities. Finally, Section 6 (Conclusion) gives some concluding remarks.

DEFINITIONS

Manufacturing

There is no consensus about the definition of the term "manufacturing". In this work we adopt a definition close to that given by Bonvoisin et al. (2017), but which expands the scope of manufacturing beyond production (i.e. the process of making goods). In the context of this work, manufacturing is defined as "the range of activities contributing to the fabrication or assembly of raw materials and components into products and solutions through systematic division of labor". This definition entails all required activities from the customer to the factory and back to the customer. Such activities include manufacturing operations as well as the different kinds of services that are connected to the manufacturing chain, such as the design and running of physical processes (e.g. machining), corresponding support processes (e.g. steam generation), and involved organizational processes (e.g. product development, manufacturing and factory planning, and supply chain management).

This definition considers not only discrete manufacturing but also process industries and services. The process industry concerns the production of continuous products (e.g. chemicals and refined petroleum products), as opposed to discrete manufacturing, whose outputs can be identified and measure in units rather than by weight or volume (Duflou et *al.*, 2012). Additionally, the definition adopted excludes handcraft activities by reference to the systemic division of labor.

Sustainability

The term "sustainability" is a concept that highlights the balanced maintenance of nature life and human society (Vehkamäki, 2005). The quest for sustainability has raised public awareness and become a very important issue for governments and industries worldwide.

Arguably, the most frequently used definitions related to sustainability is from the report known as "Our Common Future", made by the World Comission on Environment and Development established by the United Nations and published in 1987. This report states that "Humanity has the ability to make sustainable development to ensure that it meets the needs of the present without compromising the ability of the future generations to meet their own needs" (United Nations, 1987). Other definitions have expanded the concept of sustainability to stress the need to "create and maintain conditions under which humans and nature can exist in productive harmony, and fulfil the social, economic, and other requirements of present and future generations" (Environmental Protection Agency, 2019).

Based on these definitions, this work considers the broad sense of sustainability, consisting of three main pillars: economic development, protection of the environment and social justice.

Sustainable Manufacturing

As in the case of the term "manufacturing", it seems that there is no wide consensus on the definition of the term "sustainable manufacturing". The divergences shown by the literature mostly come from the explicit inclusion of the social dimension of sustainability. While some papers define sustainable manufacturing referring only to the environmental dimension of sustainability, some others include the three dimensions in their definition.

Following the definitions that consider all dimensions of sustainability, we adopt the definition stated by Garetti and Taisch (2012). The authors describe sustainable manufacturing as "the ability to smartly use natural resources for manufacturing, by creating products and solutions that, thanks to new technology, regulatory measures and coherent social behaviors, are able to satisfy economic, environmental and social objectives, thus preserving the environment, while continuing to improve the quality of human life".

The adopted definition emphasizes the inclusion of the three dimensions while however leaving their specific concretization open. In this view, manufacturing technology not only acts as a powerful tool for "meeting the needs of the present generation without compromising the ability of future generations to meet their own needs" (United Nations, 1987), but also as a fundamental enabling element of sustainability. Here, technology influences and interacts with the economic dimension (e.g. by allowing new business models), with the environment (e.g. by degrading/protecting nature) and with society (by fostering new living models). On another end, the "coherent social behaviors" pinpointed in this definition, calls for education as the prerequisite for consumers and the general population to address the sustainability goals through appropriated lifestyles and the correct use of products and technology (Garetti & Taish, 2012).

SUSTAINABILITY STRATEGIES AND INITIATIVES

This section discusses the main strategies and initiatives that factories are adopting to prepare for a challenging future ahead. Companies face an increasingly complex situation as governmental organizations issue more strict environmental regulations and standards, which have a deep impact on the sustainability attitude to business and creating compelling challenges for compliance (Garetti and Taish, 2012).

Sustainable Business Models

A business model is "a conceptual tool to help understand how a firm does business and can be used for analysis, comparison and performance assessment, management, communication, and innovation" (Osterwalder, Pigneur, & Tucci, 2005). Business models describe how a company defines its competitive strategy through the characteristics of the product or service it offers to its target market, what it charges for it, what it costs to produce, how it differentiates itself from other firms by the value proposition, and how the company integrates its own value chain with those of other companies (Rasmussen, 2007).

To be sustainable, the business models of the future need to be oriented to a global market, considering not only the development of emerging countries but also the need for more sustainable products and services. Indeed, the dramatic rise of the five countries collectively referred to as the BRICS (Brazil, Russia, India, China and South Africa) countries, is irrevocably changing the global economy. This would imply for the European industry new fast-growing markets on one hand, but intensified competition and lower prices on the other. In this context, the current business model based on the idea of maximizing product sales is neither adequate nor sustainable and needs to be revised, taking into account not only cost, service and quality but also sustainability. One possible solution lies in the increased implementation of new business models based on "Product-as-a-Service systems" (also called product-service systems), along with the use of lean manufacturing techniques.

Product-as-a-Service is a business model that provides service in areas that were traditionally sold as physical products. Here, the producer is responsible for managing and optimizing the usage phase, thus assuming a much relevant role beyond merely manufacturing, and allowing a bigger income stream by selling services instead on products only (Meier, Roy, & Selinger, 2010). This new role involves ongoing interaction with customers including support, and can be one of the solutions for addressing differentiation needs. Customers may engage to product-service models due to flexibility, enhanced support, reduced upfront costs and lower risks. For instance, in the case of joining a car sharing service, the customer is not responsible for maintenance and has lower upfront costs as compared with purchasing a car. Besides, it can help suppliers to provide both physical goods as well as complementary services to their customers, by working together with manufacturers on the configuration of product-as-a-Service systems (Legnani, Cavalieri, & Ierace, 2009).

Despite the big potential of product-as-a-Service systems from an industrial perspective, such systems are not easy to implement. This is especially true for small and medium-sized enterprises (SMEs), due to the increased complexity in the model and the need of a new organization (including new customer relationship, new contracts based on expected usage and new financial models). Because of the important contribution of SMEs to the manufacturing sector, appropriate methodologies and business models need to be developed to increase SME's sustainability, while minimizing their inefficiencies (Garetti and Taish, 2012). Additionally, the increased service activity of the new system requires a higher consumption of resources to develop this activity, thus efficiency in service is particularly needed. To this concern, lean management, which has been used in the improvement of manufacturing industries, stands out as a promising approach for service industries as well. Specifically, lean management helps businesses optimize their processes by reducing waste – activities that consume time and resources but add no value to the customer – so that the service flows efficiently. The aim is to attain maximum business performance with the fewest resources.

Sustainable Manufacturing Processes

To cope with demand uncertainties, sustainable manufacturing companies need to be able to adjust output quickly to match market demands. To this end, flexible production systems may be a solution for smoothing demand uncertainties. Two examples of flexible production systems are "mass-customization" and "build-to-order" paradigms. Such manufacturing paradigms are suitable for enhancing the performance and productivity of manufacturing systems, and at the same time are highly flexible and adaptive to both volume and variants of products. This higher flexibility and adaptability can be a powerful tool to mitigate process inefficiency in case of changing conditions, thus being advantageous for sustainable performances. To achieve such sustainable objectives, a holistic approach is needed to configure flexible and adaptive manufacturing systems and processes that integrate all aspects of the life cycle of the product, from conception and assembly to use and disposal of the product and its components.

Adaptability of manufacturing systems can be achieved in two main directions: by developing models to allow optimal performance through process control, and by developing improved equipment (e.g. cooperative communities of robots capable of collaborating among them and with human workers).

In addition, it is also important to allow the re-configurability of manufacturing systems (i.e. the ability of the system to achieve cost-effective and rapid system changes regarding complexity and variety of products), particularly in the case of the introduction of new products. Some examples of re-configurable approaches include the so-called "plug-and-interoperate" devices, interoperable manufacturing platforms, fast modeling and simulation tools, and re-configurable assembly and disassembly systems (ElMaraghy, 2005).

At the factory level, a holistic design approach is needed to achieve a breakthrough in factory sustainability. The design of the factory should integrate all its elements, from machine level up to the company level. Incremental innovation is not sufficient to reach the objective to become a near-to-zero-emissions factory and may lead only to marginal environmental improvements. According to (Fantini, Palasciano, & Taisch, 2015), to achieve sustainable manufacturing processes new developments should address the integration of many aspects from a holistic and multi-disciplinary perspective, including:

- high-efficient plant control methods.
- equipment for energy saving, recovery and harvesting.
- modeling and simulation tools for the design of flexible and adaptive manufacturing systems.
- the concurrent design of the factory building.
- the plant layout and machines configuration.
- the development of a structured set of key performance indicators (KPI) able to real-time capturing the multi-dimensional objectives of the sustainable factory.

Sustainable Product Design and Life Cycle Management

A product life cycle consists of three major stages: beginning of life (where products are designed and manufactured), middle of life (where products are distributed, used and maintained) and end of life (where products are retired by the company in order to be recycled or to be disposed) (Kiritsis, Nguyen, & Stark, 2008).

The sustainability of products during their life cycle involves three main approaches, namely:

- the development of product design activities aimed at assuring a sustainable life cycle of the product.
- the integration of information within the product life cycle for achieving sustainability.
- the development of the so-called intelligent products to allow better communication between suppliers and users.

These approaches are described next.

Firstly, from the point of view of sustainable design, different methodologies are available for designing machines, products, as well as production technologies and services. A well-known approach is Design for X, which has evolved from design for assembly and manufacture, in order to consider environmental concerns during the design stages, such as energy saving, disassembly and recycling issues (Kuo, Huang, & Zhang, 2001). Other techniques address a systematic and general life cycle thinking process, from an environmental perspective, to define improvement strategies for the whole product life cycle. An illustrative example is life cycle product modeling and simulation. By means of appropriate software, designers and product developers are able to model and simulate the virtual product behavior during its entire life cycle, including the interactions of people involved with the product itself (Garetti, Rosa, & Terzi, 2012). This can help to estimate the impacts and costs of the whole life cycle. Another way for companies to cut costs and increase customer retention is to involve customers in the development process of new products and services. By doing so, companies will be able to build up their own business-to-consumer communities and not only save money, but also meet their customer requirements for green products.

The use of integrated data management software with product sustainability can offer a support decision system for managing the product during its entire life cycle while taking into account environmental practices into their business strategy. To this end, product life cycle management software (PLM) is being used to connect people, processes and data across the product lifecycle to a central information database. This allows the management of complex product information, engineering processes, manufacturing workflows, and collaboration. For example, a PLM software can be linked to supplier management applications to identify and source from suppliers that incorporate sustainable practices.

Finally, the embedment of technologies such as radio frequency identification (RFId), sensors and wireless networks into products has led to the development of the so-called “intelligent products”. An intelligent product can be integrated into a network and exchange data with its environment, manufacturer, user, and other products and systems. Furthermore, the integration of intelligent products into a PLM system can greatly enhance its efficiency thanks to the traceability of the product, which is the ability to capture, store and transmit information along the product life cycle. The volume of accumulated data as the life cycle evolves can be significantly high, thus integration with a PLM system can allow the efficient management of life cycle data. Furthermore, product-related data arising from its life cycle can be used to extract knowledge useful to enhance actual or future products.

Scarce Resources Management

The increasing scarcity of the most important raw materials asks for reuse of materials and new recycling technologies. For instance, waste materials should be reused in supply chains as raw materials or as a source of energy. Factories will need to focus on efficient recycling technologies based on reuse of components, combined with more efficient collecting systems. In particular, factories will need to develop methodologies and tools to improve materials reuse after disposal, with especial emphasis on electronic products. Indeed, the disposal of electronic products has a high environmental impact because of their manufactured materials are being prematurely trashed because of obsolescence.

Thus, in order to achieve a sustainable factory, it is necessary to adopt a life cycle comprehensive approach, using advanced recycling technologies, with the aim to optimize the use and reuse of products, as well as their disposal impact. Packaging is an important form of waste in manufacturing. For this reason, it is crucial to mitigate its effects by developing reusable, biodegradable or even edible packaging, while considering existing standards and regulations, and optimizing packaging sustainability from both an economic and ecological perspective.

Energy- and Emissions-Efficient Manufacturing

Among the different resources consumed by factories, energy is crucial for achieving sustainable manufacturing due to the impact of fossils fuels (its dominant source) on the environment. Manufacturing represents one third of the global energy consumption (Kepplinger, Templ, & Upadhyaya, 2013) and about 40% of CO_2 emissions worldwide. The transport sector, households and the services industry represent the rest (Davis, 2012). In addition, for a factory to be sustainable, it has to fulfil increasingly strict energy consumption and emissions regulations, and reduce its environmental and health impact by exploiting renewable energy resources, co-generation and industrial synergies, as well as reusing any available source. Therefore, sustainable manufacturing requires factories to assure limited energy consumption of industrial processes, while producing limited waste and reducing the consumption amount of natural resources.

From the point of view of energy efficiency, the discrete manufacturing industry and the process industry are different. Process industry focuses on energy-intensive processes, whose efficiency has been under focus and improved for many years, as it represents the largest share of energy consumption among all industries. Today, the use of advanced computer control systems allows these processes to be highly efficient, and, while a marginal improvement would still be possible, the real possibility for reducing energy consumption in the process industry relies in the creation and development of new production processes (Garetti & Taish, 2012). The scenario is completely different in the discrete manufacturing case, where the control of energy use is very poor and the room for improvement due to a better control is very high. However, the distributed and fragmented nature of manufacturing processes has made it very difficult to address these improvements until now. On the one hand, the development of information and communication technologies (such as RFId, computer systems, and wireless networks) provides a powerful tool to solve this kind of issue. On the other hand, the application of lean approaches on energy and materials consumption can also lead to further improvement.

At the company level, the advantages of achieving energy-efficient manufacturing are manifold (European Comission, 2009):

- An improvement in the customers' perception of the company and their intention to purchase green products (i.e. products which not only are biodegradable, easy to reuse or energy efficient when used by the customer, but also which have been obtained using as little energy as possible).
- Costs savings in spite of increased energy prices.
- Fulfilment of environmental regulations.
- Reduced dependence from energy suppliers.

In this context, the European Commission adopted the environmental target of reducing global emission of greenhouse gases to 80-95% by 2050 (European Comission, 2011). According to this vision, reducing the greenhouse emissions is an urgent matter of utmost importance. Four major actions with respect to energy-efficiency for mitigating the carbon footprint can be highlighted (Garetti & Taish, 2012):

1. *New energy sources for factories*. In order to reduce both energy consumption and dependence from energy suppliers, it is necessary to develop energy autonomous factories for facilitating the use of renewable energy sources and for enabling the self-generation of energy. Another important issue concerns the development of new and innovative solutions for energy harvesting, with the aim to convert ambient energy (e.g. solar, thermal, wind, kinetic) to electricity.
2. *More efficient production processes*. A considerable share of industrial energy and resource consumption corresponds to grinding processes, a collective term for machining processes that utilize hard, abrasive particles to shape and finish components made of metal and other materials. The development of new concepts and technologies, such as pre-treatments or flexible systems, can improve the energy efficiency of these processes. Furthermore, the efficient use of raw materials in manufacturing, distribution and disposal reduces costs and energy consumption, and, with this, reduces greenhouse emissions. By encouraging the development of "zero-waste" technologies in manufacturing, it is possible to reduce substantially the total energy and resources consumption, as it is related to the quantity of materials used in the entire supply chain.
3. *Collaborative energy use*. Energy dissipation lost in the form of unwanted heat or by products can be considered as a waste output of the production process.

However, by enabling collaboration across different types of industries, a viable reuse can often be possible in another production process if we can effectively integrate the two types of processes. Mitigation actions are especially important in resource-intensive industries, as they contribute significantly to greenhouse emissions. To this end, the enhanced utilization of waste as alternative fuels can replace natural resources and, with this, reduce the environmental impact of resource-intensive industries. Additionally, collaboration between plants in process industries can also be a solution to apply waste heat recovery to identify opportunities for reuse.

4. *More efficient management and control of energy consumption.* Information about where and how energy is used is an essential prerequisite for any improvement action. By using the information of sensors with systems for in-process measurement, factories can develop effective energy control systems. This can provide valuable information that can be further used for improvement actions in energy efficiency. To do so, energy control systems must be integrated in information systems (e.g. enterprise resource planning systems, ERP) in order to conceive integrated systems that are able to optimize energy consumption with respect to manufacturing planning and control.

Sustainable Supply Chain

When considering sustainability at the supply chain level, it is necessary to adopt new organization and management models, due to the prospect of increasingly complex and efficient supply chains arising for better products and services. Such new models should consider the trend that more and more of the value-adding chain is being transformed into a knowledge-sharing network, where fruitful cooperative clusters are expected to develop. Additionally, the sustainability trade-offs of alternatives available in the global market, such as de-location, co-location or outsourcing, should be also considered.

Another relevant issue to take into account in the organization of a sustainable supply chain is the participation of SMEs. Although the individual environmental impact of small businesses may be low, their aggregate effects can exceed that of some large companies. Thus reducing the environmental impact of SMEs by achieving more sustainable patterns of production and consumption is a key factor for success in the transition to more sustainable supply chains. This is especially urgent for SMEs in the manufacturing sector, which is responsible for most of the world's consumption of resources, environmental pollution and generation of waste (Organization for Economic Cooperation and Development [OECD], 2013). On the other hand, SMEs are frequently integrated in supply chains as producers. This opens up business opportunities as suppliers of green goods and services. SMEs are in a particularly advantageous position to seize opportunities of sustainable supply chains in local clean tech markets, which may be unattractive or impenetrable for large companies (Ehst, Kesrelioglu, & Lee, 2014). However, for SMEs it is often difficult to adopt sustainable practices and green business opportunities due to size-relates resource and financial constraints. Environmentally sustainable initiatives in SMEs are often held back by perceived technical complexities and costs, as well as a lack of awareness about financial opportunities and a lack of appropriate expertise to seize business opportunities. As a result, SMEs are often more risk-averse and less willing to invest in new and green technologies than larger firms (OECD, 2018). To this concern, the collaboration on different levels along the supply chain, like the coordinated design and development of products, common purchase, and coordinated sales and services are important potential streams for innovation in the transition to more sustainable supply chains.

The trend of manufacturing companies focusing on specialized segments of the product value chain is leading to inefficient increase of stock in the supply chain. To face this inconvenience, a multi-tier view of the supply chain is necessary. This can be achieved by integrating new information tools for higher and more transparent information flows along the supply chain, as well as self-optimizing control mechanisms to harmonies the logistics of stock, optimize inventory levels and improve the overall performance of the supply chain.

Finally, when companies simultaneously take part of several production networks, collaborative planning and optimization of resources is needed to coordinate the production activities and ensure robust performance against the propagation of unexpected disruptions across the networked enterprise.

THE NEW MANUFACTURING PARADIGMS

This section describes the profound changes and the new manufacturing paradigms that factories need to undertake in order to achieve sustainability. New changing paradigms in manufacturing entail a holistic makeover of factories, from production plant to supply chain and from employee to stakeholder. The objective of sustainable factories is to enable clean and competitive manufacturing systems and create opportunities based on sustainability issues to support corporate strategy and spur innovation. The discussion around climate change has increased environmental concerns among consumers, academia, industry and policy-makers and heads for a changing manufacturing paradigm towards a holistic factory of the future as described in the following.

Table 1. Manufacturing paradigm shifts towards a holistic vision of factories of the future (Herrmann et al., 2014)

Paradigm	Today	Tomorrow	Description
1. Symbiotic flows and urban integration	Separate focus on building, process and product efficiency	Efficiency of the whole system	• Increasing consumer interest in green products • Need to understand, analyze and improve the whole system
	Separate work and life spheres	Intelligent cities where everyone comes together	• Return of factories to urban areas • People working and producing from home • Overall positive influence of the factory on its surroundings • Resource and energy efficiency of the whole city
	Unilateral view of factories	Multi-disciplinary and integrative view of factories	• No standard KPIs • No standard global approaches • A continuous cycle of design, production and sales • Collaborative product development fueled by crowd sourcing • Utilization of life-cycle view strategies to the factory
2. Adaptive factory elements	Stiff infrastructure	Flexible infrastructure	• New requirements on the factory infrastructure (e.g. smaller, expandable or mobile) • Flexible production systems
3. Production cloud and cyber-physical systems	Production concentrated on one site	Distributed manufacturing opportunities	• Production combining manufacturing and information technologies • Production networks that can be connected on demand • Flexible small scale production
4. Learning and training environments	Manufacturing factories	Learning factories	• Continuous training of highly skilled and educated employees • Factories acting as experimental and research spaces (e.g. FabLabs)

(May, Stahl, & Taisch, 2016)

The factory of the future needs to be highly adaptive and flexible to external changes while achieving environmental sustainability. More than ever, future manufacturing needs to address all three pillars of sustainability – environmental, economy and society. The economic dimension asks for a higher profitability of production. From the environmental perspective, the ecological impact of manufacturing should be mitigated, leading to zero emissions or even a positive influence of the production site on its surroundings, reusing and exploiting local waste flows, generating energy from renewable sources and improving the quality of local air and water. From the social perspective, factories act as a place for people, focusing on their collaborative learning and training. Responding to these needs and addressing the trends presented in the previous Section, a possible vision for a holistic factory of the future is summarized in Table 1 (Herrmann et al., 2014), (May, Stahl, & Taisch, 2016).

Table 1 shows the four main changing manufacturing paradigms that are leading to the creation of sustainable factories. These new manufacturing paradigms comprehend a 360-degree makeover of factories, as described next:

1. *Symbiotic flows and urban integration*. Sustainable factories of the future will have a multi-disciplinary and integrative view of their efficiency. They will share material and energy flows not only with industry, but also with households and urban infrastructure. The impact of the factory on its surroundings will be positive; waste will be used to create new products, renewable energies will be utilized with neutral local emissions. This will allow factories of the future to return to urban areas.
2. *Adaptive factory elements*. In order to link material and energy flows with the building shell and to cope with the increasing trend of personalized products, factories will need to shift towards a highly adaptive and flexible physical factory infrastructure.
3. *Production cloud and cyber-physical systems*. As information technologies and sensors will further spread at all factory levels, the collection and processing of all information on a decentralized cloud will be reasonable. The cloud can also serve as a manufacturing network of individualized products, where every factory of the network can offer its services with respect to available capacity, product quality and delivery times, allowing for flexible small-scale production.
4. *Learning and training environments*. To keep pace with changing external requirements and technological progress, future factories will have to continuously train highly skilled employees. In this context, manufacturing factories are changing into learning factories, as they provide experimental and research environments allowing the communication and testing of knowledge in practice. Factories may also contain digital fabrication laboratories (FabLabs), which allows for the creation of new products at low cost.

FUTURE RESEARCH AND INNOVATION LINES

Currently, innovation is mainly focused in the technical aspects of the factory (May, Stahl, & Taisch, 2016). However, to fully reap the benefits of sustainability it is critical to develop a philosophy of change encompassing both the managerial level and the employees, in order to provide a vision and a direction for the company. Furthermore, companies need to embrace the mindset of sustainable factories and be guided and mentored throughout the whole organization in order to engage as many people as possible and increase implementation success.

Research and innovation for sustainable factories include five main lines that are not disjoint and must be considered together to build the holistic vision of factories of the future (Terkaj & Tolio, 2019), (European Commission, 2020):

1. *Evolutionary and re-configurable factory*. Factories will be required to evolve and re-configure themselves at all levels, from the single shop floor to the global supply chain, in order to cope with complex and variable market demands together with the technological evolution of products and processes. Specific research and innovation lines in this area may cover the development of new methodologies and tools to design and implement re-configurable production systems, as well as automation and control systems. The research for the development of methodologies to support the integrated design of the evolving products-processes-systems is also needed. Finally, another key research line in this area is the optimization approaches for performance improvement (e.g. reduction of manufacturing set-up times).
2. *Sustainable factory*. The need of new generation of factories to address all three dimensions of sustainability heads for research and innovation for the development of new materials and production technologies to exploit renewable sources and waste processes. Methodologies and tools, as well as human-robot interaction and artificial intelligence, need to be developed for the modeling, design and management of processes, considering both an efficient consumption of resources, the whole life cycle of products and the de-manufacturing or disassembly of products after their use. Finally, there is a need to research and create new business models for an efficient consumption and exploitation of resources (e.g. by offering targeted services).
3. *Factory for the people*. Societal and demographic changes such as the global aging of the population and the increase in the retirement age need to be considered in the design and management of future factories. The continuous improvement of robotics and automation technologies asks for an increasingly relevant role of human-machine interactions, as well as safety and ergonomics, in the future of manufacturing. In this context, the following research and innovation lines with respect to social sustainability are required: i) the development of technologies to improve working conditions due to ergonomic studies, ii) risk reduction of dangerous processes thanks to higher automation and control, iii) more effective training using virtual reality, and iv) reduction of air, land and water pollution. Research on development of smart machines collaborating with humans is also needed to exploit human skills in changing working conditions, along with the development of adaptive human-machine

interfaces (e.g. voice recognition, autonomous moving devices) to effectively support this human-machine collaboration.

4. *Factory for customized and personalized products.* The production of products and services based on a full personalization of the specific customer needs represents an important strategic opportunity to compete in the global market with high value added-products that are difficult to replicate. This requires a shift from high production volumes to factories able to offer one-of-a-kind products. In order to do so, factories will need to address research and innovation with respect to the development of new business models and ICT tools to support product personalization, as well as new approaches for the design, management and collaboration of single factories and supply chains aimed at the production and maintenance of personalized products along their life cycle.
5. *Advanced-performance factory.* In order to meet the demand of highly personalized products, factories will need to minimize the inefficiencies of production systems and processes by properly collecting and monitoring data. Necessary enabling technologies include advanced sensors, robotic components, ICT platforms and digital twins. Combined with innovative techniques of data fusion and artificial intelligence, factories will be able to autonomously identify causes of failures, implement preventive strategies and operate production conditions to achieve high efficiency and zero defects. Specific research and innovation topics include the development of new advanced-performance production and transportation systems and processes, models and platforms for data fusion, and new methodologies and solutions to support the design and modeling of quality control systems, as well as to continuously monitor the production systems performance.

CONCLUSION

The literature on sustainable manufacturing is mature but fragmented, with review studies mainly focusing on specific areas rather than a holistic and integrated approach. In order to address this gap, this chapter provided an overall, three-dimensional (i.e., environmental, economic and social) overview on the field, and identified the new manufacturing paradigms with the aim to unify these disparate contributions on sustainable manufacturing under a common theme.

Factories will need to create new kinds of products, processes and business models to face the evolving trends and the challenges arising from sustainable manufacturing. Such trends include customers demanding personalized products, the need for green production systems reducing their carbon footprint, the advent of information technologies enabling the development of smart factories and products, an increasing relevance of social aspects related to production learning and training, as well as the need for integration of factories within their spatial context.

In order to respond efficiently to such challenges and trends, a thorough understanding of a factory system is required. This chapter presented a vision of a holistic manufacturing company comprehending four main changing paradigms, beginning with the synergic industrial flows and the integration of the factory with its urban environment, as well as the integrated and multi-disciplinary focus on the efficiency of the whole system, leading to a positive combination of ecology and economy. Concerning the production elements and infrastructure of the factory more flexible production systems are needed, allowing adaptive characteristics for the elements of the building itself and the production system. In order to properly manage this system, a production cloud paradigm together with a broad application of ICT technologies is required, to allow the decentralized gathering and usage of data. Finally, human skills and abilities being one of the key success factors of future factories, learning and training will be increasingly important to keep pace with changing external requirements and technological developments.

Ultimately, the holistic view of the factory represents a global response to the aforementioned trends and challenges by addressing all three pillars of sustainability, shifting from the current rather economic focus towards the inclusion of ecological and social dimensions. This holistic perspective of future factories based on changing manufacturing paradigms can be viewed as a starting point to broaden and unify the research agenda for sustainable manufacturing.

REFERENCES

Ahmad, S., Wong, K. Y., & Rajoo, S. (2019). Sustainability indicators for manufacturing sectors: A literature survey and maturity analysis from the triple-bottom line perspective. *Journal of Manufacturing Technology Management*, *30*(2), 312–334. doi:10.1108/JMTM-03-2018-0091

Albulescu, C. T., Artene, A. E., Luminosu, C. T., & Tămăşilă, M. (2020, September). **Arte**ne, A. E., Luminosu, C. T., & Tamaşila, M. (2019). CO 2 emissions, renewable energy, and environmental regulations in the EU countries. *Environmental Science and Pollution Research International*, *27*(27), 33615–33635. Advance online publication. doi:10.100711356-019-06155-1 PMID:31473925

Althor, G., Watson, J. E., & Fuller, R. A. (2016). Global mismatch between greenhouse gas emissions and the burden of climate change. *Scientific Reports*, *6*(1), 20281. doi:10.1038rep20281 PMID:26848052

Arnold, J. E., Graesch, A. P., Ragazzini, E., & Ochs, E. (2012). Life at home in the twenty-first century: 32 families open their doors. Los Angeles, CA: The Cotsen institute of Archaeology Press. doi:10.2307/j.ctvdjrr2c

Arrow, K., Bolin, B., Costanza, R., Dasgupta, P., Folke, C., Holling, C. S., Jansson, B., Levin, S., Mäler, K., Perrings, C., & Pimentel, D. (1995). Economic growth, carrying capacity, and the environment. *Ecological Economics, 15*(2), 91–95. doi:10.1016/0921-8009(95)00059-3 PMID:17756719

Beattie, V., & Smith, S. J. (2013). Value creation and business models: Refocusing the intellectual capital debate. *The British Accounting Review, 45*(4), 243–254. doi:10.1016/j.bar.2013.06.001

Bhanot, N., Rao, P. V., & Deshmukh, S. G. (2017). An integrated approach for analyzing the enablers and barriers of sustainable manufacturing. *Journal of Cleaner Production, 142*(4), 4412–4439. doi:10.1016/j.jclepro.2016.11.123

Bilge, P., Badurdeen, F., Seliger, G., & Jawahir, I. S. (2016). A novel manufacturing architecture for sustainable value creation. *CIRP Annals, 65*(1), 455–458. doi:10.1016/j.cirp.2016.04.114

Bocken, N. M., Short, S. W., Rana, P., & Evans, S. (2014). A literature and practice review to develop sustainable business model archetypes. *Journal of Cleaner Production, 65*, 42–56. doi:10.1016/j.jclepro.2013.11.039

Bogue, R. (2014). Sustainable manufacturing: A critical discipline for the twenty-first century. *Assembly Automation, 34*(2), 117–122. doi:10.1108/AA-01-2014-012

Bonvoisin, J., Chang, J., Neugebauer, S., Lehmann, A., Scheumann, R., & Finkbeiner, M. (2017). Glossary. In Sustainable Manufacturing. Challenges, Solutions and Implementation Perspectives (pp. 277–283). New York, NY: Springer.

Central Intelligence Agency. (2017). *The World Factbook.* Retrieved November 14, 2019 from https://www.cia.gov/library/publications/the-world-factbook/geos/xx.html

Chan, F. T. S., Li, N., Chung, S. H., & Saadat, M. (2016). Management of sustainable manufacturing systems-a review on mathematical problems. *International Journal of Production Research, 55*(4), 1210–1225. doi:10.1080/00207543.2016.1229067

Colantonio, A. (2007). Social sustainability: an exploratory analysis of its definition, assessment, methods, metrics and tools. In *Measuring social sustainability: Best practice from urban renewal in the EU 2007/01* (EIBURS Working paper series). Oxford, UK: Oxford Institute for Sustainable Development, Oxford Brookes University.

Davis, S. (2012). *Success Stories in Industrial Energy Efficiency*. Retrieved from https://ccap.org/resource/success-stories-in-industrial-energy-efficiency/

Duflou, J.R., Sutherland, J.W., Dornfeld, D., Herrmann, C., Jeswiet, J., Kara, S., Hauschild, M., & Kellens, K. (2012). Towards energy and resource efficient manufacturing: A processes and systems approach. *CIRP Annals - Manufacturing Technology*, *61*(2), 587–609. doi:10.1016/j.cirp.2012.05.002

Ehst, M., Kesrelioglu, S., & Lee, H. (2014). *Building competitive green industries: the climate and technology opportunity for developing countries*. World Bank Group. Retrieved from https://documents.worldbank.org/curated/en/837201468165876160/Building-competitive-green-industries-the-climate-and-clean-technology-opportunity-for-developing-countries

ElMaraghy, H. A. (2005). Flexible and reconfigurable manufacturing systems paradigms. *International Journal of Flexible Manufacturing Systems*, *17*(4), 261–276. doi:10.100710696-006-9028-7

Environmental Protection Agency. (2019). *Learn About Sustainability*. Retrieved from: https://www.epa.gov/sustainability/learn-about-sustainability

Esfahbodi, A., Zhang, Y., Watson, G., & Zhang, T. (2016). Governance pressures and performance outcomes of sustainable supply chain management - an empirical analysis of UK manufacturing industry. *Journal of Cleaner Production*, *155*(2), 66–78. doi:10.1016/j.jclepro.2016.07.098

European Commission. (2009). *Ecodesign requirements for energy-related products directive 2009/125/EC*. Retrieved from: https://eur-lex.europa.eu/eli/dir/2009/125/oj

European Commission. (2011). *Communication from the Commission to the European Parliament, The Council, the European Economic and Social Committee and the Committee of the Regions – Energy Roadmap 2050 (COM (2011) 885/2)*. Retrieved from: https://eur-lex.europa.eu/LexUriServ/LexUriServ.do?uri=COM:2011:0885:FIN:EN:PDF

European Commission. (2020). *IMS2020: Supporting global research for IMS 2020 Vision*. Retrieved from: https://cordis.europa.eu/project/id/233469

Fantini, P., Palasciano, C., & Taisch, M. (2015). Back to Intuition: Proposal for a Performance Indicators Framework to Facilitate Eco-factories Management and Benchmarking. *Procedia CIRP*, *26*, 1–6. doi:10.1016/j.procir.2014.07.099

Garetti, M., Rosa, P., & Terzi, S. (2012). Life cycle simulation for the design of product–service systems. *Computers in Industry*, *63*(4), 361–369. doi:10.1016/j.compind.2012.02.007

Garetti, M., & Taisch, M. (2012). Sustainable manufacturing: Trends and research challenges. *Production Planning & Control: The Management of Operations*, *23*(2-3), 83–104. doi:10.1080/09537287.2011.591619

Gbededo, M. A., Liyanage, K., & Garza-Reyes, J. A. (2018). Towards a Life Cycle Sustainability Analysis: A systematic review of approaches to sustainable manufacturing. *Journal of Cleaner Production*, *184*, 1002–1015. doi:10.1016/j.jclepro.2018.02.310

Hämäläinen, P., Takala, J., & Tan, B. K. (2017). Estimates of occupational accidents and work-related illnesses 2017. In *Proceedings of XXI World Congress on Safety and Health at Work*. Singapore: Workplace Safety and Health Institute.

Herrmann, C., Schmidt, C., Kurle, D., Blume, S., & Thiede, S. (2014). Sustainability in manufacturing and factories of the future. *International Journal of Precision Engineering and Manufacturing-Green Technology*, *1*(4), 283–292. doi:10.100740684-014-0034-z

Hu, H., Huang, T., Cheng, Y., & Lu, H. (2019). The evolution of sustainable business model innovation: Evidence from a sharing economy platform in China. *Sustainability*, *11*(15), 4207. doi:10.3390u11154207

Intergovernmental Panel on Climate Change. (2014). *Climate Change 2014: Synthesis Report. Contribution of Working Groups I, II and III to the Fifth Assessment Report of the Intergovernmental Panel on Climate Change* (R. K. Pachauri & L. A. Meyer, Eds.). IPCC.

International Labour Organization. (2017). *Global Estimates of Child Labour, Results and Trends, 2012-2016*. Retrieved from: https://www.ilo.org/global/publications/books/WCMS_575499/lang--en/index.htm

Jackson, T. (2009). *Prosperity without Growth: Economics for a Finite Planet*. Earthscan Publications. doi:10.4324/9781849774338

Jovane, F., Yoshikawa, H., Alting, L., Boer, C. R., Westkamper, E., Williams, D., Tseng, M., Seliger, G., & Paci, A. M. (n.d.). The incoming global technological and industrial revolution towards competitive sustainable manufacturing. *CIRP Annals - Manufacturing Technology*, *57*(2), 641–659. doi:10.1016/j.cirp.2008.09.010

Kepplinger, D., Templ, M., & Upadhyaya, S. (2013). Analysis of energy intensity in manufacturing industry using mixed-effects models. *Energy*, *59*(15), 754–763. doi:10.1016/j.energy.2013.07.003

Kiritsis, D., Nguyen, V. K., & Stark, J. (2008). How closed-loop PLM improves Knowledge Management over the complete product lifecycle and enables the factory of the future. *International Journal of Product Lifecycle Management*, *3*(1), 54–77. doi:10.1504/IJPLM.2008.019970

Kuo, T., Huang, S. H., & Zhang, H. C. (2001). Design for manufacture and design for 'X': Concepts, applications, and perspectives. *Computers & Industrial Engineering*, *41*(3), 241–260. doi:10.1016/S0360-8352(01)00045-6

Legnani, E., Cavalieri, S., & Ierace, S. (2009). A framework for the configuration of after-sales service processes. *Production Planning & Control: The Management of Operations*, *20*(2), 113–124. doi:10.1080/09537280802707522

Littig, B., & Griessler, E. (2005). Social sustainability: A catchword between political pragmatism and social theory. *International Journal of Sustainable Development*, *8*(1), 65–79. doi:10.1504/IJSD.2005.007375

Malek, J., & Desai, T. N. (2020). A systematic literature review to map literature focus of sustainable manufacturing. *Journal of Cleaner Production*, *256*, 120345. doi:10.1016/j.jclepro.2020.120345

Mani, V., Gunasekaran, A., & Delgado, C. (2018). Enhancing supply chain performance through supplier social sustainability: An emerging economy perspective. *International Journal of Production Economics*, *195*, 259–272. doi:10.1016/j.ijpe.2017.10.025

Marafa, L. M. (2002). Socio-ecological impact and risk assessments in the urban environment: A multidisciplinary concept from Hong Kong. *The Environmentalist*, *22*(4), 377–285. doi:10.1023/A:1020723132203

May, G., Stahl, B., & Taisch, M. (2016). Energy management in manufacturing: Towards eco-factories of the future - A focus group study. *Applied Energy*, *164*, 628–638. doi:10.1016/j.apenergy.2015.11.044

Meadows, D. H., Meadows, D. L., Jorgen, R., & Behren, W. H. (1972). *The Limits to Growth; A Report for the Club of Rome's Project on the Predicament of Mankind*. Universe Books., doi:10.1349/ddlp.1

Meadows, D. H., Randers, J., & Meadows, D. L. (2004). *Limits to growth: The 30-year update*. Chelsea Green Publishing.

Meier, H., Roy, R., & Selinger, G. (2010). Industrial Product-Service Systems – IPS^2. *CIRP Annals - Manufacturing Technology*, *59*(2), 607–627. doi:10.1016/j.cirp.2010.05.004

Moldavska, A., & Welo, T. (2017). The concept of sustainable manufacturing and its definitions: A content-analysis based literature review. *Journal of Cleaner Production*, *166*, 744–755. doi:10.1016/j.jclepro.2017.08.006

Ocampo, L. A., & Clark, E. E. (2015). A sustainable manufacturing strategy framework: The convergence of two fields. *Asian Academy of Management Journal*, *20*(2), 29–57. doi:10.1108/SCM-01-2014-0007

Organization for Economic Cooperation and Development. (2013). *Green entrepreneurship, eco-innovation and SMEs*. Retrieved from https://oecdinfo.oecd.org/info.aspx?app=OLIScoteEN&Ref=CFE/SME(2011)9/FINAL

Organization for Economic Cooperation and Development. (2018). *Environmental Policy Toolkit for SME Greening in EU Eastern Partnership Countries*. OECD Publishing.

Ortiz-Avram, D., Domnanovich, J., Kronenberg, C., & Scholz, M. (2018). Exploring the integration of corporate social responsibility into the strategies of small-and medium-sized enterprises: A systematic literature review. *Journal of Cleaner Production*, *201*, 254–271. doi:10.1016/j.jclepro.2018.08.011

Osterwalder, A., Pigneur, Y., & Tucci, C. (2005). Clarifying Business Models: Origins, Present, and Future of the Concept. *Communications of the Association for Information Systems*, *16*. Advance online publication. doi:10.17705/1CAIS.01601

Pantelic, D., Sakal, M., & Zehetner, A. (2016). Marketing and sustainability from the perspective of future decision makers. *South African Journal of Business Management*, *47*(1), 37–47. doi:10.4102ajbm.v47i1.51

Peñuelas, J., Poulter, B., Sardans, J., Ciais, P., Van Der Velde, M., Bopp, L., Boucher, O., Godderis, Y., Hissinger, P., Llusia, J., Nardin, E., Vicca, S., Obersteiner, M., & Hanssens, I. (2013). Human-induced nitrogen–phosphorus imbalances alter natural and managed ecosystems across the globe. *Nature Communications*, *4*(2934), 1–10. doi:10.1038/ncomms3934 PMID:24343268

Randers, J. (2012). *2052: A Global Forecast for the Next Forty Years*. Chelsea Green Publishing.

Rasmunssen, B. (2007). *Business Models and the Theory of the Firm* (Working Paper). Melbourne, Australia: Victoria University of Technology.

Song, Z., & Moon, Y. (2017). Assessing sustainability benefits of cybermanufacturing systems. *International Journal of Advanced Manufacturing Technology*, *90*(5-8), 1365–1382. doi:10.100700170-016-9428-0

Spieth, P., Schneckenberg, D., & Ricart, J. E. (2014). Business model innovation–state of the art and future challenges for the field. *R & D Management*, *44*(3), 237–247. doi:10.1111/radm.12071

Steckel, J. C., Edenhofer, O., & Jakob, M. (2015). Drivers for the renaissance of coal. In M.G. Morgan (Ed.), *Proceedings of the National Academy of Sciences (PNAS)* (vol. 112, pp. E3775–E378). Washington DC: PNAS. 10.1073/pnas.1422722112

Stubbs, W., & Cocklin, C. (2008). Conceptualizing a "Sustainability business model.". *Organization & Environment*, *21*(2), 103–127. doi:10.1177/1086026608318042

Terkaj, W., & Tolio, T. (2019). The Italian Flagship Project: Factories of the Future. In T. Tolio, G. Copani, & W. Terkaj (Eds.), *Factories of the Future* (pp. 3–35). Springer. doi:10.1007/978-3-319-94358-9_1

Thies, C., Kieckhäfer, K., Spengler, T. S., & Sodhi, M. S. (2019). Operations research for sustainability assessment of products: A review. *European Journal of Operational Research*, *274*(1), 1–21. doi:10.1016/j.ejor.2018.04.039

United Nations. (1987). *Report of the World Commission on Environment and Development: Our Common Future*. Oxford University Press.

Vehkamäki, S. (2005). The concept of sustainability in modern times. In *Sustainable use of renewable natural resources: From principles to practices* (pp. 23–35). University of Helsinki – Department of Forest Sciences.

World Health Organization. (2018). *Preventing disease through a healthier and safer workplace*. Retrieved from https://www.who.int/quantifying_ehimpacts/publications/healthier-workplace/en/

Xia, X., Govindan, K., & Zhu, Q. (2015). Analyzing internal barriers for automotive parts remanufacturers in China using grey-DEMATEL approach. *Journal of Cleaner Production*, *87*, 811–825. doi:10.1016/j.jclepro.2014.09.044

Zott, C., Amit, R., & Massa, L. (2011). The business model: Recent developments and future research. *Journal of Management*, *37*(4), 1019–1042. doi:10.1177/0149206311406265

Chapter 4
Pull Production Systems:
Link Between Lean Manufacturing and PPC

Eduardo Guilherme Satolo
https://orcid.org/0000-0002-8176-2423
São Paulo State University (UNESP), Brazil

Milena Estanislau Diniz Mansur dos Reis
https://orcid.org/0000-0003-1080-8735
Federal University of Rio de Janeiro (UFRJ), Brazil

Robisom Damasceno Calado
Fluminense Federal University (UFF), Brazil

ABSTRACT

This chapter aims to organize knowledge about pull production systems by presenting the underlying concepts of lean manufacturing as for its origin, principles, and relations with PPC. Pull production is one the fundamental principles of lean manufacturing, and its implementation can bring positive impacts. For such a purpose, sequential and mixed supermarket pull systems stand out in which the integration between pull production systems and PPC and its various levels is a main subject of discussion. The JIT model or Kanban method and hybrid systems, such as conwip and lung-drum-string theory, are mechanisms for managing pull production systems. Finally, a pull production system implementation is presented for illustration purposes. At the end of this chapter, it is expected that skills are developed by readers, which are going to assist them in using the tools presented to model production systems and aid decision-making processes.

DOI: 10.4018/978-1-7998-5768-6.ch004

INTRODUCTION

Lean Manufacturing Principles

The concept of Lean Manufacturing had its inception during the post-World War II period in Japan in the 1940s. At that time, the Toyota Company was experiencing difficulties due to economic and social issues that plagued the country. To rebuild itself, Toyota set out to rethink its production model and reinvent the means of production.

For such an end, it was determined that its manufacturing process would not be carried out through push production, as in mass production. In a production strategy called pull system, Toyota starts manufacturing only the number of items necessary to meet the customer's demand now of need by seeking excellence in manufactured items. At this point, to support this new way of thinking about production, principles and tools have been created over the years, which guided the creation of the Toyota Production System (STP).

Its success was not only due to economic terms, but also to quality, reliability, competitive prices, and the wide operational margin achieved by Toyota, which aroused curiosity and revealed the need for studies aimed to understand this new production method that culminated in incursions into Japan. One of the most notorious studies about it was conducted by Professors James Womack, Daniel Jones and Daniel Ross (1980s) on a research project at the Massachusetts Institute of Technology - MIT. This research aimed to study the global automobile industry by focusing on a comparison between European, American, and Japanese industries. The results obtained using the Japanese model by Toyota have been mapped through interviews involving employees, union members and government officials, which were highlighted as the best practices in the sector.

However, Womack el al. (1990s) published a book entitled "The machine that changed the world" in the 1990s, and a new philosophy has been extensively disseminated henceforth. In this book, the authors coined the term "Lean Production", which is considered synonymous with the term "Toyota Production System". This book served as reference for production management development in several other types of organizations.

As the theme started to become more deeply explored, Womack and Jones (1996) compiled a book called "Lean thinking - banish waste and create wealth in your corporation", which became a guide for the adoption of the lean philosophy by companies. Lean thinking is a management philosophy which is also described as an approach with which organizations develop skills towards gradually eliminating waste and creating value.

Since then, Lean manufacturing has been successfully disseminated and applied in different sectors of the economy and it has been named specifically according to its application area: lean education, lean public, lean healthcare, lean agrifood, and lean construction. However, its essence and principles remained unchanged.

Lean Manufacturing involves more than just eliminating waste, and it has five guiding principles, which are: value, value chain, continuos flow, pull production and perfection. The steps regarding these principles are going to be described as follows.

The 1st principle is to define value of given product in Lean Manufacturing from the customer's perspective, which is its first step. It is important that an organizational culture is established within the organization, which comprise predominant attitudes and behaviors that characterize the functioning of a group or organization, from the perspective of lean thinking associated with management commitment on strategic, tactical, and operational levels.

Understanding the current stage of the product within its life cycle, i.e., introduction, growth, maturity or decline, demand behavior, and order volumes is essential. Establishing the product value perceived by customers is associated with constant studies and a refined perception about changes in the customer's behavior that indicate the need for improvements.

The second step to determine value is to calculate the target cost of a product based on resources and efforts required to produce it by eliminating waste. When the value of a product is exactly the price a customer is willing to pay (product cost), a cost-value balance is achieved, as shown in Figure 1. Value is increased when something, such as a resource or a service, is offered and valued by the customer, e.g., when there is faster delivery. A reduction in activities that do not add value to their costs also creates value.

The value chain, which is the 2nd principle, involves actions from conception to launch, i.e., the transformation of raw material into a finished product, from sale to delivery, in short, all activities involved in the creation of a finished product. The importance of identifying the value chain of a product or product family provides a distinction between steps that add value and those that do not, in addition to allowing effective actions that focus on a subsequent elimination of waste.

The value chain is designed by value stream mapping – VSM, which, visually represents the stream of materials using symbols and the flow of information within an organization to reproduce its current state and forecast its future state to indicate the path that a particular business is taking.

Along with the value stream mapping, performance indicators associated with cost, time, quality, and flexibility are calculated, for example: cycle time, value-added and non-value-added time, setup time, availability and number of operators, scrap rate, and lead time of activities that add value and of those that do not.

Figure 1. Relation among value, cost, and waste

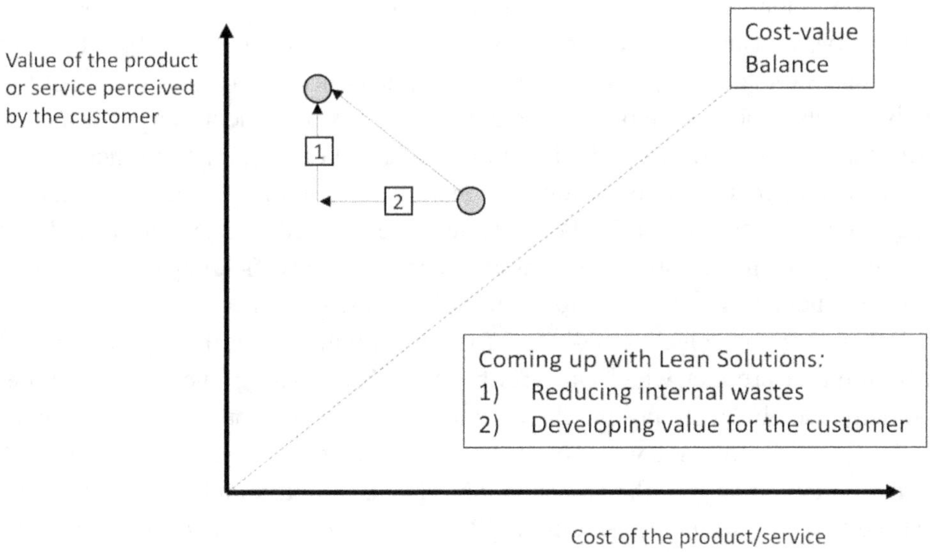

For calculating these indicators, standardization is essential, since it is only through stable processes that a study on improvements is started. The success of standardization depends on the use of visual management, which will allow employees to have easy access to production stages, tools and parts needed for manufacturing a given product or providing different services. Knowledge acquired from the entire supply chain allows acting on the flow.

The 3rd principle lies in establishing a continuous flow which is the rate at which resources necessary for the transformation of products and services are allocated. This task involves working continuously with no dawdling, interruption, or rejection, either within or between the steps of the process. Therefore, it is worth studying, designing, and adapting an organization's layout, i.e., the physical arrangement of resources in each workspace, through which an organized production space is made useful by the 5S and visual management.

To achieve a continuous flow, knowledge and application of tools such as cellular manufacturing, production leveling (*heijunka*) and line balancing are relevant towards seeking an ideal flow, i.e., a continuous flow, a single-piece flow or a one-piece flow. In a one-piece flow, instead of manufacturing many products that will then be queued and remain idle for the next stage of the process, products go through each stage one at a time, with no interruptions, thus leading to quality improvement and cost reduction. This flow is considered the opposite of batch production, where parts are processed and moved on to the next process (regardless of need) in which they remain idle for some time.

A continuous flow allows identifying problems that may be an obstacle to an effective transportation of products and provision of services. Thus, there are possibilities for achieving process improvements based on the use of tools of total quality management, total productive maintenance, simultaneous engineering, fast tool change - FTC (SMED - single minute exchange of die), among others.

The evolution of results obtained with respect to items of equipment must be monitored according to the lead time of activities that add value and those that do not, quality rate indicators, performance rate (also called efficiency rate) and defect rates for calculating overall equipment effectiveness - OEE.

Pull production, which is the 4th principle, essentially means that a process should only produce a product, which can be either a good or service, if the previous process had demanded it. When pulling the product, the customer appreciates the delivery of something desired at a given moment of need, instead of what occurs when the supplier attempts to give the customer a hard sell. Womack and Jones (2007) state that the best way to understand the logic and challenge of pull systems is to start from an actual customer expressing demand for a real product and move on to the opposite direction, i.e., by taking all steps necessary to deliver the product to the customer. Details about pull production systems are presented throughout the chapter.

According to lean thinking, perfection, i.e., the 5th principle, occurs when a process provides sheer value (defined by the customer) with no waste. Although this is an ideal scenario by performing a continuous improvement of the system, perfection must be considered goal to be achieved.

To eliminate waste and seek perfection, companies must make a radical improvement, otherwise known as *kaikaku*, *kaizen* breakthrough, flow *kaizen*, *kaizen* intervention, and continuous improvement, i.e., kaizen, point kaizen and process *kaizen*. *Kaikaku* consists in drastic improvements, usually involving an administrative process that affects the value chain flow afterwards. Kaizen, literally translated as “radical change”, in turn refers to a weeklong, rapidly developing event that adopts a focused, teamwork-based approach encompassing a small gradual improvement of an activity aimed to generate greater value with less waste.

Finally, a constant search for perfection based on concepts of continuous and systemic improvement of processes is based on the use of 5S and *poka yoke*. The latter is defined as a mistake proofing device or procedure that prevents a defect from proceeding to the next stage of operation or process. Audible or visual alarms are examples of poka yoke; or something that enables to halt the process to avoid making mistakes such as: component failure in assembly, part assembled in the wrong position, among others.

By constantly monitoring processes through statistical control and diagnosing the need for changes that allow minimizing mistakes made by employees through autonomation (*jidoka*), which uses automated machinery instead of human intelligence, machines can detect the manufacture of any defective part and can immediately stop performing activities autonomously while indicating the need for maintenance. To systematize these improvements, procedures are used through kaizen projects and the PDCA cycle (plan-do-check-action).

Figure 2 summarizes lean principles as well as their interactions. By applying the first four principles, i.e., value, value chain, continuous flow and pull production, to a system and their interaction, there is the formation of a virtuous circle that provides participants with the view that waste reduction is countless. Such view and application results foster the fifth principle, perfection, which in turn provides a growing increase in the perception of value by customers.

Figure 2. Lean manufacturing Principles

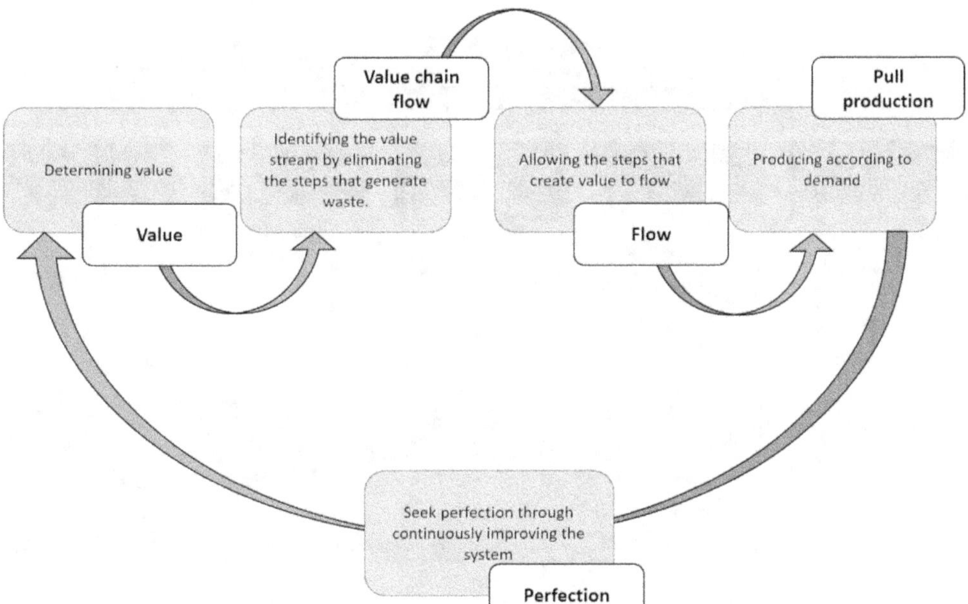

Management based on the elimination of losses and waste proposed by Lean Manufacturing plays an important role, mainly in assisting to mitigate the effects inherent to the particularities of each production system. Lean Manufacturing is implemented in organizations through a coordinated and structured use of several tools, which aids the process of minimizing and/or eliminating waste and inefficiencies from the process, and enables reduced lead time and costs, quality improvement, increased productivity, reliability, and flexibility. In addition, its implementation allows organizations to become more agile, flexible, and innovative, i.e., if achieved successfully.

For the implementation of Lean Manufacturing, a set of tools are used which requires studies that meet the organization requirements. These studies must consider available resources that satisfy and fulfill the organization's objectives. In this sense, the tool *hoshin kanri* stands out, which consists in setting goals and guidelines through a directed effort and established goals to avoid its wrong definition, thus obtaining unsatisfactory results from lean manufacturing implementation into an organization. *Hoshin kanri* is a strategic tool that assists business executives in decision-making processes that focus their resources on critical initiatives required to fulfill the company's business objectives.

Figure 3. Lean Manufacturing, its principles, and tools

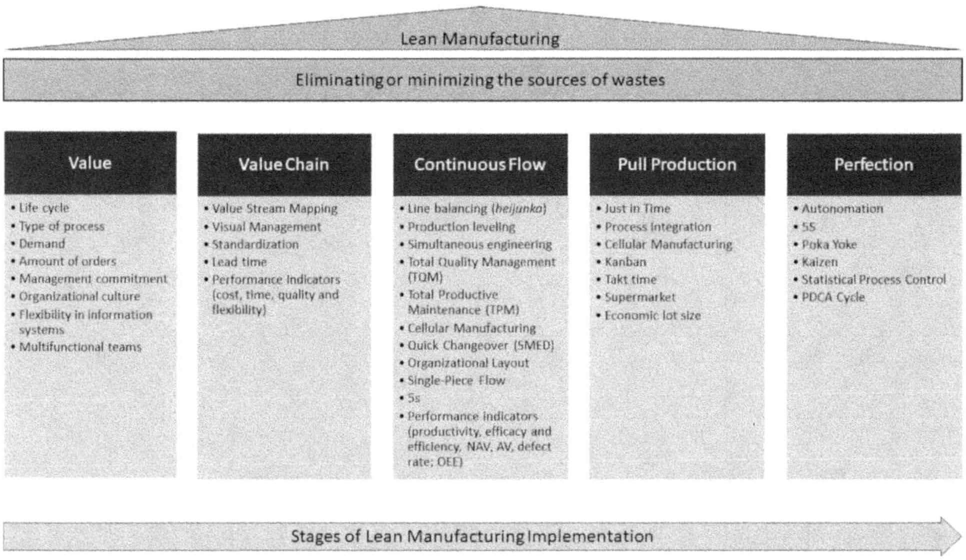

THEORETICAL BACKGROUND

Value and Waste (Muda, Mura, and Muri)

The elimination of waste, which means *muda* in Japanese, and the creation of value represent the quintessence of the Lean Manufacturing philosophy. According to Womack and Jones (2007), the underlying issue of Lean Manufacturing involves a way to do more and more with less and less - less human effort, less equipment, less time, and less space - while becoming closer and closer to providing customers with exactly what they want.

Muda comprises a classification called 3MU's used to identify waste. The 3MU's is an acronym for the words *muda*, *mura*, and *muri*. As aforementioned, changes mean waste. Muri involves purposelessness by subjecting employees or equipment to functions beyond their limits (capability). Its consequences are problems regarding safety and quality, interruptions, and defects.

Mura can be conceptualized as inconsistency due to involving production volume fluctuations or irregular production. It refers to subjecting an equipment or employee to functions beyond their capacity at times and below their capacity at others. Such unevenness can be caused, for example, by the occurrence of failures or absence of parts. Table 1 presents a practical example of *muda, mura* and muri.

Table 1. Muda, mura and muri

A worker on an assembly line must manufacture 60 products using a machine that allows working on 20 products simultaneously. ***Muda*:** assembling the products in groups of 10 (6 different cycles). ***Mura*:** assembling the products in 2 groups of 20 cycles and 2 groups of 10 cycles. ***Muri*:** asking the employee to finish production in 30 groups of 2 cycles (condition in which work capacity limits are determined for employees). **Ideal situation:** dividing the products into 3 groups of 20 cycles (a balance between capacity and workload becomes the goal to be achieved).

In most operations, the ratio of value to *muda* is 5/95. Most activities only increase costs and add no value, which makes it crucial to define waste in production processes. Seven types of *mudas* were identified, namely: defects, overproduction, transportation, waiting, inventory, motion and overprocessing. Recently, non-utilized talent was added as the eighth type of waste. Table 2 describes waste in industrial and hospital environments.

Table 2. The seven types of waste

Waste	Hospital environment	Industrial environment
Defects (Errors)	Time spent performing, inspecting, or repairing something containing errors. Example: lack of item in surgical cart or incorrect medication administration.	Manufacture of defective parts or those requiring correction, rework, disposal, or inspection.
Overproduction	Do more than the demand or produce before demand arises. E.g., performing unnecessary diagnostic procedures.	Production of items that are not in demand, which generates transportation costs and losses due to excessive personnel or inventory.
Transportation	Unnecessary movement of "products" (patients, samples, materials) in a system.	Inventory movement (of process and finished products) over long distances which leads to inefficient transportation.
Waiting	Waiting for work activity. E.g., personnel waiting due to unbalanced workload and patients waiting for examination.	Idle time. E.g., employees watching equipment operation, and waiting for work or resources.
Inventory	Excess inventory involving financial costs, storage and transport costs, waste, and damage. E.g., expired medications.	Excess of raw material, in-process stock, or finished products.
Motion	Unnecessary motion of employees in the production system. E.g., employees in each sector traveling long distances due to layout problems.	Useless movement made during work. E.g., searching for parts.
Over processing	Perform worthless work or one which does not meet the needs of the customer. E.g., insertion of data in forms that are not used.	Unnecessary steps to process parts, generating obsolete or damaged products, increased lead times, and incurring transport and storage costs.
Non-utilized talent	Ignored opinions, skills, improvement, and opportunities. E.g., disregarding the opinions of users of the system as a possibility of improvement.	Ignoring proposals for improvements by machine operators.

LINK BETWEEN LEAN MANUFACTURING AND PPC

As described above, the lean philosophy and its tools are successfully used in organizations from different sectors and aid to improve the manufacture of goods and services. When implemented in a production environment, it focuses on improving projects, processes, and products.

Production planning & control (PPC) is aimed at managing the flow of information and materials by focusing on how they go through the production process, but not on process design.

Thus, the PPC determines timing, quantity, and route of production items, while Lean Manufacturing improves these environments through continuous improvements that bring benefits to different levels of PPC.

The demand forecast established by PPC is positively affected by Lean Manufacturing which aids to reduce preparation time. It does not directly affect the development of production plans, since these are still necessary, but it reduces the complexity of the master production schedule (MPS) by allowing its elaboration time to be close or the same as that of the rate of sales.

In production planning, establishing a partnership relationship with suppliers allows the purchase process to be reorganized towards reducing lead time in the supply chain, thus favoring the creation of a synergistic work environment that promotes an organized and continuous flow of materials.

Lean Manufacturing implementation positively affects the master production schedule (MPS) by making delivery times shorter. The closer the time to manufacture a given product is to estimated delivery time calculated by the sales team, the less is the need for forecasting and, therefore, the planning horizon. The tools of lean manufacturing enable a stabilization of production processes by allowing the use of just in time (JIT), which assists in manufacturing items to meet customer demands only at a moment in which they place an order.

As regard material requirements planning (MRP), Lean Manufacturing aids to reduce inventories (raw material in the process and finished products), since a reduction in lead time allows products to be manufactured according to the customer's need to follow the customer's order as close as possible to its request date and run the production flow continuously. A reorganization of production environments by adjusting the layout and cellular manufacturing reduces the number of products to be assembled and, consequently, their component lists, thus facilitating management and determining production sequencing.

It can be observed that PPC is necessary in a lean environment, just as PPC environments require Lean Manufacturing practices if they are willing to improve competitiveness and become more effective and efficient.

PULL PRODUCTION SYSTEMS

The central principle of pull production is that there must be a continuous optimization along the value-adding chain oriented towards customer demand. This effort is aimed at increasing customer satisfaction, in which the waste of inventory is avoided, and simple and transparent processes are generated. Thus, companies are posed with the challenge of making the right product available at the time, on the spot, in quantities and at a level of quality desired by the customer, and at minimum cost.

A production system oriented towards pull production is described in literature in three different ways: supermarket pull system, sequential pull system and mixed pull system (sequential with supermarket).

Figure 4. Schematic of a supermarket pull system

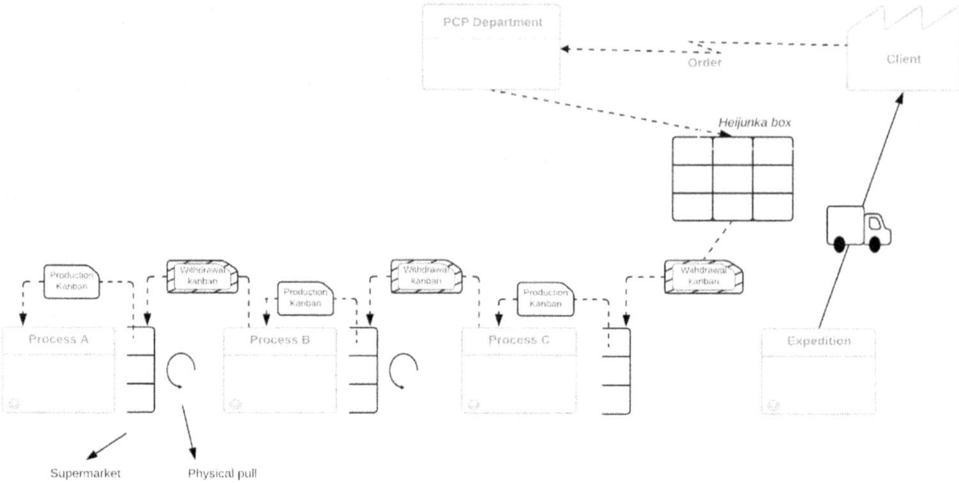

Supermarket Pull System

In this type of pull production system, the term used in its name, supermarket, is used as basis for its operation.

A supermarket represents a place where customers buy items. It has inventory control (also called a buffer) used to set up an upstream process. In such supermarkets, minimum and maximum amounts of raw materials, items and assembly groups are maintained to meet the demand of subsequent processes. At supermarkets, the customer or the next process removes the necessary items from shelves, which triggers the replenishment of items by the seller, i.e., the previous process.

Inventory control in supermarkets (minimum and maximum quantities of items and production flow) is managed with the aid of *kanban* cards which is made possible by the *heijunka* box through visual management to allow a practical and easy inventory replenishment control. The *kanban* card is normally placed in a rectangular vinyl envelope that allows visual controls and management of the flow of information to regulate the supply of materials between processes. The *heijunka* box is a visual programming tool used to promote production leveling aimed to allow a smooth flow of items production. Its appearance can be, for example, that of a rectangular cube, divided into boxes (box) that represent a given time to produce a certain set of items. At the top of the box, there is a schedule in specific time units that provides the current stage of production.

The production of new items takes place by inserting *kanbans* that are added to the *heijunka* box, thus providing a visual representation of the following items to be produced. The *heijuinka* box allows operators to easily visualize the following tasks in a production process, in addition to the time at which they are scheduled for. Visually, the concept of supermarket pull system, otherwise known as replenishment, pull system, is outlined in Figure 4.

From the moment that there is the need for producing a particular item according to the *heijunka* box, the withdrawal *kanban* is collected, which establishes the need for process C production. Process C checks the availability of items and starts their production. Depending on the need for production, process C removes items from the supermarket provided by process B. Once it is noticed that there are items requiring replenishment established by the supermarket (determined by *kanban* tags assignment), process B starts manufacturing them as needed, and the items made available by process A are removed from the supermarket for product manufacturing. Thus, each process is due to keeping its supermarket stocked with items it produces. This type of pull system is effective for companies that have a small variety of items.

Sequential Pull System

The sequential pull system differs from the previous one in such a way that it is recommended for an organization in which there is a wide variety of items. When selected, the storage of all types of items in supermarkets is not feasible.

As a solution, it is recommended to change the production environment into order placement which minimizes total inventory.

Therefore, the role of the PPC department becomes fundamental to determine the production mix and the quantity of items to be produced.

Commonly, for such cases, sequencing is guided by the concept of first in, first out - FIFO. In this type of sequencing, the objective is to respect the order in which orders are placed with items that arrive at each process being produced based on what was originally defined in the sequencing and presented by the *heijunka* box.

For a sequential pull system to operate properly, there is a need to be aware of the lead time of items, and that such lead time in the supply chain is the same between processes, or that there is already an adequate and minimum inventory at the supermarket to avoid the wastes of overproduction, inventory or waiting.

Figure 5 shows a sequential pull production system which, based on the customer's request, the PPC Department determines the production schedule to meet the customers' needs and have the best production mix for the factory floor, which is made available by means of the *heijunka* box. Once schedule is defined, Process A starts manufacturing items which are made available sequentially to process B until the final product manufacture occurs, which is then sent within the estimated delivery date.

Figure 5. Schematic of a sequential pull system

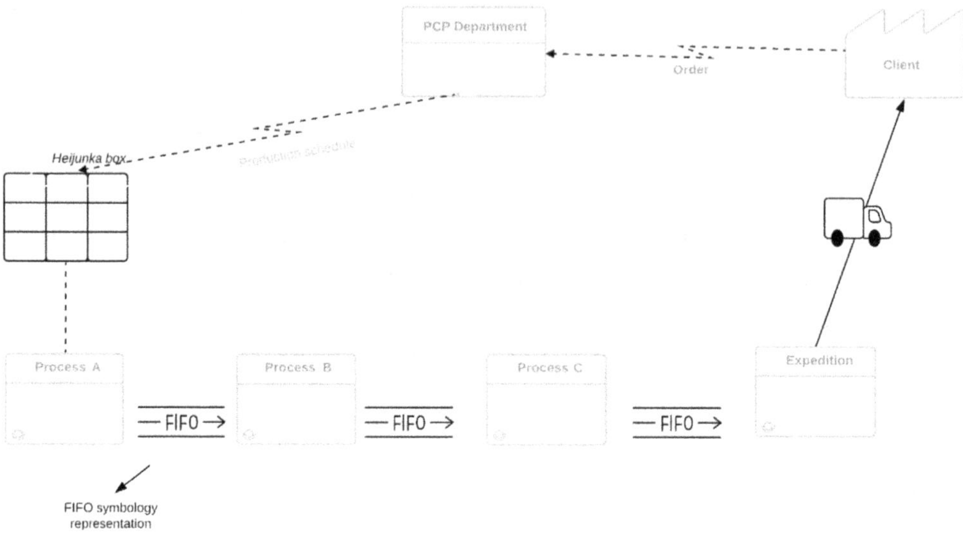

Mixed Pull System (Supermarket Sequential Pull System)

The third type is called mixed pull system, as it operates based on the concepts presented in the two previous models, i.e., the supermarket and sequential pull systems. This type of pulled system is recommended when there is a variety of items, although it is noteworthy that few items represent a large production volume.

For identifying these representative items, the ABC classification concept is used to determine 20% of A items (called high ordering rate) that represent 80% of the production volume. Items B are classified as average ordering rate and items C as low ordering rate.

Items A and B are commonly managed by the supermarket pull system due to the rate of orders and constant production. Less frequent items, such as C items or even special requests are inserted into the supply chain schedule using the sequential pull system, as pointed out by the PPC Department.

The complexity in managing the two pull system models requires extra care in assigning daily activities in the mixed pull system to identify potential improvements and waste in production processes.

It can be seen in Figure 6 that after receiving an order by the PPC department and the classification of the requested item, a supply chain schedule is established and taken to the *heijunka* box. Items A and B are removed at appropriate time by process C, which starts manufacturing the item from those available in process B at the supermarket. When it is identified that the items supplied require replenishment, their production begins by withdrawal *kanban* from the production in process A, which directs production to process B and follows the sequencing. For orders classified as C items, production is sent to the *heijunka* box and, at the time set for production, it is initiated by process A to run smoothly through the other production stages.

Figure 6. Schematic of a mixed pull system

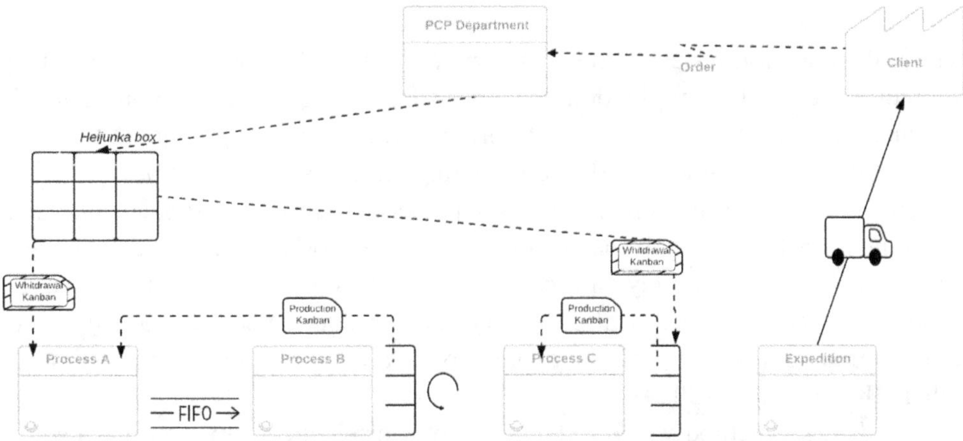

Pull Production and PPC

Lean Manufacturing is the origin of the concept of pull production, which is a primary aspect of JIT and an alternative to the concept of push production defined by PPC.

In push production, items are produced to meet future demand estimated by forecasting models or created through marketing actions. Thus, items are produced in their final state, or at least close to it.

In pull production, manufacturing orders are issued when there is a real demand, which generates lower intermediate and final product inventory to avoid the waste of overproduction. For such a purpose, the use of small manufacturing batches and production leveling recommended by Lean Manufacturing aids to avoid that organizational performance is adversely affected by changes in orders, problems with suppliers, forecast errors and product defects.

The role of PPC in productive environments in which pull production has been implemented is relevant, mainly for sequencing orders of customers and controlling inventory during production stages.

When the supermarket pull production system is adopted in a production environment, PPC is responsible for informing the needs of the final consumer, which triggers the entire pulling process along the production line. In a sequential pull production environment, the role of PPC is to establish the most effective order for production by considering the characteristics of each workstation and making it available on the *heijunka* board. Finally, in a mixed production system environment, the PPC identifies the complexity of managing the characteristics of the two previous types.

JIT Model/Kanban Method

The pull production system is an effective way of JIT production and it efficaciously promotes the level of production and operation management of companies. JIT production conceptualizes the need to produce the necessary product at the right time, in the right quantity and thus eliminating unnecessary inventory.

The JIT model is a production planning approach that replaces material requirement planning for better materials management because it works on the principle of zero inventories. It inflicts highly enforced quality control processes to ensure that the product meets quality specification as per customer requirements. It is a pull system of operation, in which the point of application of a worker represents the actions of others (Kaswan et al., 2019).

The *kanban* method was developed for pull production aimed to manage JIT production.

The term *kanban* means signboard and is currently used in manual or automated forms such as electronic *kanbans*. It allows a visible and harmonious control of production and inventory quantities in a production system. The replenishment of stocks on supermarket shelves represents the basis of the *kanban*.

The use of *kanban* enables to level production and obtain a reduction in the stock of finished products and raw materials and, consequently, lead time. Constant changes in orders by customers are no longer a problem, as the company's production can adjust the pace of items to be produced during the day, week, or month. In addition, there is an increased flexibility of response to customers, which results in a production which is closer to the actual demand.

The *kanban* plays an important role in communicating information throughout the product's production cycle and ensures manufacture of the right number of items needed at the right time. Such information communication is carried out through two basic types of *kanban*: withdrawal and production, as shown in Table 3.

Kanban cards, as a means of visual management of production control, employ three colored tags, which are green, yellow, and red, that allow the employee to identify whether items need to be replenished or not.

The green tag defines production leveling. At that time, there is no need for production. The yellow tag sets response time. While using items of the green tag, and entering the stage of the yellow tag, it is necessary to produce the item. Then, the due production process must finish its production, carry out its setup, produce a transfer batch and place it on the supermarket shelves. Finally, the red tag represents the necessary safety so that there is no halt in production at sectors, and it exists to safeguard production from any problems in the supply chain process.

Table 3. Types of kanban and their definitions

Types of *kanban*	**Other nomenclatures**	**Definition**
Withdrawal *kanban*	Move cards Conveyance Kanban	It authorizes to move products towards a downstream process (customer). This card must contain information on the type and quantity of product to be moved.
Production *kanban*	Manufacturing Kanban	It authorizes the manufacture or assembly of items in downstream processes (customer) based on the use of upstream processes (supplier).

The tags are determined according to the needs to replenish each item. Thus, there are variations in the number of kanbans, as shown in Figure 7.

Figure 7. Schematic of kanban and its colored tags

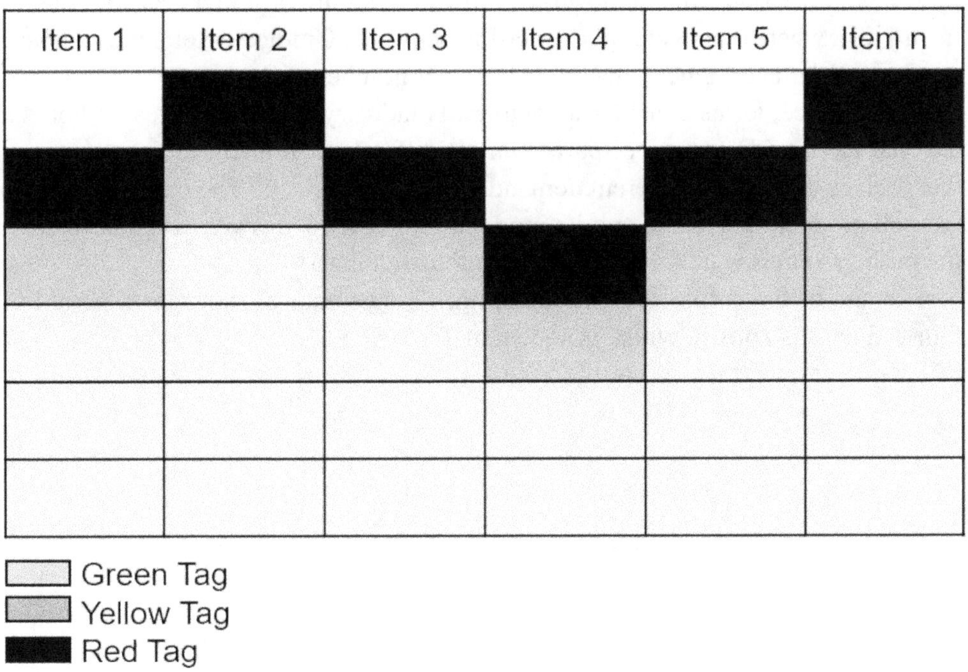

To Kanban function correctly, it is recommended to follow four rules: (1) the customer should only take items from the stock when it is in fact necessary; (2) the supplier can only produce items which have production *kanbans* and in quantities defined therein; (3) only good parts can be placed in stock; (4) *kanbans* must be in full packages or on a signboard.

Determining the Number of *Kanbans*

The number of *kanbans* is calculated using the following equation.

$$N = \frac{UxT\left(1+P\right)}{C}$$

Where:

N = total number of containers needed between the two workstations (i.e., number of *kanbans*)

U = utilization rate of the next workstation, usually measured in items produced per hour

T = average elapsed time required so that the container completes the entire cycle from the moment it leaves the previous workstation, and it is filled with products and leaves again.

P = policy variable that indicates the efficiency of the system. P can have values from 0 to 1, where 0 indicates perfect efficiency and 1.0 indicates total inefficiency.

C= Standard container capacity in terms of number of items.

For instance, let us consider an auto parts industry which operates 24 hours a day and has a wide variety of parts to meet different customer requirements, and that each customer has different demands for these products. The company adopts the pull production system, and its production is daily. However, when necessary, the push system is used for products having less demand.

For one of the company's production lines, customer demand for a period of fourteen days is known, which is shown in Table 4.

Table 4. Customer demand for five types of products

Period	Parts				
	A	B	C	D	E
1st Day	4.890	1.360	1.870	1.310	1.100
2nd Day	4.890	1.360	1.870	1.310	1.350
3rd Day	4.890	1.360	2.540	2.260	840
4th Day	4.890	0	1.870	0	0
5th Day	4.890	2.720	1.870	0	0
6th Day	0	0	0	0	0
7th Day	0	2.720	1.920	0	840
8th Day	4.890	1.360	0	0	0
9th Day	4.890	1.360	2.540	1.310	1.230
10th Day	4.890	1.360	1.870	1.310	840
11th Day	4.890	1.360	2.520	1.310	1.470
12th Day	4.890	1.360	2.540	1.310	840
13th Day	0	0	1.870	1.310	700
14th Day	0	1.360	720	0	450

Table 5. Production mix planning for the five types of products

Period	Parts				
	A	B	C	D	E
1st Day	3.750	1.370	1.970	800	750
2nd Day	3.750	1.370	1.970	800	750
3rd Day	3.750	1.370	1.970	800	750
4th Day	3.750	1.370	1.970	800	750
5th Day	3.750	1.370	1.970	800	750
6th Day	3.750	1.370	1.970	800	750
7th Day	3.750	1.370	1.970	800	750
8th Day	3.750	1.370	1.970	800	750
9th Day	3.750	1.370	1.970	800	750
10th Day	3.750	1.370	1.970	800	750
11th Day	3.750	1.370	1.970	800	750
12th Day	3.750	1.370	1.970	800	750
13th Day	3.750	1.370	1.970	800	750
14th Day	3.750	1.370	1.970	800	750

Table 6. Number of kanbans for each type of part

Item	U	T	P	C	K
A	3.750	1.020	0,30	450	8
B	1.370	1.020	0,30	150	9
C	1.970	1.020	0,30	230	8
D	800	1.020	0,30	40	19
E	750	1.020	0,30	150	5

The PPC Department carries out the process of determining the production mix according to the forecast of customer demand according to Table 5. These data represent the utilization rate (U) of each part daily.

The next step is to calculate the average elapsed time required for the product to complete the entire cycle (T), which is the result of a sum of *kanban* collection time (T1), waiting time (T2), replenishment time (T3), processing time (T4) and shipping time (T5).

$$\mathrm{T} = \Sigma\left(T_1 + T_2 + T_3 + T_4 + T_5\right)$$

For the case under study, the time for the *kanban* to be sent from the supermarket to the production line is 50 minutes (T1 = 50 minutes). Waiting time (T2) for production is 830 minutes. Replenishment time (T3) is 25 minutes, processing time (T4) is 25 minutes and shipping time (T5) from the supermarket to the warehouse is 90 minutes. Thus, the production cycle time (T) is:

$$T = 50 + 830 + 25 + 25 + 90 = 1.020\left(min\right)$$

Let us suppose that, for part A, the system efficiency is 0.30 and the capacity of a standard container is 450 items. The *kanban* quantity is then calculated as:

$$N_{(Items\ A)} = \frac{UxT\left(1+P\right)}{C} = \frac{\dfrac{3.750\,pç}{24h\,x\,60\dfrac{min}{h}}\,x1.020min\,x\left(1,30\right)}{450\,pç} = 7,67 \therefore 8\,kanbans$$

The method of calculating *kanban* production for the other items is similar and the results of calculating the *kanban* quantity of these five types of products are shown in Table 6.

Therefore, the number of *kanbans* for the entire production line has already been calculated. The PPC Department prepares and releases the *kanban* based on the calculation result, and the production department schedules production according to information provided by the *kanban*.

Determination of Tags

The calculation of the number of *kanbans* (N) ensures that the production system will have enough items in stock at the supermarket to meet its daily production. In this case, the calculated N value makes it possible to establish green and yellow tags.

Taking the previous case as an example, a set of 8 *kanbans* is required for item A, i.e. 8 containers with 450 items each, which will meet the demand of 3,750 items to be utilized daily. It is now necessary to identify the moment at which the previous process needs to start producing items to replenish the containers that will be used along the day.

To determine replenishment time, which represents the yellow tag, it is necessary to identify (TF1) average time to finish producing items in the process; (TF2) average time to prepare machinery for producing the item; (TF3) time to produce items of a container; and (TF4) time to ship the container to the supermarket in the supply chain process.

If the supply chain process for A items lasts as follows:

TF1 - average time to finish producing items in process - 45 minutes.
TF2 - average time to prepare machinery for producing the item - 15 minutes.
TF3 - time to produce A items of a container - 250 minutes.
TF4 - time to ship the container to the supermarket - 10 minutes.

The sum of these values amounts to 320 minutes required for the supplier to replenish a container at the supermarket. Assuming that there is a demand for 3,750 items to be used over the course of a working day, the utilization rate is 2.6 items per minute.

Thus, if the production of 1 container requires 320 min, 832 units of A items (320 min x 2.6 items/min) will be utilized over that period, which indicates that 02 containers are required according to the yellow tag. Given the need for 8 *kanbans* for item A, it results in the green tag, representing 06 containers.

The red tag considers average repair time in case of machinery stoppage or in the absence of products due to failures by the supplier associated with safety stock.

Table 7. Calculation of kanban tags adjusted according to type of items

Item	TF1	TF2	TF3	TF4	Maintenance Time (hour/ day)	Consumption rate (items/min)	Green *Kanban*	Yellow *Kanban*	Red *Kanban*
A	45	15	250	10	1	2,60	6	2	1
B	45	15	210	10	1	0,95	7	2	1
C	45	15	90	10	1	1,37	7	1	1
D	45	15	220	10	1	0,56	15	4	1
E	45	15	120	10	1	0,52	4	1	1

If item A were supplied internally and the average time to repair defects in machinery were 1 hour a day in the supply chain process, 156 units of A items (2.6 items/min x 60 minutes) would be utilized in this period. If each container holds 450 units, there is a need for 1 container in the red tag, i.e., 1 *kanban* in the red tag.

Thus, the number of *kanbans* adjusted for the total A items is 9:8 (*kanban* to meet daily production), being 6 *kanban* green tags and 2 *kanban* yellow tags, and 1 *kanban* red tag.

For the other items, the calculation is similar, as shown in Table 7.

Figure 8. Kanban, CONWIP and Hybrid Systems

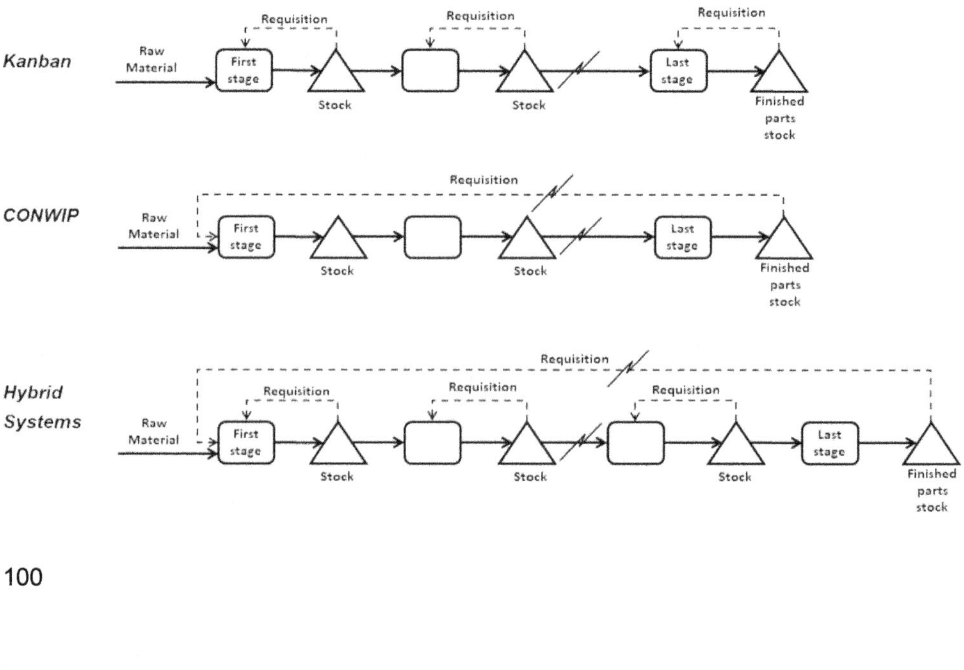

Hybrid Production Systems

CONWIP - constant work in process - is a pull system as a generalized form of *kanban* as it depends on signals. However, in the *kanban* system, each card is used to indicate the production of a specific item, but these cards are assigned to a cycle that covers the entire production chain and does not correspond to a specific item in CONWIP. The specific need for an item is represented by cards at the beginning of the production chain, e.g., it is determined according to a log of demanded items called as backlog. When it is necessary to start production, a card is removed from the production chain and marked as the first type of item on the list to be produced for which raw material or components are present in a standard container at the beginning of the production chain.

The CONWIP illustrated in Figure 8 is applicable to production environments that manufacture different types of products. The CONWIP system seeks to control the plant at a constant level of work in progress, although it does not attempt to control the location of the WIP within the system.

To identify whether a production line behaves as a CONWIP system or not, the following parameters are observed: a) number of signboards (to determine the maximum level of WIP); b) target production quota (for a given period); c) the maximum amount of work ahead. If the target production quota were added to the maximum amount of work performed during a given period, the line would come to a halt until the beginning of the next period; d) a trigger of lack of capacity as a function of effective production until a given time. By satisfying these parameters, it is possible to consistently adjust the production system to the CONWIP method, thereby avoiding non-compliance with the benefit proposed by the method.

CONWIP is like technique used in air traffic control. On days of heavy air traffic, a departure plane will sometimes be kept on the ground at the airport of origin, instead of being allowed to take off and remain in a waiting pattern at the congested destination airport. As the objective is to avoid delays at the destination airport, planes are kept in the air, even if runways are free at the airport of origin, therefore increasing safety and reducing fuel consumption without additional delays.

In this example of CONWIP application in air traffic control, a job will not start unless a location in the system has been vacated for it. The same balancing challenge that air traffic controllers face must be met in CONWIP. Excess work, such as airplanes in the air, must be placed on the ground while arriving so that the bottleneck station, which is the runway, is seldom idle, but not for the point of work, i.e., airplanes, having to wait for too long. If successful, the system will achieve a maximum throughput without excessive flow time or WIP. CONWIP will naturally follow what had been stated by Goldratt (1984), in the book "The Goal", which aimed to balance flow, not capacity. The operation of a CONWIP line is regulated according to bottleneck features.

Integration of MRP, *Kanban* and the Theory of Constraints

By constantly seeking optimizing customer service and reducing costs and inventories, systems such as MRP, *kanban* and the theory of constraints (TOC) stand out, however, to reap better benefits, one must focus on the production strategy inserted in the context of a business environment, as described in Table 8.

MRP, also known as ERP, is a forecasting system that provides answers related to deadlines for making, buying and assembling components at all levels of product structuring. It depends on reliable data from the company's infrastructure to ensure good decision making. Operations and service environments are dynamic and constantly changing, which leads to less efficiency and high costs in the case of incorrect data.

The just in time production approach and *kanban* are alternatives to simplify the see and act in a quick reaction system. Kanban assists in lead time reduction and might eliminate or reduce uncertainties in process and inventory, thus making production efficient, although it depends on a stable and predictable environment. The *kanban* system is favored by a level production with predictable demands and few processing problems, such as long setup times, low equipment availability, few product changes, and a good level of reliability of suppliers' deliveries.

Table 8. Peculiarities of Systems

System	Advantage	Disadvantage
MRP (ERP)	Resource forecasting instrument.	It depends on the reliability of data to generate assertive information.
Kanban	Simplifying the information system and corroborating to reduce lead time.	Low compliance with an uncontrolled environment involving high setups and low availability of equipment in an unpredictable and volatile environment.
Theory of constraints	Identification and improvement of bottlenecks in the operation.	Low efficiency in a product mix environment followed by very often changes in the constraint during a given day.
CONWIP	Applicable to environments with a wide variety of product types.	Stock levels within the system are not individually controlled.

The theory of constraints – TOC, is a similar system to the Kanban method in which it is adopted a principle establishing that difficulties are related to a factor called *constraint*. When the bottleneck is identified and managed, for example minimizing high setup time the bottleneck equipment which is a constraint in an operations environment, benefits are generated to process management, given that there is no significant modification in the product mix followed by changes in the constraint along the day.

Production planning using the drum-buffer-rope (DBR) tool aims to synchronize production by balancing the production flow, and not the individual capacity of each process, which is one of the tools of the TOC. The drum-lung-string tool was created based on the five TOC focusing steps, which are: 1. Identifying the system constraint; 2. Exploiting the system constraint; 3. Subordinating the entire production system to its constraint; 4. Elevating the system constraint; 5. Restarting: after action is taken, in which the constraint should still be the focus of attention, unless it is no longer a constraint.

In specific cases, hybrid systems are alternatives to improve the results of operations, such as when you have a conventional production system, and you wish to start turning it into a lean system. Rother and Shook (2003) utilizes value stream mapping to illustrate a push system by using MRP in the production system, which evolves to the *kanban* system in the future value stream mapping without MRP in the lean production system, but through a combination of *kanban* and MRP which is also rather common in literature. In a hybrid MRP and *kanban* system, MRP is used for advanced planning and the *kanban* acts as an execution system on account of its characteristics of quick response to customer orders and reduced levels of inventory along the process. In a hybrid system that integrates Lean Production with the TOC, the TOC prioritizes areas for improvement and the principles of lean production, in addition to promoting process improvements from the elimination and/or reduction of waste.

By allowing data on the WIP (work in process) to be collected further up ahead in the process or at the bottleneck, CONWIP can operate with less WIP than *kanban*. CONWIP uses the number of pending items, which can allow task sequencing to be performed by production planning & control personnel. The backlog sequence can be a way to ensure adequate capacity when there are significant configurations and optimize production synchronization of item components.

IMPLEMENTATION

In this section, we present the case of a production system in the sector of pad printing of injected items which is responsible for the manufacturing process of subsets used in a domestic refrigerator. The sector has three macro processes considered as departments: (i) plastic transformation, where the study has been carried out; (ii) metal transformation and (iii) product assembly. In this case, there is an internal customer, which is part of the company's logistics policy, its distribution center structure and external customers who are distributors through their stores, such as Casas Bahia and Magazine Luiza (in Brazil), and the final consumer, i.e., the consumer.

In this manufacturing process, the main indicators for managing safety, quality, production, and costs used in this production system are products per operator, service to the product mix, sales price per hour of direct work, number of employees, amount of overtime, level of consumer satisfaction, indicator of production errors, number of field complaints, number of accidents, cost of refuse and rework, material cost and stock days.

In its initial stage, the sector of pad printing of injected items classifies its production flow as a functional layout. This sector almost always served the customer in terms of production volume, despite constant variations in the product mix. This layout favored the company by partially protecting customer service, even when there was machinery breakdown, long setups, lack of employees and other types of waste. However, areas of improvement were identified, such as: waiting time, disorganization, refuse, stock in process, improper use of resources and non-conformities in terms of safety, ergonomics, and quality.

In view of a consensus on the need to improve the layout of the sector of pad printing injected items, a multifunctional team, trained in the concepts of lean thinking, starts introducing good practices and tools from the pull production system. It was proposed by envisaging a possible elimination of the safety problem, reduction of losses and waste, and an increase in quality and productivity indicators.

To guide the improvement project, short-term goals (6 months) were agreed upon aimed at an enhancement of safety, quality, productivity, and strategic production planning guidelines, which are: to reduce 25% of the stock in process, increase 10% productivity, gain 15% of utilized area, study the possibility of improving production flow, and eliminate the safety problem in rack/corridor movement.

Figure 9. Outline of the current layout

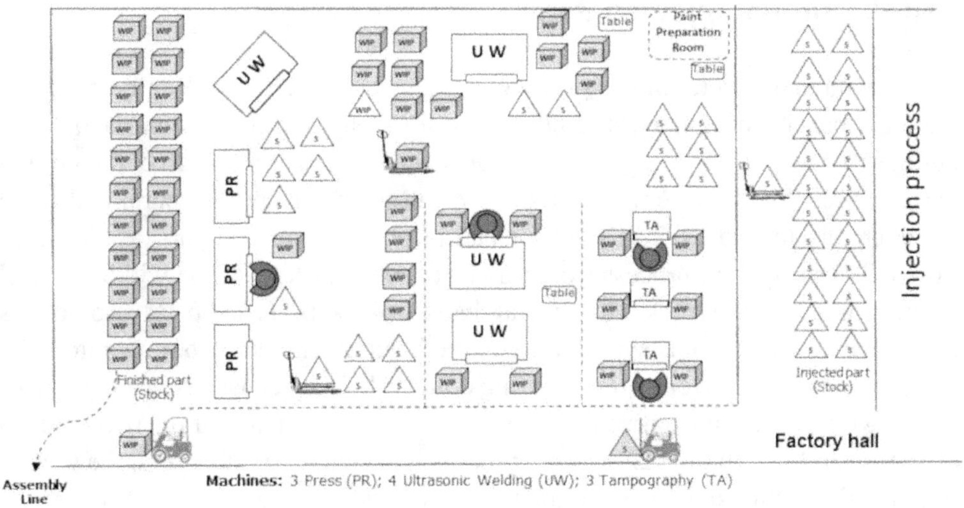

Figure 10. VSM of a Push System involving a refrigerator defrosting tray

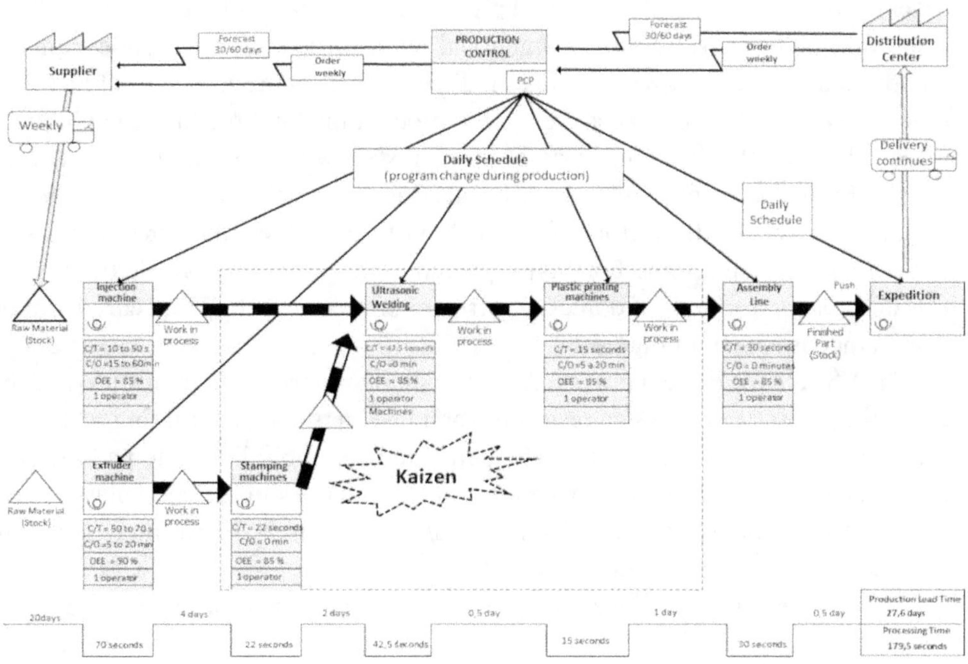

Visualizing the Flow of the Sector of Pad Printing of Injected Items

To understand the sector of pad printing of injected items, it was initially measured the area, bench, machinery, and packaging for designing the current layout. An outline of the current layout demonstrates three presses (PR) used for cutting plastic sheet, four ultrasonic welders (UW) used to join plastic parts and three tampography machines (TA) used for pad printing the plastic part, as shown in Figure 9. The stock in process stored in a steel wire rack (WIP) is within the pad printing area, and the stock with several ready-made subsets and parts also utilizes a steel wire rack, which is symbolized by the letter S. There are four operators per shift and a total of nine operators working double shifts by considering a supervisor working during administrative hours. The production sequence of the two families of sub-assemblies studied is: the first process is with a hydraulic press, the second one is with an ultrasound welder, the third one is with a pad printer (tampography machines) and the last one is to send final products to stock. It is worth mentioning that steel wire racks are moved and stored using forklifts, and packages are moved and repositioned with a hydraulic hand pallet truck in the pad printing area. In this case, the flow is not continuous, and work is carried out according to manufacturing orders controlled by the planning and production control - PPC sector.

In this study phase, the multifunctional team started out from understanding that the push production system makes it difficult to meet the company's functional requirements, such as daily changes in the product mix and production volume. It is also observed that stock in process stored in steel wire racks contributes to area misuse. From that point on, the stage of analyzing the initial situation and projecting a future scenario of the sector was carried out through the value stream mapping (VSM) of pad-printed items. For this purpose, the family of products called defrosting tray and evaporator door of domestic refrigerators was selected. The current value stream map is shown in Figure 10.

The VSM of the current process for the evaporator port subset family (A) and the defrosting tray (B) involves a press, ultrasound welders and pad printing machines. Subsequently, the station supplies the automatic assembly line. The pad printing sector depends on the injection machine sector and has a single shared plastic sheet extrusion machine as its internal supplier. It is also observed that there is a high level of inventory before, during and after each process.

As examples, it is mentioned the raw material received at least once a week and stored in a warehouse for 20 days or so depending on deliveries. After the injection and extrusion processes carried out by different and distant departments in the plant, 4 days of stocking manufactured parts have been mapped and allocated to a warehouse of stock in process. In the pad printing sector alone between three different processes containing ten machines, which are three presses, four ultrasound welders and three pad printing machines, 3.5 days of in-process stock have been mapped, i.e., involving parts stored in a steel wire rack and moved several times.

Figure 11. Outline of cell layout of the pad printing process

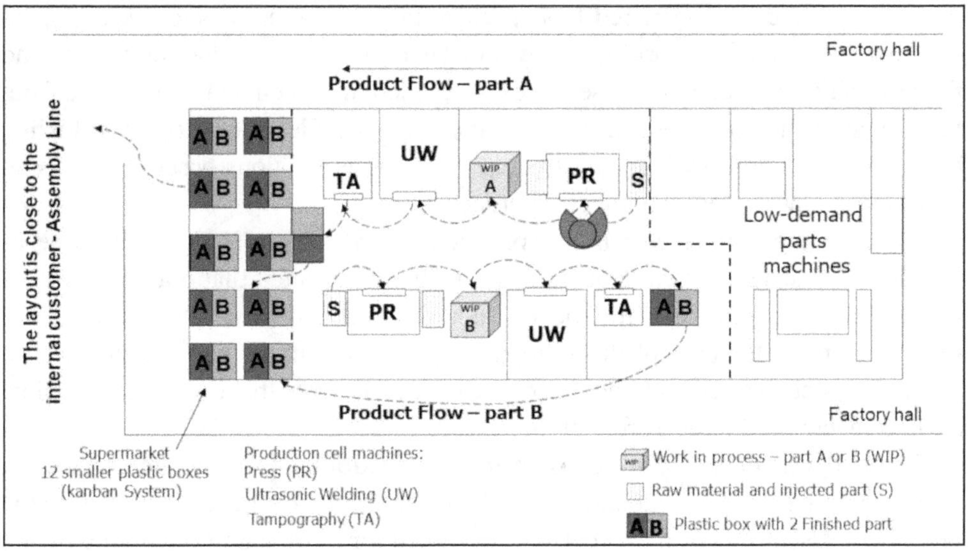

The assembly lines take 1 day to stock evaporator door subsets (A) and defrosting trays (B), which is why the flow of final product assembly is hardly ever interrupted. Subsequently, while assembling the product, these are sent in batches directly to the logistics distribution center, which in turn equates deliveries to customers and informs demand to the production planning and control department.

Then, the production planning and control sector assumes the role of operator of the push production system through production orders for what to produce, how much and when to proceed to the following process. Through the value stream mapping, production rate, i.e., takt time for manufacturing subsets is defined. The cycle time of the press is 22s, that of the ultrasound welding machine is 42.5s and the pad printing lasts 15s. The sector works double shifts whose work hours equal 8.4h/shift with 15% of the time spent at lunch and other necessary stops. Therefore, the time available is equal to 7.14h/shift or 25.704 s/shift considering a demand for 300 subsets a shift, which results in a takt time of 85s for both subsets.

In this configuration, the PPC department issues the order of production and delivery to those responsible for each sector and/or equipment based on the priority of daily delivery scheduled monthly and reviewed weekly.

FUTURE IMPROVEMENTS TO THE CURRENT PROCESS

The outline of a future layout shown in Figure 11 comprises the change of layout through the creation of a cell composed of three types of equipment added to a supermarket at the end of the process to meet the customer demand level in any circumstance. It has not been demonstrated, but it is worth mentioning that it was necessary to reduce the wired-type packaging by smaller plastic packaging to eliminate the need for a forklift in the cell. There was also a reduction in area and the possibility of accidents ceased to exist, as there has been a layout change from functional to mobile. The cell was designed and installed close to the following process (main assembly line), thus being close to the previous process (extrusion process) and away from the injection process.

Another important factor is overproduction that occurs at the same time as the lack of unscheduled components, but it cannot be stated that only one type of production system is ideal. It depends on several variables, mainly on functional requirements. In the case of this company, specifically in the pad printing sector, the push system was considered one of the causes of waste that does not add value, such as inventory and overproduction.

The cell was designed by considering the flexibility to meet the changes in customer demand without generating high inventory in process and considering the priority of assembly lines. The cell was designed for a collaborator by considering takt time of 85s for the two previously mentioned subsets, however, two or more collaborators must be added to work simultaneously in case of a larger volume.

The finished set packaging has changed from standard steel wire racks to smaller plastic boxes. In each plastic box, the defrosting tray subset and the evaporator door subset were stored in the same proportion. The mixing of different parts in the same packaging occurred due to the point of use in the same final product on the assembly line. Therefore, the packaging was shared by two subsets.

To ensure sales cadence on the assembly line, the operator's routine was standardized in which the method was described in an orderly fashion. Another relevant aspect was the visualization of limits by cycle time versus takt time.

The routine of standardized operations, in addition to establishing the sequence of activities that each employee performs within a given cycle time in a pull system, is relevant to reach the preset takt time and meet customer demand. Such a routine has two purposes: the first one is to provide the operator with a sequence order or routine to pick up the part, place it in the machine and remove it after the process, and the second one is to determine the sequence of operations that a multifunctional employee must perform on multiple machines during a cycle time.

Despite the proposed improvements, such as the establishment of production cells and standardized work, possibilities for future improvements in this cell were envisaged. It is observed that there is a level of utilization of 20% of machines (17s/85s = 20%), and that the operator works 86% of the time and rests 14% of the period. It is also noted that the cell productivity can improve because that the operator travels 30% (25s / 85s = 30%) of time inside the cell and 56% (47.5s / 85s = 56%) of time is spent on handwork. This enables to conduct future studies on the implementation of autonomation.

The study on cell layout and the development of standardization allowed redesigning the company's value stream mapping and establishing its future state, as shown in Figure 12.

Figure 12. VSM of a supermarket pull system

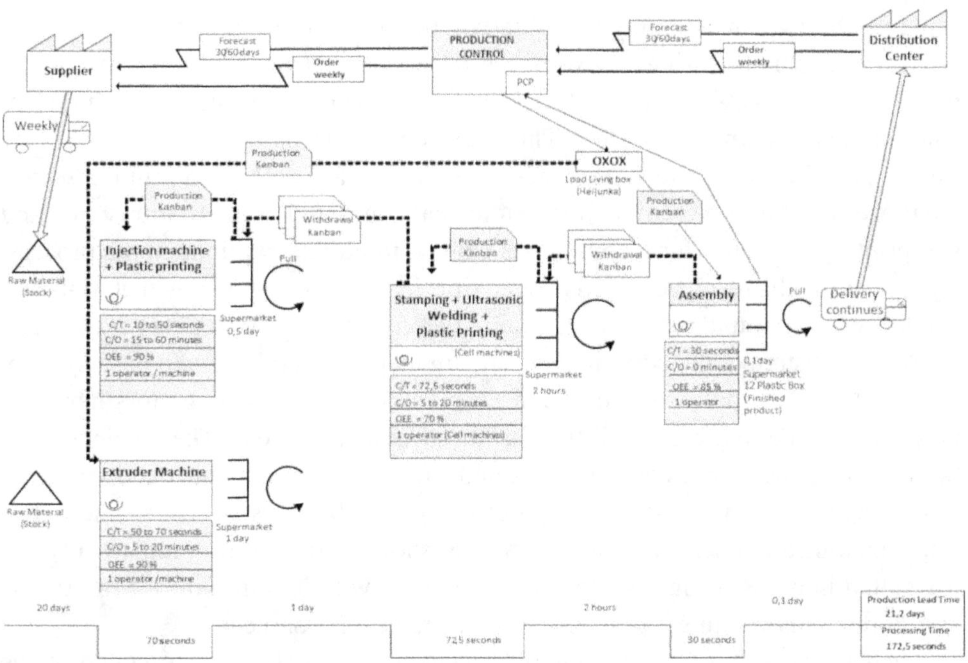

CONCLUSION

Measurable results were satisfactory, since greater savings were obtained than those estimated by the initial plan. For example, the stock in process went from 9,784 items to 2,596 items with a reduction in the size of packages and the *kanban* implementation. There was an improvement of 73%, while the initial estimate was a 25% reduction in inventory in process. By implementing the pad printing cell and packaging changes, it was possible to account for savings of 30% of the area used in processes and stocks once it was estimated an improvement of 15% as target. There was a reduction from four to three direct employees and the production capacity improved 19%, i.e., higher than the 10% desired at the beginning of the project.

The implementation of the pull production system improved response time for the assembly line by 92%, i.e., from 5.3 hours to 0.38 hours, thus making it possible to change a stock in process involving steel wire racks into smaller supermarket plastic boxes. These box-type plastic packages are in the pad printing cell, which is an area next to the assembly line, and are ready to be pulled by the assembly line. The cell works to supply the 12 boxes with items, totaling a maximum of 228 products in the established supermarket. Therefore, it was only after the implementation of basic tools of pull production, for example, a supermarket with families of products, i.e. subsets comprising an evaporator door and defrosting tray that customer service was achieved through an assembly line based on the Just in time concept.

The total manufacturing cycle time has been reduced from 79.5 s to 72.5 s. It was noted that it is possible to eliminate activities that do not add value in a push system when compared to the pull system. This occurred, for example, from eliminating the time to pack items and move packages since, in the case of the pad printing sector, the process was grouped in a cell layout format composed of press, ultrasound and pad printing machines for a continuous flow with no interruptions in the process. By measuring the result achieved on lead time by considering no raw materials and suppliers, it was found an improvement of 72%, i.e., from 8.03 hours to 2.24 hours.

As demonstrated, the current value stream mapping is related to a push system, where a subset of defrosting tray and evaporator door for refrigerators has been produced and there are possibilities for improvement (kaizen). The conduction of these improvements between mistakes and successes by the multifunctional team has changed the way of thinking about the production system over time. After implementing the supermarket pull system, as shown in the future value stream map, it resulted in items being dispatched in less time, with less inventories in process and stored, thus creating a productive and flexible environment.

Such a change from the creation of a manufacturing cell and, consequently, the elimination of the pad printing sector, generated a cultural change in the injection sector, not only as regards the implementation of *kanban* to manage and avoid the lack of materials, but also through the provision of a subsequent implementation of total productive maintenance to avoid machine downtime and enhance the overall equipment efficiency aimed to improve results. There was also a change in the occupation of operators since they started doing activities of pad printing of other injected items by respecting the available time of the machine during which the operator was idle. Such an improvement was estimated and leveraged through the value stream mapping of the supermarket pull system.

Therefore, concepts of Lean Production are linked to the Toyota Production System and aim to constantly improve its organizational systems by eliminating or minimizing the eight types of *muda* (waste). For such a purpose, it is based on five principles (value, value chain, continuous flow, pull production and perfection) that assist in determining activities that add value and those that do not add value which can be identified, analyzed, and improved by a wide set of tools.

The pull production principle not only reduces the stock of in-process and finished products, but it also reduces production costs, in addition to reducing the entire product production cycle, which improves the company's responsiveness to fluctuating customer demands and lowers the time to deliver products.

ACKNOWLEDGMENT

Brazilian Ministry of Health, Fluminense Federal University and Euclides da Cunha Foundation [TED 125/2019, grant numbers: 25000191682201908]; and the National Council for Scientific and Technological – CNPq [grant number 424722/2018-6].

REFERENCES

Aghajani, M., Keramati, A., & Javadi, B. (2012). Determination of number of kanban in a cellular manufacturing system with considering rework process. *International Journal of Advanced Manufacturing Technology*, *63*(9-12), 1177–1189. doi:10.100700170-012-3973-y

Arlbjørn, J. S., Freytag, P. V., & de Haas, H. (2011). Service supply chain management: A survey of lean application in the municipal sector. *International Journal of Physical Distribution & Logistics Management*, *41*(3), 277–295. doi:10.1108/09600031111123796

Chiarini, A. (2013). *Lean Organization: from the Tools of the Toyota Production System to Lean Office*. Springer. doi:10.1007/978-88-470-2510-3

Dennis, P. (2008). *Produção Lean Simplificada - Um Guia para Entender o Sistema de Produção mais Poderoso do Mundo*. Bookman.

Graban, M. (2016). *Lean Hospitals Improving Quality, Patient Safety, and Employee Engagement*. Productivity Press.

Hines, P., Holweg, M., & Rich, N. (2004). Learning to evolve: A review of contemporary lean thinking. *International Journal of Operations & Production Management, 24*(10), 994–1011. doi:10.1108/01443570410558049

Kaswan, M. S., Rathi, R., & Singh, M. (2019). Just in time elements extraction and prioritization for health care unit using decision making approach. *International Journal of Quality & Reliability Management, 36*(7), 1243–1263. doi:10.1108/IJQRM-08-2018-0208

Liker, J. (2003). *The Toyota Way: 14 Management principles from the world's greatest manufacturer*. McGraw Hill Professional.

Ohno, T. (1998). *Toyota production system: beyond large-scale production*. CRC Press.

Rother, M., & Shook, J. (2013). *Learning to see: value stream mapping to add value and eliminate muda*. Lean Enterprise Institute.

Smalley, A. (2004). *Creating level pull: a lean production-system improvement guide for production-control, operations, and engineering professionals*. Lean Enterprise Institute.

Spearman, M., Woodruff, D., & Hopp, W. (1990). CONWIP: A pull alternative to kanban. *International Journal of Production Research, 28*(5), 879–894. doi:10.1080/00207549008942761

Womack, J. P., & Jones, D. T. (2007). Lean thinking: banish waste and create wealth in your corporation. New York: Simon and Schuster.

Chapter 5
Seru Production System:
A Conceptual Model Proposal for the Clothing Industry

Emre Bilgin Sarı
Faculty of Economics and Administrative Sciences, Dokuz Eylul University, Turkey

Sabri Erdem
Dokuz Eylul University, Turkey

ABSTRACT

Seru production system is a flexible, cost-effective, workforce competence-oriented manufacturing management system that provides the opportunity to respond quickly to customer demand. As in parallel to technology and physical improvements, customer demands are also effective for development of production systems. The impact of change in demand has been seen on changeover from job shop to mass production, flexible, and lean manufacturing systems. Seru production system is more appropriate for targeting work both cost-effectively like mass production and maximum diversification like job shop production. This chapter clarifies the Seru production system and explain its use and benefits in the clothing industry. In the application, a shirt production is illustrated according to the principles of mass production, lean production, and Seru production. Thus, different types of production systems have been benchmarked. There will be potential study areas for proving the efficiency of Seru soon.

DOI: 10.4018/978-1-7998-5768-6.ch005

INTRODUCTION

The historical development of production systems, which are in parallel with technological developments, has reached a different dimension when it comes to today's competitive world, both due to the ever-changing demands of the customers and the presence of technologies aimed at meeting these demands. Production Systems have gone through very important stages since the 18th century. Before the Industrial Revolution, production had been traditionally done in job shops. Production had been carried out by the craftsmen in the job shops. The craftsmen used his mastery skills to dominate the production of the product. Handcrafts and craftsmanship were used in job shop and artisan works. In this production period, which was based on qualified labor more than capital, the transition from manufacturing into production was made primarily with the development of the division of labor.

Manufacturing refers to manual production from the combination of the Latin "manu & factum'' words. Today, manufacture can be defined as "the process of producing large numbers and quantities of goods by using machinery especially in factory" (Combley, 2011: 520). The difference that emerges when compared with the first meaning of the word shows the change of the concept over time. Nowadays, the manufacturing word has gained the meaning of machine-assisted production of any good.

The first phase seen as a transformation in the industry was the discovery of steam engine by James Watt in the 1760s and the use of steam power in the service of humanity. This was a great revolution for humanity and a giant step in the transition to mechanization. In the 18th century, Adam Smith, while arguing that each worker would work more effectively when he did only one job, divided the work into smaller pieces with his theory of specialization. These parts were suitable for the desired function where used. Thus, everyone got rid of producing their own piece. Standard parts can be produced in large quantities in certain places and therefore much cheaper. The production units who need these parts, have purchased these parts and started using them directly in their production. At that time, these replaceable parts have accelerated production tremendously and provided a transition from a skilled craftsman to a cheaper labor force. And then the concept of division of labor has developed according to specialization.

At the beginning of the 19th century, the concepts of standardization and interchangeability was developed and implemented. In this period, Frederick W. Taylor developed the concept of Production Management, which is another important stage for manufacturing. Taylor, by examining the workers in the finest detail at work, formed the foundations of scientific management. Frank and Lillian Gilbreth have introduced the principles of motion economics and have defined micro-movements (Therblig). Henry Lee Gantt developed the Gantt diagram to be used in business scheduling. Following these developments, the transition to mass production was made with the assembly lines developed by Henry Ford in the early years of the 20th century. This development further reduced the price of products in the market, accelerated production and enabled the products to be obtained by the wider community.

With mass production, the made of a product is 8-10 times faster, and even in some sectors 50-100 times faster and the labor effort spent on the product has decreased at the similar rate. In the beginning, this meant that the enterprises could produce more, thus adding wealth to their ownership. At the same time, entry barriers to sectors have fallen due to lower costs. Thus, small and medium enterprises were competing with large companies in the market. Together with this competition, a market environment has emerged in which not only production is sufficient, but at the same time it is necessary to sell what it produces. In the first years of mass production, it meant "More production = More sales" but this equation was broken with this competition.

Businesses that succeeded as long as they could sell the products, they produced turned to establishing cooperation with their customers based on the loyalty and partnership. The basis of this relationship lies in the fact that the company cannot sell a product or service without a customer, while the customer desire to meet their needs in the best way. On the other hand, in changing world, each customer has different and customized needs that should be taken into account by the companies. The changing demands of the customers and the failure of the companies to able to sell each product they produce to different customers caused the production systems to gain flexibility. The Flexible Manufacturing System (FMS), which first came to the agenda in the 1960s, refers to a production system with different characteristics in terms of operational and control features. In the FMS, standardized products are moved away, and a system is adopted that can better respond to continuous development. The FMS has accelerated the introduction of new products to the market and increases product diversification.

In the late 1980s, mass customization, that became more flexible as a result of quality and cost-optimized processes, has changed the scale of the production. This understanding has created an environment for developing customized products or services for the customer and thus achieving their goals with the differentiated production and more competitive understanding with low costs. In the competitive environment, this dynamic relationship between the enterprise and its customers is explained by the concept of mass customization. Mass customization practices are primarily concerned with the creation of combinations the products produced by mass production in accordance with customer requirements. The features of the parts designed in most of the customization process are then converted into production information by entering computer aided design and computer aided manufacturing

(CAD/CAM) systems. Computerized design systems enable customers to implement design changes and incorporate them into the production chain in a timely manner, while computerized production systems maximize machine utilization with allowing a wide range of products to be processed. The standardization of products and processes is the most important part of mass customization initiatives in order to enhance the flexibility.

Nowadays, instead of mass customization, the concept of mass personalization is on the agenda because customers desire individualized products and services. In this regard, mass personalization could provide a significant competitive advantage to meet the needs of every customer in the market that is primary concern for all businesses. This concept is widely accepted and frequently discussed by marketing and customer relationship management. However, the field of mass personalization has many question marks about production system.

Mass production has enabled the production of a large quantity of the same product after job shop-type production. As a result of the economies of scale, fixed costs in production have been reduced and products have been produced and sold at low cost. This has created customer satisfaction with the low price manner. Then, product differentiation has become essential for people upon customer needs and preferences came into play instead of low price. At that time, the adoption of just-in-time production system reducing the amount of inventory facilitated the flexible production and the production technologies used by the enterprises. FMS provided the flexibility and efficiency conditions of the production and mass customization has been realized. This transformation of the production systems is shown in Figure 1.

Figure 1. Transformation of the production systems

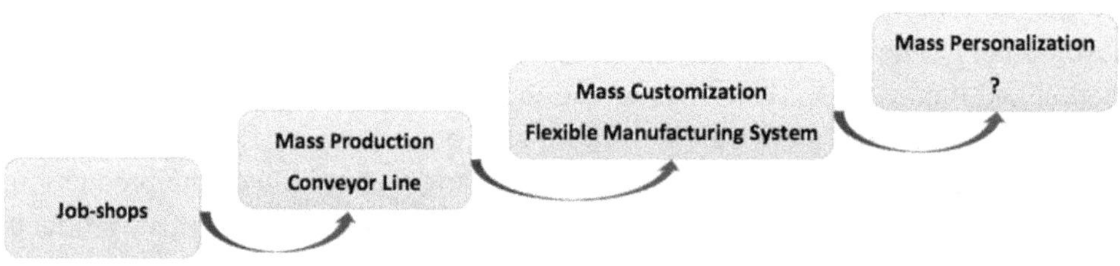

For the mass personalization, a production system does not exist like in the previous periods. In the current situation, works that require individualized product design and production are done in hot couture job shops and at high prices. However, customers expect these personalized products to reach lower prices as before. On the other hand, enterprises have owned their systems, have investments made and have established production facilities in this regard suitable for mass production. In general view, personalized products are seen more suitable for boutique business and are considered niche markets. However, the customer perceptions of the product have changed every passing day and insist on personalization. In order to respond to this insistence, the method that businesses should apply leads to a search that will both protect the effectiveness and provide personalization. When considering whether businesses will close their existing facilities and return to job shop production, or if there is anything to do to turn their facilities into job shops, Seru Production System comes into play.

Seru production system is an innovative approach over mass customization and lean production in a combined manner. This production system is based on the principle of converting the conveyor line system into cellular production. Figure 2 depicts this converting and emphasizes the Seru – cell system. In the Seru production system, more advanced workforce competence is needed in production cells instead of the line system because in the Seru system workers in the production cells aims to do work finish completely. Seru Production System has combined the flexibility of job shop production and the efficiency of mass production. For advanced production companies to use their investments for more efficient and flexible production, Seru production system could be recommended. This system, which makes it possible to respond to special customer requests especially for mass personalization, has lean, mass and flexible production characteristics as shown in Figure 2 that depicts this integration of these different production systems.

Figure 2. Seru Production System

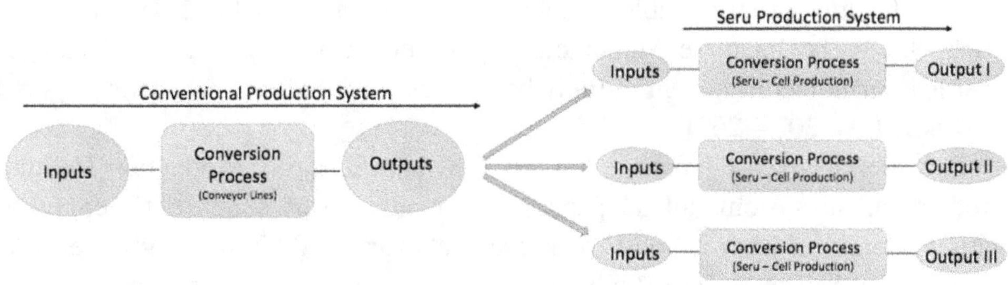

In this book chapter, it is aimed to introduce Seru Production System and its application in the clothing industry in a comparative manner with the other appropriate production systems that have been used so far. It is thought that the clothing industry, especially with a changing fashion perception, should follow the trends closely and still have an opportunity with this production system for a labor-intensive sector.

Additionally, since it has been emerging recently as an alternative to the existing production systems such as lean, flexible and mass customization, there is not any remarkable application of Seru Production Systems in textile industry in the world and not any applications in most of the countries like Turkey.

The rest of the chapter is organized as follows: in the background section Seru production system is introduced, the next section introduces the clothing industry and characteristics of producing clothes, problems and solutions. Solutions and recommendations section illustrates the Seru with an application from the clothing industry as evolving from previous production systems. Conclusion section concludes the chapter and future research directions recommends the potential research area regarding Seru production system in the near future.

BACKGROUND

Seru Production System

Rising of people's living standards and the transformation of consumption perceptions cause the fluctuations in quantity and the demand variability in many industries. Flexibility in production systems is used to respond to market mobility in this regard. Seru Production system is also called Beyond Lean in the Japanese manufacturing industry, and seen as next generation of lean, like a production management innovation (Stecke, et al., 2014: 46; Liu, et al., 2014: 4). The Seru Production System, which first appeared in Sony Japan in 1992, combines the flexibility of job shops, the efficiency of mass production and the environmentally friendly features of sustainable production (Zhang, et al., 2017: 291; Yin, et al., 2018: 851). The Japanese Manufacturing System, which has many advantages over lean manufacturing, can initially be seen as a kind of cell production system (Suzuki, 2004: 202 – 203).

Seru Seisan means Japanese Seru = Cell and Seisan = Production Cellular Production. This system, called Japanese Cell Manufacturing, should not be confused with conventional known Cell Production (Sakazume, 2005: 341). However, it is used as Seru Seisan or Seru Production System in order not to interfere with Cell Production, which is common in USA and Europe. Cell Production is based on the application of group technology. In group technology, similar parts or products are grouped in a part / product family regarding the specifications of the parts / products, the similarity of the processes and the production methods. In addition, the equipment used is grouped according to the similarity of the products rather than the function of the machines. In cell production, traditional job shops are clustered into several production cells with cellular layouts based on the principles of Group Technology in theory (Liu, et al., 2010: 95 – 96; Villa, & Taurino, 2013: 963; Liu, et al., 2014: 1 – 2). Distributing the conveyor lines to the cells forms the cells mentioned by Seru Seisan. Mass production, depicted by the conveyor line, is a type of generic production management that is widely preferred by many manufacturing companies around the world (Liu, et al., 2010: 90). Recently, Japanese manufacturing companies dismantle the conveyor line and distribute it to the cells, and implement Seru Seisan. Figure 3 shows Cell Production vs. Seru Seisan.

Figure 3. Conversion of Seru and GT Cell

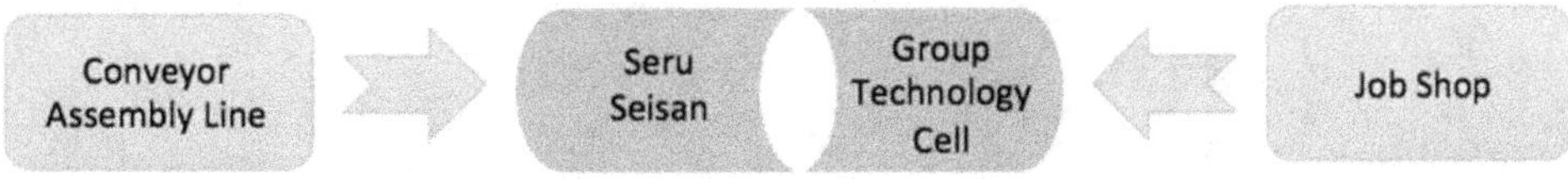

With the help of Seru Production System, flexibility is provided to the conveyor line and solutions are provided for capacity utilization problems (Miyake, 2006: 5). First in 1992, at one of Sony's factories, the long multi-product conveyor line was divided into several short conveyor lines dedicated to a particular product. After dismantling the long conveyor line, each product was completed on a short line as in the original conveyor line (Yin, et al., 2008:). Figure 4 shows how a conveyor line is dismantled.

With the continuous improvement, these short lines were gradually shortened and continuous innovations were created in the layout. This change facilitated adaptation to rapid and frequent changes in the production system, enabling it to meet dynamic market demands. These production cells, which can be modified and restructured, increase flexibility, are seen as an innovation in production management (Liu, et a., 2010: 91). The Seru production system is a type of cellular production of configurable cells instead of a fixed line. Cells can be formed as I - L or U. The formed cells are used not only for production but also for assembly, packaging and testing. Cells are identified in order of priority and are sensitive to cost reduction in determining the operation strategy (Yin, et al., 2017: 1).

Seru is considered as a human-centered production system and consists of one or more multi-skilled workers (Liu, et al., 2013: 938). Kono (2006: 52), explained Japanese Cell Manufacturing System according to number of workers and their assignments to works. And Then these are referred three basic types of Seru: Divisional Seru, Rotating Seru and Yatai (Liu, et a., 2010: 104 – 106). In Divisional Seru, each worker is assigned to single one task. A worker visits several specific workstations. Every worker in Rotating Seru manages all tasks from start to finish.

Figure 4. Dismantling the conveyor line

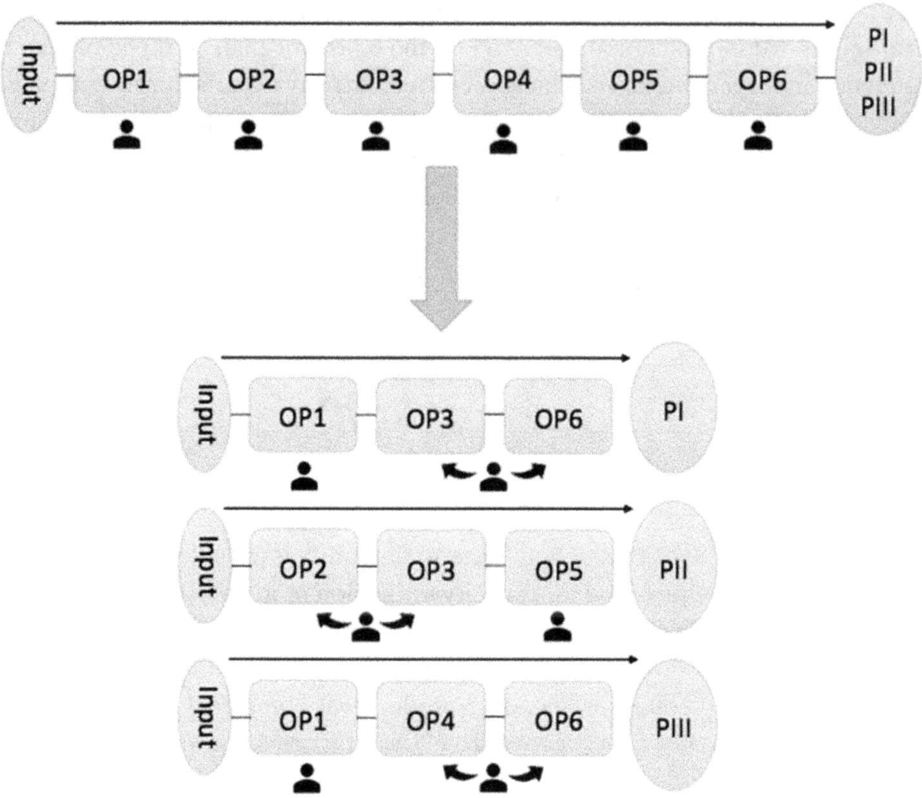

Figure 5. Divisional – Rotating and Yatai Serus

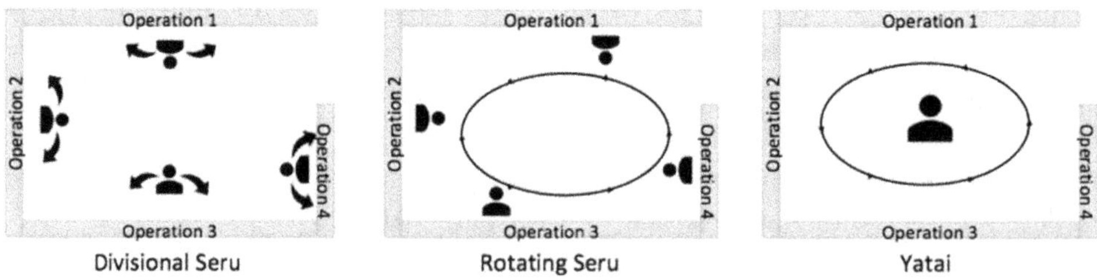

These workers, in turn (in a fixed order), complete their work by visiting them from one workstation to another. In a Yatai Seru, the worker's cycle time should be set to an acceptable level to achieve the best production efficiency and to prevent workers from slowing down. Yatai, a type of seru, is regarded as the most perfect human /

business organization (Stecke, et al., 2012: 107). Figure 5 exhibits the Divisional – Rotating and Yatai Serus.

Yatai is seen as the ideal seru mode to be achieved in the Seru production system. Stecke, et al. (2012: 108 – 109), describes the characteristics of Seru with Japanese words: kanketsu, majime, and jiritsu. Kanketsu means completeness and explains the completion of all operations necessary to produce the product in one Seru. Majime is to place all Seru inputs such as people, equipment, machinery and tools closely each other. Jiritsu explains the ability of Seru to be self-sufficient, to learn and to provide its own development (Singh, 2017: 83; Shao, et al., 2017: 1166).

The primary factor to implement Seru Seisan is to increase the flexibility of the conveyor line to produce products on time. In Seru Production System, production processes of a product usually are manufactured in a cell. In addition, in some cases, each cell may be dedicated to one or more specific product types that cooperate with other cells. Such a production unit requires that all production resources, such as equipment, equipment and tools, be located within or near the production unit (Liu, et al., 2010:91). Since Seru Seisan is known as a low automation system, it is not rational duplicating automation equipment with high cost when distributing conveyor lines to cells. To avoid equipment conflict and reduce costs, it is recommended that expensive automation equipment be replaced with simple equipment with similar functions that can be easily reproduced and replaced at low cost (Zhang, et al., 2017: 291). Furthermore, in the Seru production system, the workers are expected to be cross-trained with many skills. Compared to the worker on a conventional conveyor line carrying out a particular task, a Seru worker must always be able to perform a variety of non-consecutive tasks. As a result, the automation level of the entire production system is reduced, and the supports of highly skilled employees are required.

The Seru production system is suitable for tracking more agile and flexible production systems. The cells consist of mobile workstations, light equipment and highly skilled workers. The forming and reconstruction of the cells in the Seru Production System is faster and easier. The demand fluctuations can be handled by changing workforce and workstations in the cells. Short delivery times, low inventory and high employee motivation are other strengths of Seru Seisan (Isa and Tsuru, 2002: 551).

Seru production is regarding as an ideal mode of production for mass customization. However, the application of Seru production won't be same in various market conditions as well as changing conditions amongst factories. So that, it would be appropriate to draw up a general framework of implementation Given the market conditions as well as the peculiarities of each company, the process of implementing Seru does not follow a single formula. However, some general methods or guidelines are discussed in the literature and may serve as a guide for the study and implementation of such a production system.

According to Sakazume (2006) as cited in Zhang, et al., (2017: 291), one can outline eleven conditions for the successful implementation of Seru production, in which they were categorized into three criteria:

a) **market conditions**: variable product mix, variable demand, need to produce multiple products;
b) **product conditions**: total assembly hours, low number of components, products and small components and;
c) **process conditions**: operators with various skills, few difficult operations, no need for expensive equipment, high use of shared equipment, small equipment.

In this sense, Yagyuu (2003) as cited in Liu et al. (2014: 3) discusses the procedure of building a Seru system in eight steps, including manufacturing system selection and product type, investigation and improvement of current manufacturing conditions, manufacturing system engineering design, operation planning, multifunctional training of operators, production balancing, redesign for low automation equipment and production stabilization work. In addition, Liu et al. (2014: 4 - 15) propose a framework for implementation that divides the process into five main steps, starting with the selection of the product and process type, and unfolding in the planning of the type of layout, process flow, supply forms, operations balancing and multifunctional training. The framework also provides information about the stabilization phase of a Seru operation as well as ways of performance evaluation and the search for continuous improvement.

Seru Seisan, is mostly applied in the electronics industry in Japan (Isa and Tsuru, 2002: 552). Suitable for the processing of small products, such as electrical goods produced by hand-crafted or simple equipment. Seru Seisan is difficult to apply to heavy products with complex operations. In addition to them, the transport equipment industry has intention to implement Seru Seisan. However, Seru Seisan has several successful applications in complex manufacturing industries such as Toyota. Unlike other production systems used in the mechanical industry, Seru is suitable for hand-crafted industries that produce a very small number of special types of equipment in assembly. It is also possible to use Seru widely in the service sector, which has switched to personal operations with several types of equipment (Wang, et al., 2013: 4). This chapter aims to focus on the clothing industry. Therefore, firstly production in garment is given and then, Seru implementation in clothing is explained.

Clothing Industry

In the clothing industry, products are manufactured by confection method. Confection means the production, which is brought out quickly, in a short time. The goods produced in clothing industry protect people from external influences, as well as dressing adds a decorative quality and indicates that it belongs to a certain socio-economic community (Weller, 2004: 1–2). The production of serial garments according to standard sizes is called ready-to-wear. Ready-to-wear is an industrial branch that meets the clothing needs of people with the fabricated production method.

Ready-to-wear includes all the garments produced in series, which are produced in large numbers and sold according to the buyer's measurements, based on the average measurements obtained by statistical data. Ready-to-wear is a cheap and easy form of clothing that carries on fashion trends. It provides the pleasure and comfort of wearing the garment without waiting to produce the garment, without the problems of finding a model, choosing the fabric, deciding who and how to sew the garment (Waddell, 2013: 23). In addition to be a universal and social clothing style that provides economic gain and convenience to the working people, it has a positive effect on daily life.

The clothing industry has emerged to meet an important human need, such as the protection and beautification of the human body. However, the roots of the sector have gone back to much older times. Archaeological excavations revealed that during the Stone Age in Northern Europe, people covered their needs to cover their bodies with clothing made of leather. Archeological studies in the southern part of Europe reveal that the pointed and hook-shaped bone needles are an indication that people were simply trying to sew clothes at that time. Progress in sewing apparel was achieved by making iron needles in the Middle Ages (Harris, 2009: 104 – 105).

With the sewing machine found by French Barthelemy Rhimonnier, garment sewing was saved from being handcrafted, and with a needle and a shuttle cross-stitch a two-threaded machine developed by American Elias Howe a big step was taken in terms of industrialization in this field. Later, the industry became more widespread with the sewing machine developed by Isaac M. Singer. The discovery of Hoffman irons, which made ironing easier with the buttonhole machine found in the late 19th century, led to a more rational operation of the clothing industry in industrialized countries (Niwa, 2002: 238 – 239).

Developments throughout history have led to the rapid growth of the sector. The materials used to protect, close and show the body beautiful, vary widely among ethnic groups around the world. Nevertheless, so that these materials will adapt comfortably to the shape of the human body and the movement of "aesthetic" has called a common feature. As a result, these attributes affect almost every stage of the entire production and distribution process in the clothing industry and this is the biggest obstacle to the automation of the production process. Therefore, most of the garment production is based on human labor. The most effective means of holding two pieces of fabric, joining the corners and sewing them in the machine are still human hands. Soft materials also affect packaging and dispensing methods.

In the retail sector, hangers, shelves and stationary stands are designed for attractive and beautiful display of clothes made of flexible materials. For a long time, apparel has been defined as the most intensely changing consumer product category. The combination of fashion and seasonal influences in developed countries has led to major changes in product offerings at both wholesale and retail levels many times a year. As economies develop, fashion has become an important factor in apparel if people can afford it.

The effects of fashion are reflected in design changes such as color, line, silhouette, pattern, shape and body. In addition, weather conditions, annual holidays and the opening of schools, such as calendar events in seasonal changes are important in product change (Jackson, 2007: 194). There is still dependency on people during the garment production phase. With the development of technology in recent years, automation has started to find a place in this field to decrease this dependency gradually.

Workflow and Production Systems for Clothing Industry

The clothing industry is a dynamic sector due to its structure, it has high export opportunities, and the product type and production type can be easily changed and has the advantage of adapting to market conditions. However, apparel is the branch where the most errors and problems arise in the textile industry because it is the last production chain where all raw and auxiliary materials come together in a model and gain meaning.

In the textile job shops and garments where mass production is made, the division of labor is among the most important details. It is important to make this distribution correctly and to have experts in the field in order to keep things moving smoothly. Clothing process generally consists of raw material uptake and control, pattern and mold preparation, cutting, sewing, pressing and packaging.

Production Systems Used in the Clothing Industry

An apparel manufacturing system consists of a combination of material handling, production process, labor and other equipment that directs the workflow. Various garment manufacturing systems will be described in this section.

- **Straight line system:** Although it is a system that is rarely used today, it is accepted as the first attempt to regulate the workflow in serial garment production. The production unit passes through operations that have a certain sequence using carriers or channels. When necessary, one operator can even support multiple operators by shifting the channels (Senanayake, 2004: 11 – 12 / Babu, 2012).
- **Straight Bundle System:** The production unit consists of pieces cut into bundles. Usually the amount of the bundle is determined by the amount of the layers of the part. The bundle is opened by the operator, processing is performed on each part, cuts the label on the bundle and reconnects the bundle to the next operator. The bundle waits until the next operator is ready (Senanayake, 2004: 12 / Babu, 2012).
- **Progressive Bundle System:** This system, which is derived from the straight bunddle system, reduces the time lost by opening and reconnecting the beam. The bundles coming to the sewing section are given to the operator who is planned to perform the operation. The operator performs the operation for each part in the bundle and reconnects the bundle and processes it on the coupon or electronic bundle ticket. The bundle flow defines the basic operation sequence of production and the workstations on which the operations are performed (Senanayake, 2004: 12 / Babu, 2012).
- **Unit Production System:** This system, which has been used as a mechanical system for many years, has made significant progress with the development of computer systems. In the unit production system, the production unit is not a bundle but a single part. The garment pieces are transferred from one workstation to another workstation from the predetermined sequence. The system is positioned as close to the left hand of the operator as possible so that the operator can move less during pickup and placement of the workstation. All parts that form a garment are loaded into the carrier designed for this purpose. Carriers feed the production line by constantly navigating between the machines. When the operator completes the work on the carriage, he presses a button on the sewing machine and activates the mechanism. When the carrier leaves the workstation, the other carrier feeds a new job to the station (Senanayake, 2004: 12 / Babu, 2012).
- **Quick Responsive Production System:** Developed to provide rapid response to market changes; this system is suitable for small volume and individual models. Each workstation consists of two or four machines and an operator performs the desired operations on the part. In this system, operators work standing up to move easily from one workstation to another (Babu, 2012).

Flexibility in Clothing Manufacturing Systems

The demand for flexible clothing manufacturing systems has increased in the fashion industry with the development of fast fashion changes and customer-specific production concepts (Tomastik, et. al., 1995: 3152). Production flexibility in the garment industry can range from stores that produce bespoke garments to firms that produce uniform products (Wong, et al., 2001: 2547). Multi-product flexibility and the need for transformation to ready-to-wear companies that enable faster response to the market have become inevitable. In the past, even with a slight change in the model, switching from one model product to another is seen as disruptive; companies would like to continue production without making changes for a long time. However, the number of ready-made clothing companies producing uniform products is now decreasing (Wang, et. al., 1991: 443 – 444). Multi-product flexibility and the transformation to ready-to-wear companies that enable faster response to the market has become the key to rapid fashion production and customer-specific production. Therefore, it is necessary to create ready-made clothing companies that can produce more flexible production. What is needed is that ready-to-wear companies increase the number of product varieties both in terms of quantity and number of varieties and that they can respond quickly to the market with the right product. It is not economically feasible to expand the existing stock level to provide a fast response. Therefore, flexible clothing manufacturing system with fast production capability will be the solution (Tomastik, et. al., 1995: 3152).

Specifications of Flexibility in Clothing Production

In order to develop a flexible production system, materials, machines, equipment and workforce must be planned to ensure the correct number, time and availability. Preliminary engineering studies related to the workplace, machine, skill requirement and time to decide the production method are very important to achieve the goal. Training needs where possible should be reduced. Clearly written work instructions covering all business steps and motivating the worker to learn quickly are very important. Multi-talent training for employees will enable easier problem solving in bottlenecks. It is important that workers have continuous communication with the support units such as production line chief, repair maintenance personnel, etc. The flexibility of the process should include all cutting, sewing and finishing operations as well as all other functions (Lee, et al., 2000: 282).

It is not enough for garment companies to focus only on the factors described above in order to be successful. Companies should start to follow technological developments and investigate how they can adapt to existing or new clothing production systems. In doing so, smaller order quantities, quality at source, lower inventory levels, wider fabric and model facilities, multiple capability of the employees, commitment to more employees, developing problem prevention techniques, continuous product and quality improvement, information systems which are the requirements of the flexible production system, are should be considered effectively.

Productivity in Clothing Manufacturing Systems

Productivity in enterprises is an indicator that determines how effectively the various resources in production are used. If efficiency is high, costs are low or total production is high. This means more profit. Enterprises should pay attention to increase the quality as well as the amount of production while increasing the productivity rate. The increase in productivity means the increase in output per input unit. However, productivity increases not only as an increase in production, but also to produce a product with the least expense, using less input (Lin, et al., 1994:21). High productivity has always been the main purpose of working subject for every manufacturer, anywhere. It is desirable to increase productivity without reducing material and product quality, without forcing the worker over, damaging the operating means and increasing the cost.

Nowadays, apparel companies are expected to high flexibility, low manufacturing costs, high level of quality security, contemporary modern and technological breakthrough and reliability against the customer. Increasing importance of fashion, product design and product variety in garment demand requires more flexible production structures. Factors affecting productivity in clothing industry are qualification of workers, Working methods, Workflow organization, Wage system, Product model, Maintenance of machinery and equipment, Absenteeism and worker turnover rate, Supply of raw materials and equipment, Opportunities of imports, exports and incentives (Bheda, et al., 2003: 14).

In fact, productivity and efficiency measurement have a dynamic structure. For the measurement of productivity and the effective use of these measured values, it is necessary to know what affects productivity, to collect and test appropriate data, and to use experience and judgment locally. In addition to this, education is an integral part of the transition to a new system. Learning quickly increases productivity first. However, this impact slows down as it approaches saturation. The way to delay saturation is continuous training. Productivity in clothing industry, can be raised with controlling, reorganizing the missing areas and improving.

SOLUTIONS AND RECOMMENDATIONS

Conceptual Seru Model for Shirt Production

The characteristics of production systems are superior to each other (Hayes, and Pisano, 1994: 77-78). Seru Production System as mentioned above, is a Japanese production system which has the advantages of previous production systems. Table 1 is intended for comparison purposes in order to better understand the characteristics of the Seru production system and compare it with other production systems.

Table 1. Characteristics of production systems

Charecteristics of System	Assembly Line System	Lean Production	Divisional SERU	Rotating SERU	Yatai SERU
Skilled Workforce	Very Low	Low	Low	Moderate	High
Customization	Very Low	Moderate	High	High	High
Personalization	Very Low	Low	High	High	High
Continous Flow	Yes	Yes	Yes	Yes	Yes
Quick Response to Changes in Market	Very Low	Moderate	High	High	High
Single Part Flow	No	Yes	Yes	Yes	Yes
Need for Concurrency/ Integration of Process of Design/Manufacture	Low	Low	High	High	High
Production Volume	High	Moderate	Moderate	Moderate	Moderate
Tangible/Intangible	Both	Both	Both	Both	Both
Need for Rotation	Low	Moderate	Moderate	High	High
Need for Job Enrichment	Low	Moderate	Moderate	High	High
Layout	Line System	Cellular Production	L or U Shaped	L or U Shaped	L or U Shaped
Main Argument	Productivity	Zero Waste	Made-for-me	Made-for-me	Made-for-me

Seru production system, especially with the high competence of the workforce, allows for personalization and quick response to market demand. While providing the motivation with the enrichment of the work of the employees, the infrastructure is created for tailored production. Seru production system is also shaped with different features according to its developmental stages. Following these developments, the ideal production mode should be determined and used for production purposes. In this study, as a result of the cases encountered in men's shirt production factory, production is carried out with different production systems. With the help of these cases, the application of the enterprise against the demand is explained with the reasons of the production system.

Figure 6. Men's shirt production in assembly line system

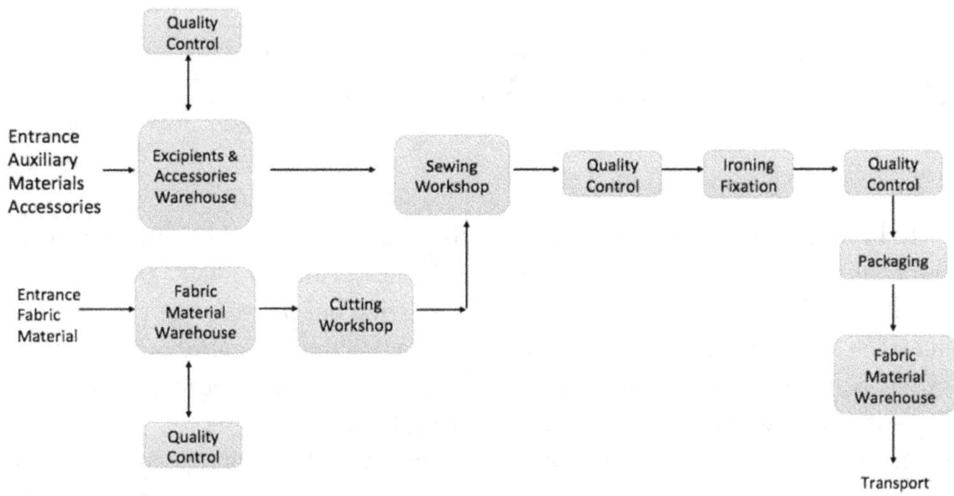

Production Systems Before Seru

Assembly Line System

Men's Shirt production factory will produce thousand of pieces in a lot, medium size, regular fit mold, long sleeve, double cuff, buttoned collar, white cotton fabric, white buttons and white stitched men's shirts. With the assembly line system installed in the factory, mass production is made in bunches. In order to complete this order, fabrics, auxiliary materials (yarn, buttons, interlining) and accessories were supplied from suppliers. All products coming from suppliers are stocked in material warehouses and controlled in quality process. Then the fabrics go to the job shop for cutting to the desired mold and standard for thousands of men's shirts. The cut fabrics pieces are transported in bundles to the sewing job shop. Shirt fabrics cuts are sewed, and arm, cuff and collar characteristics are created in the sewing job shop.

After sewing process shirts that are manufactured, is gone through to the quality control process. Sample shirts are selected and analyzed with statistical quality control. After the quality control stage, the shirts are carried in bundles to the ironing area.

Ironing and fixing process is done. After ironing, the final control phase is passed, and approved products are packaged and sent to the product warehouse. When the product order arrives, the finished products are transferred from the warehouse. Figure 6 shows the steps and flow of the production process in assembly line system.

Figure 7. Men's shirt production with lean manufacturing

In the shirt factory, the production of the shirts in assembly line system leads to lots of inventory (material - semi-finished and finished product). In addition to this, the quality control in the line system is carried out between the process steps and if the error / defect is encountered; there is a risk that the entire product bundle will be defective. There is also a lot of carrying for fabric bunches between the job shops in this system. This creates another workload density. In order to overcome all these negativities, a lean production system would be appropriate instead of assembly line system and unnecessary inventories, control steps and carryings would be prevented, and waste would be eliminated.

Lean – Group Technology

Men's shirt production factory was ordered thousands of pieces shirts in various different types of size, color, sleeve length, collar feature. These are like these; hundreds of pieces in a batch, medium size - regular fit mold, long sleeve, double cuff, buttoned collar, white cotton fabric, white buttons and white stitched men's shirt; hundreds of pieces in a batch, large size - slim fit mold, short sleeve, single cuff, straight collar, green poplin fabric, white button and green stitched men's shirt, etc. The cellular production method used in the factory and group technology machines were brought together. In the production area managed by the lean production system, after the order is received, fabric, auxiliary materials (yarn, button, interlining (wadding)) and accessories are ordered from the suppliers in accordance with the specifications of the shirts to be produced and so that the pull system operates. All materials coming from suppliers are taken into production area without stocking and quality control. The fabrics are cut according to the order specifications in the cutting cell and the cut pieces are delivered to the sewing cell without waiting. Products are sewn in the sewing cell, and also collar, cuff, buttonhole and button sewn operations are completed and delivered to the ironing cell in one-piece flow and packaged after ironing. In the one-piece flow system, inventories do not occur. In addition, all the operators involved in the cutting - sewing and ironing stages act as quality controllers and the failures is eliminated as soon as it is noticed. Thus, lean system does not require extra the quality control stages. Figure 7 shows lean manufacturing in cellular production.

Figure 8. Men's shirt production in Seru production system

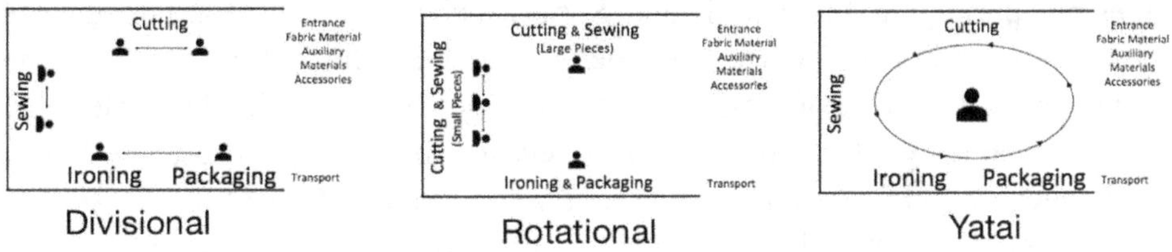

In the shirt factory, different products in the cellular production order are made easily with the help of group technology. In the system, the inventories are not needed and the view of quality at the source is dominant. In addition, the products are transmitted between cells by one-piece flow. However, today's consumer market demands far more than ordering five different types of shirt for one thousand of orders. Every consumer wants to wear the shirt he preferred or even he designed. In this case, millions of different products arise. It is necessary to perceive these requests of customers as tailored-production and to design the shirt production factory in this direction. The Seru production system is suitable for mass personalization and is presented as a choice in this study. With Seru production system, both the efficiency of the line system and the flexibility of the lean system are provided simultaneously. The way to realize this, especially in a labor-intensive work area such as the clothing sector, is through workers. By increasing the productivity of the employees with educating and encouraging them to do more than one job, diversity will reach unlimited dimensions.

Seru Production System

Consider that, men's shirt production factory got an order for one type of shirt by the customer named "Efe Deniz Arat (EDA)". This shirt is medium size - slim fit mold, long sleeve, single cuff, shoulder width (47 cm), straight collar, blue cotton fabric, white button and blue stitched. On the cuffs "DAE" and "MES" characters will be embroidered with navy blue shiny yarn. Shirt production cells are formed with Seru production system in the factory. As soon as the customer's order is received at the store, the order form is also included in the sales system of the suppliers and the necessary materials and fabrics are shipped to the factory. The materials arriving at the factory come directly to the Seru which is responsible for the shirt production order of by EDA. Production in this cell can be carried out in three levels, which show how effectively the Seru application is used in the shirt factory. Figure 8 exhibits this development levels.

The Seru is constructed as shown in Level 1 for the purpose of producing only a specific product. Shirt productions stages are basically aligned as cutting-sewing-ironing and packaging. In the divisional Seru, which is also seen as a starting level, the line layout was established for the ordered product. In this order, two workers are assigned for cutting. One of these workers is responsible for cutting large pieces for shirts and the other worker is responsible for cutting small pieces. These workers are interchangeable and able to do each other's work and have trained on this subject. Two workers are also working in the sewing stage, the first one is sewing large parts and also responsible for buttonhole and button sewing operations. The second worker is responsible for sewing small pieces such as cuffs, collars, processing the letters "DAE" and "MES" in the cuffs that the customer wants, and combining the parts with the shirt body. The two-employee performing the cutting operation are also cross-workers. After the sewing process, there are two employees for the ironing and packaging works, and these workers are also able to do each other's work. In the packaging step, a label with EDA is printed on the package and the product is shipped to the ordered address.

The Seru production system requires an application that to improve employee competencies and to do production with workers who can do each other's work. As Seru production system deepens its practice, it is aimed to increase employee qualifications. In the rotating Seru, which is also seen as an advanced level, cutting and sewing operations are combined. An employee is responsible for cutting and sewing large parts. Three workers were assigned to cut and sew small pieces. One of these workers cuts the cuffs and sewing, and sewing the cuffs' button and the buttonhole, the second is responsible for buttonhole and button sewing operations on the body, also processing the letters "DAE" and "MES" in the cuffs, the third one is assigned for cutting and sewing the collar and assembly the cuff and collar to the body. All three workers are inter-cross worker. One worker is responsible for the ironing and packaging of the shirt. In the packaging step, a label with EDA is affixed on the package and the product is shipped to the ordered address. In the level 2 Seru, which is an improved Seru compared to the starting, it is possible to work with fewer workers and the competencies of the employees have increased. Employees complete both cutting and sewing at the same time.

The Seru production system aims to improve employee competencies to make production that can be done by the single worker from start to finish. When the application of Seru production system in the enterprise reaches upper level, it is intended to use Yatai Seru. As shown by Level 3, the cutting, a single worker carries out sewing, ironing and packaging stages of the EDA's shirt. This worker is liable for EDA's shirt and is doing tailored - production. Similarly, another worker employed in a parallel Yatai Seru, produces another shirt ordered by "Mr. Brown". In this way, each Seru works like a tailor's job shop in the men's shirt factory. After completing the order and labeling it for delivery to the customer, each employee continues to carry out the personalization of a new customer demand. When this work is occurred with as many Seru as the number of employees in the factory at the same time, the goal of mass personalization is achieved.

It is clear that the Seru production system is a production system capable of providing products that are desirable to the consumer market with assembly line efficiency and lean manufacturing flexibility, and as a result, mass production-mass customization can provide advanced mass personalization. With continuous improvement methods, Seru is developed and changing customer demands supports the competencies of the workers, so that every order is fulfilled in a complete and personalized way. Tailored - production is done by implementing Seru in shirt factory. This makes each employee work as if they were producing hot couture products in a tailor's job shop. In this way, the production system, where thousands of workers produce personalized products, can only be formed with the Seru production system and mass personalization can take place.

CONCLUSION

The change in customer demands progresses in parallel with the developments in the production systems. Nowadays, customers prefer different and personalized products in a reasonable time and cost. This leads businesses to seek a production system where they can produce personalized products at low costs.

Seru Production System, as originally Japanese cell manufacturing system, could be used to respond quickly to changing customer requests. Seru Production System, which is based primarily on the division of the long assembly line for the purpose of providing product variety, has been frequently used in the electronics industry. In order to reflect the differentiation on the products, assembly lines are divided into Serus. However, when these implementations were examined, it was seen that instead of being shaped by individual customer demands, they were used to adapt to the product diversity targets of the companies.

In this study, it has matched the clothing industry with the Seru Production System which has not been associated before. It is known that there is a rapidly changing demand in the clothing industry with the perception of fashion. However, in addition to fashion, customers expect the personalization performance, which they receive from a hot couture shop, from the apparel industry. Besides the purpose of meeting this expectation, factories in the apparel industry is willing to use their current production capacities.

In the application part, in accordance with the purpose in the study, a case from textile industry was investigated in that the order of products were coming to a shirt production facility and they were produced according to the principles of mass production, lean production and Seru production. It has been seen that Seru has many advantages in terms of managing many attributes, organizing the production environment, inventories, equipment and employee well. Seru production was expressed in detailed as divisional, rotating and yatai Seru types according to number of employee and their duties. The purpose of the study has been achieved by completing a specific order with a single employee in each Seru cell. In this case, in order to produce products with the desired dimensions and features for each customer, ready-made factories could be transformed into mega tailor job shops.

FUTURE RESEARCH DIRECTIONS

Seru Production System is proposed as a method for producing products with the desired dimensions and features for each customer, ready-made factories have to be transformed into mega tailor job shops. Long assembly lines are divided into small cells as a working principle of the production system. In the clothing industry, it will be possible to produce different products by making production in these cells. The aim of this study is to give an idea about the usage areas of Seru Production System for future studies.

Hereby, an alternative model is shown for production systems that have to be changed according to customer requirements. For industries that are more labor intensive than equipment, Seru Production System can be considered as a suitable model and examples can be increased with new researches. For the next decade, simulation studies and applications of the Seru could be widened both in national and international scale for various types of products. In the long run, it will be the efficient alternative against to existing mass customization based production environments.

Additionally, during nCovid-19 pandemic period, it can be seen that many textile companies were stopped producing except for those where responsive, agile, flexible and lean production systems adapted ones. In these companies, there were fast transformation in production from regular ones into mask, protector overalls clothes since there were huge demand across the world and existing companies are not ready for such an unexpected big demand. Seru production systems could be a better solution than others for such unexpected cases for shifting the product spectrum for both textile and other relevant medical good producers. In the near future, there could be research studies in depth regarding these transformational processes in case of such unexpected situations.

REFERENCES

Babu, V. R. (2012). *Industrial engineering in apparel production*. Woodhead Publishing Limited. doi:10.1533/9780857095541

Bheda, R., Narag, A. S., & Singla, M. L. (2003). Apparel manufacturing: A strategy for productivity improvement. *Journal of Fashion Marketing and Management*, *7*(1), 12–22. doi:10.1108/13612020310464331

Combley, R. (Ed.). (2011). *Cambridge business English dictionary*. Cambridge University Press.

Harris, S. M. (2009). *Smooth and cool or warm and soft, investigating the properties of cloth in prehistory*. Oxbow Books.

Hayes, R. H., & Pisano, G. P. (1994). Beyond world-class: The new manufacturing strategy. *Harvard Business Review*, *72*(1), 77–86.

Isa, K., & Tsuru, T. (2002). Cell production and workplace innovation in Japan: Toward a new model for Japanese manufacturing? *Industrial Relations*, *41*(4), 548–578. doi:10.1111/1468-232X.00264

Iwamuro, H. (2004). *An Easy Book About Seru Production*. Nikkan Kogyo Shimbun. (in Japanese)

Jackson, T. (2007). The process of trend development leading to a fashion season. In *Fashion Marketing* (pp. 192–211). Routledge.

Kono, H. (2006). Fundamental Principles and Viewpoints of Manufacturing Revolution Activities,«. *IEMS*, *5*, 48–56.

Lee, L. H., Abernathy, F. H., & Ho, Y. C. (2000). Production scheduling for apparel manufacturing systems. *Production Planning and Control*, *11*(3), 281–290. doi:10.1080/095372800232243

Lin, S. H., Kincade, D. H., & Warfield, C. (1994). Productivity and production in the apparel industry. *International Journal of Clothing Science and Technology*, *6*(1), 20–27. doi:10.1108/09556229410054477

Liu, C., Lian, J., Yin, Y., & Li, W. (2010). Seru Seisan-an innovation of the production management Mode in Japan. *Asian Journal of Technology Innovation*, *18*(2), 89–113. doi:10.1080/19761597.2010.9668694

Liu, C., Stecke, K. E., Lian, J., & Yin, Y. (2014). An implementation framework for seru production. *International Transactions in Operational Research*, *21*(1), 1–19. doi:10.1111/itor.12014

Liu, C., Yang, N., Li, W., Lian, J., Evans, S., & Yin, Y. (2013). Training and assignment of multi-skilled workers for implementing seru production systems. *International Journal of Advanced Manufacturing Technology*, *69*(5-8), 937–959. doi:10.100700170-013-5027-5

Miyake, D. I. (2006). The shift from belt conveyor line to work-cell based assembly systems to cope with increasing demand variation in Japanese industries. *International Journal of Automotive Technology and Management*, *6*(4), 419–439. doi:10.1504/IJATM.2006.012234

Niwa, M. (2002). The importance of clothing science and prospects for the future. *International Journal of Clothing Science and Technology*, *14*(3/4), 238–246. doi:10.1108/09556220210437202

Sakazume, Y. (2005). Is Japanese Cell Manufacturing a New System?: A Comparative Study between Japanese Cell Manufacturing and Cellular Manufacturing (< Special English Issue> Production and Logistics). *Journal of Japan Industrial Management Association*, *55*(6), 341–349.

Sakazume, Y. (2006). Conditions for successful implementation of assembly cells. *Industrial Engineering and Management Systems*, *5*(2), 67–73.

Senanayake, M. M. (2004). *Mixed Mass Production and Mass Customization: Best Practices for Apparel* (PhD Thesis). NCSU.

Shao, L., Zhang, Z., & Yin, Y. (2017). Production System Performance Improvement by Assembly Line-seru Conversion. In *Proceedings of the Tenth International Conference on Management Science and Engineering Management* (pp. 1165-1180). Springer. 10.1007/978-981-10-1837-4_95

Singh, S. (2017, July). A study on seru production system. In *3rd International Conference on Emerging Technologies in Engineering, Biomedical, Management and Science [3rd ETEBMS-2017]* (*Vol. 9*). Academic Press.

Stecke, K. E., Yin, Y., & Kaku, I. (2014). Seru production: an extension of just-in-time approach for volatile business environments. In Analytical Approaches to Strategic Decision-Making: Interdisciplinary Considerations (pp. 45-58). IGI Global. doi:10.4018/978-1-4666-5958-2.ch003

Stecke, K. E., Yin, Y., Kaku, I., & Murase, Y. (2012). Seru: The organizational extension of JIT for a super-talent factory. *International Journal of Strategic Decision Sciences*, *3*(1), 106–119. doi:10.4018/jsds.2012010104

Suzuki, Y. (2004). Structure of the Japanese production system: Elusiveness and reality. *Asian Business & Management*, *3*(2), 201–219. doi:10.1057/palgrave.abm.9200088

Takeuchi, N. (2006). *Seru Seisan* [Seru Production System]. JMA Management Center.

Tomastik, R. N., Luh, P. B., & Liu, G. (1995, December). Optimization-based scheduling of flexible manufacturing systems for apparel production. In *Proceedings of 1995 34th IEEE Conference on Decision and Control* (Vol. 3, pp. 3152-3157). IEEE. 10.1109/CDC.1995.478632

Villa, A., & Taurino, T. (2013). From JIT to Seru, for a production as lean as possible. *Procedia Engineering*, *63*, 956–965. doi:10.1016/j.proeng.2013.08.172

Waddell, G. (2013). *How fashion works: Couture, ready-to-wear and mass production*. John Wiley & Sons.

Wang, J., Liu, H., Qu, P., & Yin, Y. (2013). Design and operations of seru manufacturing: case study. In *The 6th PSU-UNS International Conference on Engineering and Technology (ICET-2013)* (pp. 15-17). Academic Press.

Wang, J., Schroer, B. J., & Ziemke, M. C. (1991). Understanding modular manufacturing in the apparel industry using simulation. In *1991 Winter Simulation Conference Proceedings*. (pp. 441-447). IEEE. 10.1109/WSC.1991.185645

Weller, S. A. (2004). *Fashion's influence on garment mass production: Knowledge, commodities and the capture of value* (Doctoral dissertation). Victoria University.

Wong, W. K., Chan, C. K., & Ip, W. H. (2001, July). Application of fuzzy concept in the apparel industry. In *Proceedings Joint 9th IFSA World Congress and 20th NAFIPS International Conference (Cat. No. 01TH8569)* (pp. 2547-2550). IEEE. 10.1109/NAFIPS.2001.943623

Yagyuu, S. (2003). *Synchronized seru seisan mode: beginning from one*. Nikkan Kogyo Shimbun.

Yin, Y., Kaku, I., & Stecke, K. E. (2008). *The evolution of seru production systems throughout Canon*. Neilson Journals Publishing. doi:10.4135/9781526462060

Yin, Y., Stecke, K. E., & Li, D. (2018). The evolution of production systems from Industry 2.0 through Industry 4.0. *International Journal of Production Research*, *56*(1-2), 848–861. doi:10.1080/00207543.2017.1403664

Yin, Y., Stecke, K. E., Swink, M., & Kaku, I. (2017). Lessons from seru production on manufacturing competitively in a high cost environment. *Journal of Operations Management*, *49*(1), 67–76. doi:10.1016/j.jom.2017.01.003

Zhang, X., Liu, C., Li, W., Evans, S., & Yin, Y. (2017). Effects of key enabling technologies for seru production on sustainable performance. *Omega*, *66*, 290–307. doi:10.1016/j.omega.2016.01.013

KEY TERMS AND DEFINITIONS

Divisional Seru: A type of Seru where each worker is assigned to single one task.

Flexible Manufacturing: Production system with different characteristics in terms of operational and control features.

Group Technology: It is a manufacturing on a specific layout of job-shops as workstations of similar operations.

Lean Production: It is a production system that targets operational excellence by eliminating the wastes in production environment.

Mass Customization: It is the creation of combinations the products produced by mass production in accordance with customer requirements.

Mass Personalization: It is to produce specific needs of any customer as product-by-product in a mass customization system.

Rotating Seru: A type of Seru where every worker manages all tasks from start to finish.

Seru Seisan: A novel production system based on combining the flexibility of job shops, the efficiency of mass production and the environmentally friendly features of sustainable production.

Yatai Seru: A type of Seru, is regarded as the most perfect human/business organization.

Chapter 6
Human Resources Management in the Portuguese Education System:
A Strategic Contribution to an Intelligent and Flexible Management

Diana Fernandes
University of Minho, Portugal

Carolina Feliciana Machado
https://orcid.org/0000-0002-9685-1576
Interdisciplinary Centre of Social Sciences (CICS.NOVA.UMinho), University of Minho, Portugal

ABSTRACT

Acknowledging the revival of the debate around school management models, following the criticism neoliberalism has brought into discussion, this work focusses on the human resources management paradigm in the Portuguese compulsory public education system. Analysing secondary quantitative and qualitative data, it describes the observed performance comparing it to the expected one (strategic human resources management). The system's main coordination mechanism is the standardization of competencies, centrally delegated from the Ministry of Education and Science, as so, "mobilization" is the vector in which the system is more constrained towards the expected performance, harming its potential for flexibilization and adaptation. This work concludes there is no strategic human resources management in the Portuguese compulsory public education, despite some positive efforts already being taken to overcome such situation. These insights can be used as an analytical guide to the implementation of changes towards the expected performance in the Portuguese education system.

DOI: 10.4018/978-1-7998-5768-6.ch006

INTRODUCTION

This work analysis the Human Resources Management (HRM) model applied in the Portuguese compulsory public education system, aiming to determine if there is a Strategic Human Resource Management (SHRM) paradigm. It is argued that this system's incorrect investment in its human capital, which could provide it with a source of sustained competitive advantage, is one of the grounds for its inefficient performance: this is the main working hypothesis.

This research is relevant once the general tendency in conceptualizing the education system's essence has been questioned with the triumph of Neoliberalism, raising a strong criticism to the Public School (Santos, 1997). The importance of rethinking the school management model stems from a set of trends that currently affect most European countries: the decentralization of educational policies, and the definition of national educational objectives and school achievement levels (Eurydice, 2004). Also noteworthy is the recent discontent of users and staff about the management deficiencies in the Portuguese education system, especially the public one, because being it a public service focused on welfare, such objectives would be undermined by the excessively economist vision of its functions, blindly subordinated to the prerogatives of austerity. The acuity of this research is also underscored by the scarcity of empirical work linking SHRM to organizational performance, especially those conducted in the Portuguese compulsory public education system.

This work divides into four parts. Firstly, we review literature about SHRM, referring to the factors that have encouraged the evolution of Human Resource Management Policies and Practices (HRMPP) towards this paradigm and its main characteristics. Later, we discuss the potentialities of applying SHRM models in the education sector. The theoretical model and the research hypotheses are, then, presented. On the second part we address the methods; on the third part we present the results, discussing them on the fourth part: the HRM model observed in the Portuguese compulsory public education system is described, comparing the observed and the expected performances, verifying if a SHRM model is being implemented. Conclusions and limitations of the study follow.

It was concluded that the Portuguese compulsory public education system does not present a SHRM model, as its concerns with saving assets underestimate the conceptualisation of Human Resources for Education (HRE) as the main source of this system's sustained competitive advantage. This work simultaneously configures an analytical document and a guide for the subsequent design and implementation of strategies aiming to achieve a SHRM model in the aforementioned system.

BACKGROUND

The Strategic Dimension in Human Resources Management

Storey (1995), quoted by Tavares and Caetano (1999: 39), defines HRM as "a distinctive approach to worker management", aiming to "achieve competitive advantage through the strategic development of a highly capable and committed workforce, using an integrated set of cultural, structural and personal techniques". This vision of HRM has been outlined since the 1990s, as organizations started facing a kaleidoscope of technological, economic, social, political and legal challenges, so needed to reorganise at structural and management levels to survive in the market(s). As capital and technology become increasingly available anywhere in the world, organizations seek sources of competitive advantage in their own organizational capabilities, guided by a logic of sustainability (Dyer, 1993). Thus, the Human Resources (HR) function develops, conciliating economic options, the aspiration to profit and the evolution of HR, who crystalize as true business partners (Lawler III and Boudreau, 2009), main strategic resources of the organization.

According to Wright and MacMahan (2011: 93), HR will, then, configure as "human capital", acting as mediators between organizational performance and HR practices. Boulding (1971) develops such concept, arguing the impossibility of separating people from the human capital they possess, becoming their exclusive owners: they create value because they are rare and difficult to imitate, leading to the achievement of sustained competitive advantages for the organization (Gerhart, 2005). So, the organization results from the decisions and actions taken by the individuals that compose it, in a decentralized dynamic, being its rationality conditioned by the perception they have of their role, the limits and the rules of the "game" (Ulrich, 1998). It opens the way to new approaches on human behaviour in organizations, emerging new interpretations of power relations (Carvalho, 1995). As so, HR shall assume central not in the organization's traditional activities (hiring, remuneration or administrative support), but in the setting of solutions enriching the organization's value, because they constitute an invisible asset, capable of bringing sustained competitive advantage (Becker and Gerhart, 1996; Pfeffer, 1994). Implementing a SHRM model, we will witness a dynamic of "mediation" and "game" between the capacities, interests and objectives of all organizational actors, in a dynamic and flexible field, based on fragmentation and dispersion of power and collective negotiation (Vizant and Vizant, 1996), architecting the organization as a "coalition of interests" (Carvalho, 1995: 939).

The HR function adopts a dynamic view of its resources so that organizations focus on "Personalization", "Adaptation", "Mobilization", "Sharing" and "Anticipation" as HRMPP (Peretti, 1998: 34). So, this approach forces organizations to rationalize structures, policies and practices (Pfeffer, 1994; Wright and McMahan, 1992). Hence, the paradigm metamorphosis requires the establishment of a different psychological contract, based on mutual involvement and high levels of reciprocal trust and satisfaction. Since SHRM presupposes a workforce highly involved with the organization, it relies, as key elements, on the selection, socialization, training and development of reward systems (Becker and Huselid, 2006).

The Application of SHRM Models in Education Institutions

The evolution of public organisations in the European Union (EU) vis-à-vis private ones is considerably slower, even though a Public Administration (PA) reform has been started, focusing on the adoption, in public organs, of management instruments and techniques, characteristics of private companies. However, idiosyncrasies shall be taken into account: "it is in an extremely restrictive environment that activities related to the management of workers in public functions are carried out, creating a barrier to the adoption of policies and practices, especially when it comes to budget restriction programs and public expenditure containment" (Mendes, 2012: 32).

The administrative reform in the public sector in Portugal (a long and slow process) began in 2003, with guidelines defined in the Resolução do Conselho de Ministros 95/2003, whose main objectives are to prestige to the PA; to rationalise and modernise its structures; to reassess the State's functions; to promote a culture of evaluation and responsibility, distinguishing merit and excellence.

After a traditional perspective, privileging the classroom's micro level and the education system's macro level, it has emerged the Sociology of School Organizations' approach to management models in the education system. It opts for an intermediate level in the analysis, advocating that the processes of change and educational innovation will go through the understanding of schools as institutions in all their complexity (technical, scientific and human). According to Magalhães (2010: 19), education, "more than a technical-scientific area or domain, (...) is a multifaceted and categorical transformative, scalar, interdisciplinary complex, with a variable geography and directly or indirectly affected by ideologies". Its reform has been spreading mainly in the so-called developed world, in a process sustained by various actors (academics, national and international political entities). It has been characterized as a political epidemic, reorienting education systems with distinct pathways in diverse social and political situations (Levin, 1998). The origins of such institutions stem greatly from Globalization: we are witnessing a movement of Neoliberal policies aimed at promoting an "education market", which end up distorting the true essence of the education system and its institutions (Ball, 2000: 5).

With particular vigour over the last thirty years, the attempt to produce deliberate and large-scale changes in school management systems has been on the agenda of decision-makers, researchers and education professionals, although the results have been generally disappointing (Canário, 2000). Since late 1980s, there has been a significant change in the State's role in education administration, a process which aims to transfer powers and functions from the national and regional to the local level, "recognising the school as a central place of management and the local community (...) as an essential partner in decision making", strengthening the school's autonomy, being this one of its most significant expressions and achievements (Barroso, 1996: 2).

As a consequence of the "post-entrepreneurial revolution", which "provides the possibility for each member of an organisation to express individual initiative and fully develop their potential in the service of the enterprise" (Du Gay, 1996: 62), we can, following Ball (2002: 4), schematise the main "political elements or technologies" architecting the need and modus operandi of education reform - the market, management capacity, performance and changes in the State's nature in political discourse. These will take on different emphasis depending on the context (social, geographical and temporal) they impact, although they are closely interrelated in the reform processes. These are characterised by their ability to give new shape to the organisations they monitor, because they offer a politically attractive alternative to the education tradition centred on welfare, advantages that stand out when compared "to the old technologies of professionalism and bureaucracy" (Ball, 2002: 4).

Such "political elements or technologies" combine to produce "a replacement of centralised management organs by a global structure at the expense of micro-management", devolving authority and providing flexibility to schools, requiring "changes in attitudes and behaviour on both sides". These reform processes should not be conceived as "simply a strategy of deregulation": in fact, they are "processes of re-regulation", an improvised mixture of physical, textual and moral elements that will make it possible to govern in a liberal way, adapted to current challenges (Ball, 2002: 5). As per Du Gay (1996), we will see the setting of a "controlled uncontrol" model.

Currently, the school is seen as an institution endowed with relative autonomy, an intermediate territory for decision making in the educational field, not only reproducing the macro-system's norms and values, but also not being exclusively perceived as a micro-universe, totally dependent on the social actors in presence. Based on the thinking so far, it is easily possible to suggest the acuity of inserting models of Strategic Management of Human Resources as a tool to achieve such objectives in the education system. Nevertheless, the difference between the application of SHRM models in the business and in the education context is evident. It is necessary a theoretical and methodological break with the scientific-rational paradigm, which traditionally dominated the analysis of such organisations and its administration. These institutions are now seen as social constructions, shedding prominence on the role of individuals and the socio-historical context of their action.

Then, in the management of education organizations, the principles of administrative rationality are questioned, instead of them triumphing the "primacy of values over facts" (Barroso, 1996: 1).

Schools deal with values determined by political processes, aiming at implementing responses to a changing environment (Stewart and Ranson, 1994). These institutions are settled in an attempt to balance various power games, and these (especially their influence) can be more or less visible in their policies and practices. It will, thus, be easily understandable that they need to adapt to different political purposes, changing according to internal electoral cycles and international configurations. Hence, these institutions face dynamics of protest, given the concentration within them of diverse (and sometimes conflicting) values and interests.

So, the strategy to be implemented in schools will present itself as the result of compromises negotiated between multiple actors. In fact, in these institutions the accountability is achieved above all in face of the higher political and administrative organs, according to previously outlined goals, in the name of public interest, because evaluation and control follow political criteria. Faced with such constraints, the schools' reduced autonomy and flexibility is easily perceptible, as their strategic choices are particularly hardened by the administrative and bureaucratic weight of a centralist administration, in Portugal represented by the Ministry of Education and Science (MES). Therefore, there will be a lost (or not even reached) relevant in loco knowledge, decisive for a comprehensive approach to the national education system, in order to obtain a greater commitment from all actors to the organizational performance. These are precisely the policies, procedures and objectives targeted by a strategic management model.

It shall also be noted that the these institutions' aims are subject to sometimes ambiguous formulations, making univocal interpretations impossible. Lima (1992: 45) indeed argues that "the school will not only be a locus of reproduction, but also a locus of production", that is, a (self)organized instance for the production of rules, in line with Nóvoa (1992). Then, each school must find and communicate the characteristics of its own identity. By integrating its specific structures (physical, administrative and social), each school will assume itself as a dynamic, evolving and proactive agent, building its own identity.

So, we can see that there are obstacles in implementing a SRHM in the public sector, since this type of model is oriented towards the creation of factors linked to increasing the organizations' competitiveness and productivity, dealing with combinations of markets and products, based on profit margins. It is clear that such objectives and procedures are distinct from those of the education sector, harming the transposition of such models into the field of education (Estevão, 1998).

In fact, Portuguese public schools have never been granted a planning department, with people actively and specifically involved in the architecture of strategic management systems. Portuguese public schools have never really had a strategy of their own, having been developed in the face of the mission, political project and State's strategic decisions in education (centrally derived from MES).

Although "the literature on strategic management in the public sector is limited and inconclusive" (Vinzant and Vinzant, 1996: 204), with this work we suggest that it is technically feasible, with due caution, to transfer "the arsenal of strategic management to schools", as per Estevão's perspective (1998: 16). The usefulness of these tools would consist in facilitating the capacity of their multiple actors to deal with organizational and environmental complexity, proper to the context where the education system is inserted. We suggest, then, that the application of SHRM models in the Portuguese public schools would lead to an improvement in the national education system's performance, once it would be possible, at least, to provide elements for its more efficient management, more autonomous from bureaucratic and political dependencies.

Bearing in mind that one of the main objectives of SHRM is to ensure that organizations as a whole are well articulated with their environment, schools could also gain from this type of management, once they are affected by a set of environmental factors (legislation, technological evolution, changes in working conditions/policies) that autonomy could help to overcome, providing positive impacts on the ideologies and political attitudes towards education.

MODEL AND RESEARCH HYPOTHESES

We conceive HRE in its biunivocal understanding, because as a workforce of double dimension (pupils and teaching/non-teaching staff), such capital is simultaneously subject to practices of education in a given education system, previously socially, culturally and temporally defined. Such system is based on three capitals: structural (educational institutions), professional (HRE) and citizens (children and young people of school age, served by such system). It is from the interaction between these vertices that knowledge is created, based on the information stored and shared in the structural system and on the previous knowledge of the organizations' users and professionals, instigating an increase in the organization's intellectual capital, which requires a coordinated action by all actors involved in this system's policies and management practices (Dussault and Souza, 1999). These characteristics simultaneously result from and highlight the public nature of the services provided by such system. Thus, according to Lima (1992: 45) and Nóvoa (1992: 1), it will be through a dynamic that holds, as core, HRE that it will be possible to upright schools as spaces of pedagogical, curricular and professional autonomy, which implies an effort in understanding the role of education establishments as organizations, functioning in a dynamic tension between production and reproduction, between freedom and responsibility.

It was composed a model in order to achieve this investigations' objectives, based on the contributions of three authors (Barroso, 1996; Estevão, 1998; Nóvoa, 1992), in Appendix 1. This option is due to the gaps on the literature, as it was not possible to find neither specific SHRM theoretical models for the education system, nor empirical evidence, as this subject is of recent study. These have been particularly noticed if analysing Portugal. It was only possible to find general standards that should govern SHRM, not applied to a specific sector, besides idiosyncrasies to which the education system management obeys, applied, however, to particular national contexts.

Our model, below described, develops in a chained and sequential logic, because although it can be dissected in several segments, they are articulated in interconnection and mutual influence.

The Portuguese education system is influenced by a set of inputs, divided into two categories - external and internal. The first regards to the "School's physical structure", including its size, material resources, number of classes and organization of school spaces. In the second category two vectors can be distinguished: related to the "School's administrative structure", such as its management, direction, control, inspection, processes and decision making agents, teaching and auxiliary staff, community participation, relations with central and local authorities; and related to the "School's social structure", from which stand out the accountability and involvement of parents, the organizational culture of the school, internal democracy, social climate and the relations between pupils/parents/education staff.

These inputs correlate once it is by considering the external configurations and resulting constraints that the internal environment is configured, providing more or less favourable conditions to the development of the SHRM processes. All these inputs will impact the SHRMPP: to achieve this, it will be necessary a logic that encompasses the three major areas of intervention in which any education system is structured - "Pedagogical" (Education System), "Professional" (Professional legitimation system) and "School" (System of work relations). Aiming at a SHRM model, the Portuguese compulsory public education system shall implement SRHMPP in each of its intervention areas, through three fundamental processes. The "Development of Human Resources Skills" refers to the process of building human capital structures, focusing on the knowledge production and vital skills in achieving the performance levels required by the system' objectives, in each professional category. The "Human Resources Distribution" focuses on the process of allocating workforce in the system's different services, by country's regions and sub-regions, ensuring equity of access for the entire population to all services. "Human Resources Management" aims to ensure an appropriate working environment and an adequate level of HR performance. To insert these elements into our model, we adapted the contributions of Dussault and Souza (1999), since these were defined specifically for the HRM in the health system, in the Brazilian context.

Those three processes are translated into SHRMPP, through the omnipresent awareness of the five Peretti (1998) vectors. Through "Adaptation", the organization will face internal and external requests, developing flexibility and creative thinking. Through "Mobilization", it will gather the whole hierarchy's involvement in motivating HR, including the generation and diffusion of information, opening options of choice and means of solving problems. "Personalization" will encompass global remuneration, training and information, in the management of careers, skills and time, reconciling staff's expectations with the company's constraints. Through "Sharing", it is certified that all managers participate in HRM in collaboration with the directors of the respective function, who will be the promoters of the new concept of shared function. Focusing on "Anticipation", the organisation will strengthen the awareness of the need for medium/long-term predictive management of employment in the face of the environment's unpredictability, thus being guided by the sustainability imperative.

Implementing such model, it will be possible to achieve a SHRM in the Portuguese compulsory public education system. This will result in schools' autonomy, organizational leadership, curricular articulation, time optimization, professional stability, staff training, parental participation, public recognition and support from authorities. It will be possible to think and act the school and at school from a perspective of sustainability. Only as so will it be possible to achieve an efficient organizational performance in the Portuguese compulsory public education system. Thus, we will be able to discover whether a SHRM model is currently being applied in such system, investigating the reasons that underlie its performance. Hence, we stipulate the below research hypotheses, mutually exclusive:

H0: The Portuguese compulsory public education system has a SHRM model, which justifies that its performance, in terms of the work performance of its HRE, will be efficient.

H1: The Portuguese compulsory public education system does not have a SHRM model, which justifies that its performance, in terms of the work performance of its HRE, will be inefficient.

The observed performance in the Portuguese compulsory public education system, in terms of the work performance of its HRE, will be the dependent variable; the existence and application of a SHRM model in it will be the independent variable.

Method

Performed in the year 2016, this investigation is developed from a conceptual and theoretical structure later tested by empirical observation. It has a descriptive purpose, aiming to unveil the pattern of HRM in the Portuguese compulsory public education system. Thus, describing the HRE performance in such system, we will reveal the observed performance in its management, verifying if it matches the expected performance, which is SHRM. We conducted a content analysis based on a comparative methodology and interpreting data deductively, presenting a longitudinal study – the analysis considered a time frame between the school year 2000/2001 and 2014/2015, once it is relevant to have a reasonable but also relatively recent time frame, providing data through whose analysis would we be able to derive reliable trends in the analysed variables, in order to conclude on the performance (and its evolution) of the HRM in the Portuguese compulsory public education system.

Statistical and documental analysis, as well as literature review, were the means by which data was collected, with no primary data being achieved. We accessed both national as well as international sources to collect quantitative and secondary data. As national ones, we found useful evidence on statistics from Base de Dados Portugal Contemporâneo (PORDATA), Instituto Nacional de Estatística (INE), Direção-Geral de Estatísticas da Educação e Ciência (DGEEC), Direção de Serviços de Estatísticas da Educação (DSEE), Divisão de Estatísticas do Ensino Básico e Secundário (DEEBS), Conselho Nacional de Educação (CNE) and Direção-Geral da Administração Escolar (DGAE). We also relied on data released by the MES, Portuguese legal documents and other reports available in the literature. As international sources, we focussed on the Organisation for Economic Cooperation and Development (OECD) reports on international tests in Education - TALIS (the OECD Teaching and Learning International Survey), PIRLS (the Progress in International Reading Literacy Study), TIMSS (Trends in International Mathematics and Science Study) and PISA (The Programme for International Student Assessment). TALIS presents on a cross-country analysis helping the surveyed countries to recognize others facing analogous challenges and learn about their policies (OECD, 2016); TIMSS and PIRLS are conducted by the IEA (International Association for the Evaluation of Educational Achievement), an organ of the OECD in charge of developing international tests to assess the state of education in its member countries (OECD, 2014). PISA is a triennial international survey ranking education systems worldwide by testing the skills and knowledge of 15-year-old students in Reading, Mathematics and Science, concentrating at the students' ability to apply knowledge and skills in order to analyse, reason and communicate effectively in interpreting and solving problems (OECD, 2016).

To test our hypothesis, data was needed on the HRE performance in the Portuguese compulsory public education system in each of the processes composing the theoretical model. To analyse the "Development of Human Resources Skills", evidence was collected apropos the HRE's training (quality, frequency, content and duration), the evolution of HRE's academic qualifications, and the limitations imposed by the Ministry to career entry and progression. To study the "Human Resources Distribution", we gathered data on the excess/shortage of HRE, as well as the geographical concentration of HRE in Portugal. The ratio of HRE by professional category and the ratio providers/users by education subsystem were also relevant. In regards to the "Human Resources Management", we examined criteria for HRE allocation, working hours, hiring requisites and (in)existence of incentives to advance the distribution and permanence of such staff.

RESULTS

Inputs to the Portuguese Compulsory Public Education System

First of all, it is important to bear in mind the Portuguese education system structure. It articulates into three sequential levels beginning in Pre-school, an optional cycle of attendance to children from 3 to 6 years old, continuing with Ensino Básico, which encompasses three consecutive cycles: the 1st CEB (Primeiro Ciclo de Ensino Básico) of 4 years, for children from 6 to 10 years old; the 2nd CEB (Segundo Ciclo de Ensino Básico) of 2 years, for children from 10 to 12 years old; and the 3rd CEB (Terceiro Ciclo de Ensino Básico) of 3 years, for children from 12 to 15 years old. It is followed by a three-year cycle, Ensino Secundário, for young people from 15 to 18 years old. This comprises several types of courses: Scientific-Humanistic, Professional, Specialized Artistic, Courses with own curricula plans (Scientific-Technological Courses), and Teaching and Training Courses. Ensino Superior, ministered at Universities or Polytechnical Institutes, is structured in accordance to the Bologna principles and is aimed at students who have successfully completed Ensino Secundário or who have a legally equivalent qualification.

When examining the external inputs, it shall at first be noted that, on the last decade, we see an increase on average by 2.4% on the population in EU28, nonetheless Portugal is one of its 11 countries whose population has decreased (-1.3%). This situation, by itself, impacts the physical structure of the school, but this is more severely constrained if also considering the dramatic birth rate decrease in Portugal (from a total of 120.008 live births in the year 2000 to only 85.500 live births in 2015), according to PORDATA (2016). As so, there is a decrease of about 63.000 pupils, translating into a reduction of approximately 4.4% in the total of enrolments in the national education system (DGEEC, DSEE and DEEBS, 2016), as per data in Annexes 1 and 2.

Furthermore, also Despacho Normativo 7-B/2015 impacted the physical structure of the school, once it established the rules concerning the constitution of classes and the division of students by the various facilities of the school's space, highlighting, for example, a maximum of 26 students per class in the 1st CEB, 30 in the 2nd and 3rd CEB and 30 in Ensino Secundário, with exceptions duly framed. We can, then, see an over dimension on class size, resulting in a workload for the HRE.

Besides that, we can see that Decreto Lei 299/2007 had a major impact on the school's physical structure, as it redefined the public schools' typologies, articulating the different teaching cycles – preferably consolidated in the same school building – now in school groupings. The technical requisites are defined in the Despacho Conjunto 258/97 and Despacho Conjunto 268/97, amended by the Despacho 13096/2008, in regards to Pre-School; in Decreto Lei 314/97 for establishments where the 1st CEB is inserted; and in the Despacho annexed to Decreto Lei 299/2007 in regards to 2nd and 3rd CEB schools, the latter may or not include Ensino Secundário.

In regards to internal inputs, we can see strong challenges being posed to the Portuguese education system in first instance because in Portugal there are, for children aged 3-4 years old, enrolment rates in the education system above the OECD average, as well as the universality of enrolment for children and young people aged 5-14 years old. Moreover, we see the recently extended compulsory schooling in Portugal to 12 years (from children since 6 to 18 years old) – this is indeed one of the longest periods of compulsory schooling in OECD (CNE, 2015). Truth is, the Portuguese education system has now more young people, for longer and disposing of a greater access to the qualifications system (DGEEC, 2016). So, if more people are taking part of the education system, it also makes it necessary to ensure schools are well prepared to response such social challenges in the best way possible. In relation to this, it is noteworthy to also shed light on the evaluation process of schools (Formosinho, Ferreira and Machado, 2000). It is urgent and complex because it grounds on the confrontation of contradictory political perspectives, threatening the conception of the school as a public service (Afonso, 1998). Liberals indeed conceive evaluation as an instrument of selection by merit, fixing on goals on productivity and accountability; Conservatives perceive so as an instrument of administrative and pedagogical control; and Progressives see it as a tool to exercise participatory democracy and to foster a personal and collective development. In such debates, the difficulty of reconciling accountability with performance improvements arises, not having in mind that these two vectors should be apprehended as complementary (indeed, this evaluation focuses mainly on the final result and not on the process conducting to such output).

The Portuguese school social structure has been turning more solid, on one hand by the sustained improvement in the results obtained by Portuguese students in PISA, PIRLS and TIMSS, which proves that schools are 'producing' increasingly skilled people. Portugal has indeed achieved major improvements in PISA cycles in Mathematics, Reading and Science: in PISA 2015, Portuguese students have improved results in all areas, firming the consistency of the positive evolution since the year 2000 (first edition of the test). Out of the 72 participants, Portugal is currently above the OECD average in all areas, due to the enhancement in the socio-economic conditions of these students' parents, the commitment to more clear and result-oriented education policies, and to the upgraded training of teachers (OECD, 2016). Very important to mention, these results in fact seem to transfigure into more positive attitudes towards school: according to PISA 2012 (OECD, 2014 b) Portuguese students were the ones who most considered as having good relationships with teachers (86%), around 25% feeling happy at school. Nevertheless, this profile is somewhat contradictory to the opinion of school's directors, also revealed by PISA 2012 (OECD, 2014 b), about the "indiscipline" and "lack of respect" situations. Going further on this logic, it is interesting to see that public opinion passes on that the Portuguese education system has an unfavourable public image, curtailing from a general and rushed media evaluation expressed in the public resonance of events/incidents and in the comparison of international rankings (Azevedo, 2005). Yet, the same opinion polls also show that people like 'their' school more than 'the system': as a matter of fact, the massification of education ended up generating a crisis of confidence in the school, as the gigantism of education administration is often blurred with the massification of the daily life at each school, phenomenon that results in the drawing of abusive conclusions about school life, especially given the current too much pressure for the homogenization of schools' characteristics, not recognizing its inherent diversity (Azevedo, 2005). We also see that the criteria for such assessments occasionally is not the homogeneous, so we denote a lack of rigour in the acceptance of these concepts.

Following the above thinking and deepening it, we can see that the Portuguese education system has a remarkable capacity to build human capital, which, however, does not materialize in a complete and efficient way, given its two structural problems: school dropout and retention. It still accounts for one of the highest school retention rates in Europe: for example, more than 40% of Ensino Secundário pupils face retention, in an average of more than two school years, this problem being so severe that CNE (2015: 8) classifies this level of schooling in Portugal as the "antechamber of abandonment". Another factor shall not be neglected: the opportunities for qualified insertion in the labour market have been decreasing – the rate of high qualifications' production that the education system has sustained, has not been corresponded in equal proportion by the economy in the labour market's potential creating work opportunities. So, the students often question, in a trade-off logic, which outcome will they achieve if fulfilling continuous cycles of study. This will further the students' dropout rate. Still, a positive evolution is being progressively achieved – scrutinize data for Ensino Secundário, where this problem is sharper: the retention and dropout rate recorded 39.4% (school year 2000/2001), decreasing to 18.5% (school year 2013/2014). As shown in Annex 3, the commitment to reach 10% in 2020 seems to be perfectly achievable, and there is no country in the EU that has experienced such a development in 25 years.

The school's administrative structure is right away harmed by the fact that, even though in the country we see a process of progressive expansion of decision-making capacity to local and regional levels, committed to improving the efficiency of public services (Decreto-Lei 75/2008; Decreto-Lei 144/2008; Decreto-Lei 137/2012), we also see that MES is the responsible for Education and Science policies in Portugal, in a logic of centralized control, allowing very parsimonious range for adaptation and personalization. The administrative structure of the school is also harmed once the share of Gross Domestic Product (GDP) devoted to the investment in education has decreased. There is a higher proportion of public funding in education in Portugal than the OECD average, however, the Portuguese State's expenditure on education has been decreasing, as a result of the austerity resulting from the economic and financial crisis that began in 2008. As so, managing the school's life faces critical challenges. Nonetheless, teachers in Portugal have lifelong training (Decreto-Lei 22/2014) and benefit from high salaries compared to the country's population with identical complete levels of education. If here we see good points, truth is that teachers' evaluation process is still a recent issue, mainly focusing on the provision of results for career progression and not so much on improving education and its professionals' skills. Self-evaluation of schools and external evaluation are also recent, and it is also important to notice that not all schools carry out self-evaluation (CNE, 2015; Fialho, 2009). As so, important challenges are addressed to the management of schools and its HRE.

Processes in HRM in the Portuguese Public Compulsory Education System

In what refers to the "Development of Human Resources Skills", we can see that teachers in Portugal must complete a training programme, which has an average duration of five years, including a pedagogical traineeship and a written exam performed at national level. This academical qualification may be ministered at Universities (Public and Private) and/or Polytechnic Institutes. There is also the possibility of its continuation at more complex levels (Masters/Doctoral Programmes). The certification of such study cycles' is carried out by the A3ES. Most of these study programmes are certified, however, in Private Education Institutions (and Polytechnic Institutes, although to a lesser extent), a small portion of them are not or were only partially certified, having been discontinued in the last two to three years. This is more frequent in Bachelor or Doctoral Programmes (A3ES, 2016).

A process of improvement has been initiated to promote the leadership skills and academical qualifications of HRE, introducing, in the year 2014, important changes in the teaching profession (before the start of the function, during initial training and throughout professional careers), so that teachers could perform their job more adequately, by updating their knowledge in face of the social, political and economic context needs (CNE, 2015).

Similar to the average reported in TALIS 2013 (87.7%), the vast majority of teachers in Portugal (84.8%) have participated in professional development activities in the last 12 or 18 months. The area of predominant need has been teaching students with Special Educational Needs (SEN), ranking 31.7% (OECD, 2014 a).

The professionalization of school leadership is a relatively recent phenomenon in Portugal, as schools were collegially run by teachers before the reforms introduced since 2008, revised and consolidated in 2012 (Decreto-Lei 75/2008; Decreto-Lei 144/2008; Decreto-Lei 137/2012). As these gave back more autonomy to the administration and management of public education establishments, schools have now been run by directors, with specific competences to perform such task. TALIS 2013 states that, even though the development of leadership in schools is still recent in Portugal, school leaders are well prepared to carry out their tasks, having invested in specific training. OECD also places these HRE above the average, Portugal occupying the fifth place in the rank (OECD, 2014 a).

As for academic qualification, we note an increase in the total of PhD and Master teachers in the Portuguese education system (DGEEC, 2016), situation that, however, does not portray the need for the Master as standard qualification, since 2007, to entry into the career, given the scarcity of new entries in the System. Since the school year 2011/2012, the decrease in the number of graduates may be explained by the departure of teachers, both for retirement or to acquire a higher qualification.

According to data for the school year 2014/2015, in Annex 4, the vast majority of kindergarten teachers are graduates or equivalent (12.074). The total number of teachers in 1st CEB with a doctorate/master degree increased (315 to 1.588), while the number of teachers with other academic qualifications decreased. It can also be seen that the majority of teachers in 2nd CEB (around 83%) are graduated, with a gradual increase in the number of doctorate/master degrees. In fact, in the school year 2005/2006, teachers in this category who held a doctorate/master degree counted 707, reaching in the school year 2014/2015 an increase of around 161%. Teachers on 3rd CEB and Ensino Secundário are mostly graduated, with an increase between the school years 2005/2006 and 2014/2015 of around 5.000 doctorate/master degree professionals. The decrease in the number of graduates is not gradual, revealing oscillations possibly due to the changes on the Retirement Statute, especially in its decreases between the school years 2005/2006 and 2006/2007.

According to DGEEC (2016), the majority of teachers with SEN functions (320) had specific academic qualification, mainly them (216) being Kindergarten Teachers.

In Annex 5 we notice an improvement in non-teaching staff academic qualification, specially by the increase of graduate professionals in this category (DGEEC, DSEE and DEEBS, 2016). In fact, while in the school year 2000/2001 there was a predominance of professionals with non-higher education qualifications (96.2%), in the school year 2013/2014 this category continues to be predominant, registering, however, a lower value – only 87.9% of the employees.

In what refers to the "Human Resources Management", it is necessary to take into account that the elaboration of teachers and educators' schedules obeys rules established in the Estatuto da Carreira Docente (ECD), Código do Trabalho and Regime de Contrato de Trabalho em Funções Públicas. The ECD (point 1, article 76) refers to the length of service (35 hours/week), dividing teachers' working hours in "5 working days" (point 2, article 76), separated in teaching and non-teaching components - the latter covers the performance of work at individual level and the provision of work at the level of the education establishment, concerning, for example, tasks executed at management positions. According to TALIS 2013, the teaching component is in Portugal higher (on average) than in most OECD countries: 756 hours in the 1st and 2nd CEB, compared to the OECD average of 782 hours; 616 hours in the 3rd CEB and Ensino Secundário, compared to the OECD average of 694 hours in the 3rd CEB and 655 hours in Ensino Secundário (OECD, 2014). Nevertheless, there has been a progressive decrease in the weekly workload of the teaching component for teachers of 2nd CEB, 3rd CEB and of students with SEN, according to Article 79 of the ECD, and especially for those whose age is over 50 years old, a reduction that reaches its peak with a maximum of 8 hours of classes for teachers at the age of 60 years old (Decreto-Lei 15/2007; Decreto-Lei 1/98).

In terms of career incentives, when analysing the distribution of wage compensation between the years 2000 and 2011 we see an increase of around 26% in teachers' salaries (one of the largest increases in the OECD). As so, these salaries reach the OECD average, being, in 2012, higher than those of an average worker with qualifications of a superior level. As these are not the only incentives to such career, it is relevant to note that, compared to the average of TALIS 2013, a lower percentage of teachers in Portugal considers their job is valued in society. From this we can suggest that teachers in Portugal face motivation difficulties to pursue their careers (OECD, 2014 a).

In Annex 4, we see that the vast majority of educators in the MES public system has a permanent employment contract, although this total has decreased since the school year 2007/2008. However, there has been a constant decrease in the number of hired professionals since the school year 2009/2010, the highest rank for the decade. The majority of teachers in the 1st CEB also has a permanent employment contract, with a decrease in the number of hired professionals between the school years 2005/2006 and 2014/2015 (although this evolution is not consistent). Between the school year 2005/2006 and 2014/2015, the 3rd CEB and Ensino Secundário lost 21.4% of teachers with a permanent employment contract and 12.1% hired in fixed-term. The decrease in the total number of teachers in the first category was constant, but in the second one it has been registered with greater vigour since the school year 2010/2011 (DGEEEC, 2016).

The governance of the education system in Portugal is based on a centralist logic – MES is responsible for defining the curriculum, national exams, recruitment and placement of teachers, as well as the budget in all levels of schooling. The Ministry's services are organised according to a number of other organs that assist it in defining education policies, including the Secretaria Geral (SG), Direções-Gerais (Administração Escolar, Educação, Planeamento e Gestão Financeira, Estatísticas da Educação e Ciência, Ensino Superior e Estabelecimentos Escolares), the Instituto de Avaliação Educativa (IAVE), the Agência Nacional para a Qualificação e o Ensino Profissional (ANQEP), the Conselho Científico-Pedagógico de Formação Contínua (CCPFC) and the Inspeção-Geral da Educação e Ciência (IGEC); as well as other consultative organs – the Conselho Nacional de Educação (CNE), which provides the Government with advice on national education issues; the Conselho das Escolas, Associações de Pais, the Associação Nacional de Municípios Portugueses and the Associação Nacional de Professores (including disciplinary organs); the Conselho Coordenador do Ensino Superior (CRUP), the Conselho Coordenador dos Institutos Superiores Politécnicos (CCISP), and the Agência de Avaliação e Acreditação do Ensino Superior (A3ES).

Portugal is gradually extending its decision-making capacity to the local and regional levels, investing in improving the public services' efficiency, launching a major reform in school management in 2008 (Decreto-Lei 75/2008; Decreto-Lei 144/2008), strengthening the authority of its director's leadership at management and curricular definition (Decreto-Lei 137/2012). Since then, municipalities have greater responsibilities, from pre-school to the 3rd CEB, both at social and financial levels: offering extracurricular activities at the 1st CEB, providing School Social Action, managing school facilities, hiring and firing non-teaching staff. However, local or regional authorities do not yet dispose of decision-making power in what concerns to Ensino Secundário. Although this is already a positive development for the Portuguese education system, the result remains modest when compared to what has been achieved in other OECD countries, where the index of schools' autonomy already registers values of around 41% (CNE, 2015).

The State is the main funding source for education in Portugal, in the figure of the MES and directly through the State's budget – being this allocated budget bigger than the OECD average. In addition, through the decentralised approach, municipalities can finance the costs for education establishments' management, transport and extracurricular activities. According to Annexes 6 and 7, the Portuguese State's expenditure on education has been decreasing: if, for example, in the year 2005 it totalled around 11.271 million euros, according to the INE (apud CNE, 2015); by 2014 it had decreased to 6.945 million euros, a trend that tends to perpetuate. This trend is also expressed if we analyse the Portuguese State's expenditure on education as a percentage of GDP: in the year 2005 it stood at 7.1% and in 2014 reached around 5.83%, also according to INE (apud CNE, 2015). Also important to mention is that, in the Portuguese education system, school materials are paid by families and no fees are charged for compulsory education. After the 2008 financial crisis and the consequent decrease in the education budget, measures were taken to rationalize spending and optimize the allocation of resources at all levels of the national education system. Dissecting MES' expenditure by category, and according to Annex 8, we see there is no constant evolution in personnel expenditure, this recording the highest value in the year 2010 and in 2012 the lowest. This category is indelibly the most intensive burden on the system's budget. Between the years 2013 and 2014, current expenditure fell by 19%, increasing again in 2015 to values closer to those recorded in 2009. As for capital expenditure, in 2015 there was an increase of around 50% compared to the previous year (IGeFE, IP, 2016 apud CNE, 2015).

Regarding the "Human Resources Distribution", according to TALIS 2013, the average age of teachers in Portugal is 44.7 years old, slightly higher than the recorded by TALIS (42.9 years old). The proportion of teachers in Portugal aged 50 years old or over is similar to or lower than the OECD average: 31% in the 1st and 2nd CEB (compared with the OECD average of 31%); 28% in the 3rd CEB (compared with the OECD average of 34%) and 25% in Ensino Secundário (compared with the OECD average of 38%) (OECD, 2014). In fact, since beginning 2010 we see the progressive ageing of teachers in this system, reflected in the gradual increase in the percentage of teachers aged 50 years old or over, together with the constant decrease of teachers under 30 years old. According to Annex 9, this trend has been particularly intense in public education in Portugal (OECD, 2014 a).

Also important to notice, as shown in Annex 10, the teaching profession in Portugal is clearly dominated by women (DGEEC, 2015).

For teachers, the procedures to enrol in the school year start at beginning/mid of July, through prior notice from the DGAE, and at beginning of September it is scheduled the hired teachers' definitive seriation. Also, the rules to the school year start are published, annually, by the Government, through a Despacho. As a rule, the school year always starts by September, divides into three periods of activities, having as well three vacation times (Christmas, Easter and Summer), ending by June (in case students do not have exams to perform) or later in June or July (in case exams shall be performed by students). However, there is often a delay in these processes for hiring teachers and placing them in schools, once they are highly bureaucratic. It shall also be noted that the MES presents specific clauses in such procedures, personalizing them considering the specificities of some professionals (such as the regime of mobility due to sickness). Recently, the MES has opted for the online procedure, but there have been some blockages in the informatic system due to the overload of information.

Annex 11 indicates that between the school year 2000/2001 and 2014/2015 there has been a clear decrease in the number of teachers – from 166.376 in 2000/2001 to 130.367 in 2014/2015 – and in the number of staff in the Portuguese Education System – from 85.204 in 2000/2001 to 81.132 in 2014/2015. The fall in the number of pupils, the more restrictive budgetary guidelines and the amendments to the retirement regulations could be seen as the rationale behind these phenomena (DGEEC, DSEE and DEEBS, 2016).

The country's Northern region has the highest number of HRE in all levels of schooling, while the Algarve and the autonomous regions of Madeira and the Azores rank the lowest values (DGEEC, DSEE and DEEBS, 2016).

In Portugal, according to PORDATA (2016 apud CNE, 2015), the population of school age is mostly concentrated in the Metropolitan Areas of Lisbon (725.507 children/youths) and Porto (428.754 children/youths), followed by region of Tâmega e Sousa (119.628 children/youths), the Algarve (112.024 children/youths), Cávado (110.211 children/youths) and Ave (109.739 children/youths). The less populated areas include, in the mainland, Beira Baixa (16.621 children/youths), Alto Tâmega (18.182 children/youths), Alentejo Litoral (20.298 children/youths) and Trás-os-Montes (21.994 children/youths), as evidenced in Annex 12.

In turn, and as presented in Annex 13, most education establishments in Portugal are located on the mainland, more specifically in the Northern region, as opposed to the Algarve and the Autonomous Regions of Madeira and the Azores. There are also 12 establishments under the Ministry of the Economy's management, where professional education is provided (linked to the Tourism sector): three in the North, two in the Centre, three in the Lisbon Metropolitan Area, one in Alentejo and three in the Algarve. In the last decade there has been a reduction of 48.9% in the number of compulsory public education establishments in Portugal, as a result of the system reorganisation: we see in Annex 14 that, if in the school year 2000/2001 there were 16.454 establishments, in 15 years that mumber has fallen to around half (DGEEC, DSEE and DEEBS, 2016).

According to the school reorganisation process, initiated by the Portuguese Government from 2005 onwards, schools considered to be small (with less than 21 pupils) or performing below average were closed, in return a policy of school groupings was implemented. These processes aim to facilitate the transition between different levels of education and overcome geographical isolation and social exclusion, also promoting greater collaboration between teachers, non-teaching staff and students, improving the organisation of work and offering broader learning opportunities (Despacho 13096/2008; Decreto-Lei 299/2007). Also relevant to notice, the curricular reform, launched with Decreto-Lei 75/2008; Decreto-Lei 144/2008 and Decreto-Lei 137/2012, provided schools with greater autonomy in curriculum management, teaching time and flexibility in teacher training planning.

SOLUTIONS AND RECOMMENDATIONS

With such data, HR functions were analysed to conclude on the HRM performance in the Portuguese compulsory public education system, between the school years 2000/2001 and 2014/2015. This analysis took into account the three HRM processes necessary to achieve the education objectives in a given context and moment, according to our model (see Appendix 1). As we advocate that it will only be possible for the Portuguese compulsory public education system to achieve an efficient organizational performance by a SHRM model, we have tried to verify, in the aforementioned description of such processes, the possibility of its orientation according to the five Peretti's (1998) vectors, thus concluding the existence (or not) of a strategic orientation in its HRM pattern.

Regarding the "Development of Human Resources Skills", there has been a positive evolution in the HRE's professional and academic capabilities: they are now more endowed with the knowledge and skills necessary for a more efficient performance, highlighting a very positive evolution in the investment in continuous training and higher education (Masters and Doctorates). However, it often happens that this investment does not really translate into the insertion of this workforce in the career, the achievement of more favourable working conditions or even the guarantee of employment in the system, which is assumed as worrying and demotivating for the exercise of this career. This supports H1 once we see evidence that there is no SHRM standard in this process: we highlight the existence of a positive path towards a SHRM model, whose potential, however, is not effectively materialized.

In regards to the "Human Resources Management" process, it is necessary to be aware of the specific aspects of the Portuguese compulsory public education system, as this is the only way to design and implement effective SHRMPP. Therefore, the reform of its structures, policies and procedures will not be enough to undertake effective processes of change: it is necessary to take into account the specificity of the education system as a public service aimed at social welfare, and not as a company that can be managed only by economic criteria. The very characteristics of this system seem to be the basis of its problems, once the main coordination mechanism (standardization of competencies, centrally delegated from the MES) will not be sufficient to respond to all the needs addressed to this system (Democracy vs. Autonomy dilemma is at its core). Its management is oriented mainly towards output, especially of a quantifiable nature capable of comparison in international rankings, with no real sustained concern with the process leading to them. In fact, in this process we again find evidence supporting H1, as we note the non-existence of a SHRM model in the Portuguese compulsory public education system once its management does not, in the true meaning of the term, take into account the specificity of the capital with which it operates (human capital), nor the true mission of public service it shall orient towards, articulating it only according to a paradigm of cost contraction, with concerns more focused on austerity rather than social service.

In regards to the "Human Resources Distribution", there is a general correspondence between the geographical concentration of education establishments, teachers and non-teaching staff. However, this match is more diluted when including the geographical concentration of pupils: despite the predominance of the geographical concentration of educational establishments and of HRE in the Northern region of mainland Portugal, as opposed to what happens in the Algarve and the Autonomous Regions, the concentration of the system's "users" obeys another logic, with the Algarve region being the fourth in the national territory where a large portion of the school age population is located (around 112.024 children/youths). Concerning the regions with the lowest concentration of school age population, there is a correspondence with the geographical concentration of education establishments and HRE. At the same time, there is a general and chronic underestimation of the interior region of mainland Portugal, poorly served in terms of public education services. Although a small part of the national population of school age is also located there, as well as a small fraction of the education establishments and of the available HRE in the country, the truth is that the logics of the system organisation into school groupings has severely affected this region, threatening the system's cohesion and homogeneity throughout the Portuguese territory. Once students are the "object" of HRE's work, as they are the users of this public service, any effort directed to and by the HRE should take into account the satisfaction of their needs, which implies conceptualizing the school as a social construct, able to reconfigure itself to face environmental challenges. So, implementing a SHRM model in the school's pedagogical and professional components shall have, as its core, adequacy towards its public. Thus, we again find evidence supporting H1, as there are challenges to the implementation of a SHRM model in the Portuguese compulsory public education system regarding the distribution of HRE, once the principle of equality is harmed in the capacity to access this public service to all system's "users", as there are fragilities in the system's distribution of HRE harming its consistency.

We argue, then, that the observed performance (HMR in the Portuguese compulsory public education system) does not correspond to the expected one (SHRM), so this system presents an inefficient organizational performance, confirming H1. The reasons are shown below, presenting and discussing the poor organizational performance observed in each of the Peretti's (1998) vectors.

In what refers to "Mobilisation", there is a difficulty in implementing a collective, efficient, cohesive and homogeneous image in this system, as well as a dynamic and efficient information sharing to conceive the organisation as a single group. Although the already launched devolution of autonomy to schools at the local level, it can be seen that the tradition of defining educational policies and guidelines has persisted as MES' prerogative, from where directives flow from a central authority, often displaced from the real context of operation. Thus, schools are only left to the operational phase of drawing up activity plans, understood above all as collections of activities proposed by the disciplinary groups and approved by the Pedagogical Council. In fact, in the Portuguese compulsory public education system we are faced with the existence of a "culture of dependence", with strategic decisions continuing to be imposed top-down (West-Burnham, 1994). Furthermore, and alluding to Azevedo (2005: 22), it should be noted that "institutional evaluation" is not yet as valued as it should be, also because it is a recent phenomenon. So, mechanisms for the tracking and accountability of crucial practices are lost, which could be very useful in the transmission of more rigorous information to all the HRE involved in the school daily life. These two cases provide evidence reinforcing H1. Although the HRE's full identification with the organisation (Portuguese compulsory public education system) is not achieved, we see the public and free availability of information in the field of education (Education web portal), either by specialised MES' organs (DGEEC, DSEE and DEEBS) or auxiliary entities such as the CNE. Information on career procedures is also publicly and freely available, via the Diário da República. Some categories of HRE (e.g. teachers) have their own labour unions, which also help them in this regard. These other two situations provide evidence towards H0.

In this Peretti's vector, it is also relevant to analyse the opportunities to motivate HRE in the exercise of their jobs in the Portuguese compulsory public education system, because only by holding highly motivated HR will it be possible to obtain a cohesive group able to develop different roles and skills to achieve a more efficient performance. According to PISA 2012, Portuguese students report better relations with their teachers in comparison to the OECD average, also recognizing that teachers have a good perception of their performance, trusting on their evaluation methods (OECD, 2014 b). If this is a point to positively highlight, nevertheless teachers report a less positive behaviour climate in the classroom, ranking the second lowest score among the countries surveyed in TALIS 2013 (OECD, 2014 a). This provides evidence supporting H1 as well, proving the difficulty in motivating this HRE to perform their jobs and invest in their career. The so mentioned problems in dealing with students behaviour, the difficult working conditions (marked by labour precarity) and the various changes to the ECD contribute to this, as these HRE will not be able to have a systematic and coherent guidance that provides them with stability and security in regards to their professional career. It is needed an effort to conceive the role of education institutions as organizations, functioning in a dynamic tension between "production" and "reproduction", between freedom and responsibility, being their action performed as in a space of pedagogical, curricular and professional autonomy (Lima, 1992; Nóvoa, 1992). This would be achieved if in the presence of a SHRM model. Therefore, H1 is confirmed for this Peretti's vector, as there is no real identification of HRE with the organisation they serve, once it seems to be at stake the real public service orientation that shall sustain this system. As the main rationale, we highlight the reduced autonomy the system enjoys (given its centralised management at MES). With more autonomy, intrinsic characteristic of SHRM, schools could design a clearly expressed education project of their own, more accurately outlining strategic needs, which would provide them a more efficient mapping of the served and to be served population, developing a more appropriate agenda.

In what concerns to "Adaptation", it is important to notice that in the school year 2012/2013 Portugal revised the curriculum of the 2nd and 3rd CEB, establishing standards for basic skills to be achieved by all students in Portuguese, Mathematics, Science and Foreign Languages, also allowing schools greater flexibility in curriculum management. This means, HRE acknowledge the need to review the skills and knowledge that were being ministered to students, as well as the time and method teachers had been passing those onto them. This means, the political agenda recognized the goal to conceive the management of education as a matter of strategic relevance, aiming to adapt it to the constant environmental challenges, thinking in a frame of sustainability. If a positive effort towards H0 is seen in here, it is also noteworthy is that, if, on the one hand, the Portuguese compulsory public education system has more prepared and experienced HRE, their progressive aging may decrease their efficiency in terms of their ability to adapt to environmental challenges: in here, we now find evidence supporting H1. In regards to this vector, light shall also be shed on the Technological Plan for Education (2007-2010), aimed at modernising the schools' technological infrastructure and improving the knowledge and use of Information and Communication Technologies by students and teachers. In fact, an evaluation carried out in 2010 showed that, although the general objectives had been achieved, those relating to teacher's training and the distribution of educational content in digital format had not been fully achieved (CNE, 2015).

So, the performance of this system's organizations reveals inefficiencies and the efforts towards a SHRM model are constrained, evidence supporting H1. Another reason to harm this system's capacity of adaptation to the constant environmental challenges resides on the direction of such system, which still belongs to the MES' central jurisdiction, notwithstanding the progressive concessions of autonomy to local and regional levels. If given more autonomy and flexibility, such institutions would be able to better track, think and actively react to challenges. In line with Lengnick-Hall, Beck and Lengnick-Hall (2001), we argue that this system will only achieve organisational efficiency and resilience through the implementation of a SHRM model, therefore, once again in this Peretti's vector we found evidence supporting H1.

"Personalisation" furthers the logic above. There has been a gradual progress in giving more autonomy and responsibility to local authorities (municipalities) in educational and management tasks. So, and as a result of the changes introduced with Decreto-Lei 75/2008 and Decreto-Lei 144/2008, reinforced by Decreto-Lei 137/2012, schools for 1st, 2nd, 3rd CEB and Ensino Secundário can now sign autonomy agreements with the MES: the conditions include a self-evaluation of the school and a positive external evaluation. These allow greater autonomy in areas such as pedagogical and curricular organisation, HRE management, school social support and financial management. The number of schools with autonomy contracts has grown from 22 in the year 2010 to 212 in the year 2013, also accounting for 26% of school groupings. If we see positive efforts towards a SHRM, supporting H0, these are still residual when taking into consideration that the paradigm in the Portuguese education system is a MES's centralized management model: as so, we confirm H1 for this Peretti's vector.

In regards to "Sharing", Portugal has launched a major reform in school leadership (Decreto-Lei 75/2008; Decreto-Lei 144/2008), establishing three councils: General (with representatives of school staff, teachers, parents and local authorities), responsible for planning, operational and strategic follow-up and for selecting the school director; Pedagogical, which supervises and coordinates pedagogical activities; and Administrative, responsible for administrative and financial matters. This allowed a faster and more accurate information sharing base, duly delimited in its functions but operating in articulation, promoting the propagation of values, ideas, policies and procedures in an homogeneous flow. Relevant organs were created to foster active and shared leadership, involving the various agents in the decision-making processes concerning school and education life, positively fostering schools' democratic and social character. Nevertheless, and following the conclusion when analysing "Personalisation", also in regards to this vector we confirm H1, because if mechanisms of greater and better exchange of information between the HRE were implemented, democratising the processes of educational decision-making, this process would, nevertheless, end up constrained by the parallel proliferation of bureaucratic organs/procedures, delegated from the MES, creating obstacles or multiplying mediators in the relationship. Education is not being thought of as a true social policy, oriented to public service: we confirm H1 in this vector once education in the Portuguese compulsory public system is only being conceived as a vocational policy. As so, we see an inefficient organizational performance: it is essential to foster a more systematic confluence between policies, practices and quality standards established for the work unit (HRE), the served public (students) and the organizational unit (Schools). This would be achieved through a SHRM: as so, regarding "Sharing", it would be possible to achieve the group's commitment.

The analysis of the "Anticipation" vector is related to the one in "Adaptation". The reform of school management has changed the selection processes and responsibilities of school directors from primus inter pares (teachers elected by their peers, acting as administrators) to professionally selected school leaders with clearly identified authority and responsibilities. Their compulsory specialised training was reinforced, especially since the changes introduced by Decreto-Lei 137/2012. Several initiatives were aimed at strengthening the teaching profession: stricter admission conditions and reinforcement of the scientific curriculum in teacher training programmes; introduction of a teacher assessment examination with professional qualification and/ or a fixed-term contract for teachers with less than five years of practice, as well as the creation of a lifelong training framework for teachers, linking continuing professional training to career progression. As so, it was possible to have more prepared HRE, able to respond more efficiently to the environment. These factors demonstrate positive efforts towards the confirmation of H0, but it is also useful to discuss the school evaluation, increasingly pertinent given its relevance in the development of a more efficient education system (Fialho, 2009). There are positive efforts launched with the new model of performance evaluation of teachers and school leaders (school year 2012/2013), outlined in three target dimensions (scientific and pedagogical; continuous training and professional development; participation in school life and relationship with the community), from here emerging a new hope, as evaluation interrelates the external and internal aspects, obtaining a more holistic picture of the education situation. But school's evaluation is still a recent phenomenon: there is no established general approach, so schools have, as by now, a variable capacity to carry out institutional self-evaluations and scarce information on how they can contribute to improve their results. On the same logic, also the teacher's evaluation is a recent process in Portugal, essentially based on peer evaluation and influencing career progression (Decreto-Lei 15/2007). Even though the positive efforts towards the confirmation of H0, those bow in face of the shortcomings this evaluation process presents, once it is still not firmly stablished and homogeneous, and focuses essentially on the output and not on the process leading to it, as TALIS 2013 reports (OECD, 2014 a). Therefore, we again end by confirming H1.

FUTURE RESEARCH DIRECTIONS

This work may act as a fuse for future research as it can be developed in a more applied nature, as it assumes not only as an analytical document but also as an auxiliary guide in the design and implementation of strategies to improve the HRM in the Portuguese compulsory public education system. Nevertheless, it also presents limitations, as in some cases it has not been possible to draw comparisons and patterns for the evolution, in equal circumstances, of variables for the same time period. As well, it was sometimes not possible to find data for certain categories of analysis (the case of non-teaching staff), focusing the study mainly on teacher's characteristics.

CONCLUSION

The performance observed in the Portuguese compulsory public education system does not correspond to the expected one (SHRM), confirming H1. There are no structured, consistent and coherent SHRMPP in such system, hence its inefficient organizational performance. If positive efforts are detected for the implementation of SHRM, they bow to bureaucratism, cost contraction and to a culture of "outcome" instead of "process". Although literature on strategic management in the public sector is limited and inconclusive, providing anecdotal empirical evidence, with this work we suggest it is technically feasible, with due caution, to transfer "the arsenal of strategic management to schools" (Estevão, 1998: 16). As so, this work offers useful contributions both in practical and theoretical terms.

In theoretical terms, it is a useful contribution as it elaborates a theoretical model enabling the application, to the public sector, of a typical private sector HRM model, and provides empirical evidence on how the performance of a subsystem on the public sector is and can be enhanced if managed by such model. Thus, it boosts new approaches in understanding human behaviour and reinterpreting power relations in organizations, metamorphosing the paradigm by grounding it on the establishment of a different psychological contract, based on mutual involvement and high levels of reciprocal trust and satisfaction among all organizational actors. So, it questions the principles of administrative rationality in the management of education. Crucial to highlight, it contributes to fill in the gaps in literature, as it hasn't so far been possible to find models to analyse HRM in the education system, since this subject is of recent study. It has only been possible to find general standards that should govern SHRM, not applied to a specific sector of activity, besides idiosyncrasies to which the management of the education system obeys, applied, however, to particular national contexts.

In practical terms, so far as we know this work is the first to present a theoretical model and empirical evidence for Portugal on the interconnection between the application of HRM models and organisational performance in the education system. By empirical stating that the application of SHRM leads to an improvement in the such system's performance, it offers an attractive alternative to micro-management (the state-centred management today ruling in education), providing a refined work basis in which such institutions act as the result of multiple actors' negotiated compromises, generating a very relevant in loco knowledge, decisive to obtain a firmer and broader commitment from all actors involved in the organizational performance, seeking sustainability. HRM changes from a police character (executed only by the managers) to a proactive, strategic, integrated and long-term dynamic (involving all organizational actors). The paradigm of self-control collapses, triumphing relationships on a basis of trust and not of intimidation/retaliation. Flexibility and versatility of roles and skills, as well as the use of investment, conquests over a monolithic HRM concerned only with cost reduction.

REFERENCES

A3ES. (n.d.). *Acreditação e Auditoria. Acreditação de Ciclos de Estudos*. Retrieved from http://www.a3es.pt/pt/acreditacao-e-auditoria/resultados-dos-processos-de acreditacao/acreditacao-de-ciclos-de-estudos

Afonso, A. J. (1998). Políticas Educativas e Avaliação Educacional. Para uma Análise Sociológica da Reforma Educativa em Portugal (1985-1995). Braga, University of Minho: Instituto de Educação e Psicologia, Centro de Estudos em Educação e Psicologia.

Azevedo, J. M. (2005). Avaliação das escolas: Fundamentar modelos e operacionalizar processos. In M. I. Miguéns (Ed.), *Avaliação das escolas. Modelos e processos*. Conselho Nacional de Educação.

Ball, S. J. (2002). Reformar escolas/reformar professores e os terrores da performatividade. *Revista Portuguesa de Educação, 15*(2), 3–23.

Barroso, J. (1996). O estudo da autonomia da escola: da autonomia decretada à autonomia construída. In J. Barroso (Ed.), *O estudo da Escola*. Porto Editora.

Becker, B., & Gerhart, B. (1996). The impact of human resource management on organizational performance: Progress and prospects. *Academy of Management Journal, 39*(4), 779–801.

Becker, B., & Huselid, M. A. (2006). Strategic human resources management: Where do we go from here? *Journal of Management, 32*(6), 898–925. doi:10.1177/0149206306293668

Canário, R. (2000). Gestão da escola: Como elaborar o plano de Formação. In *Cadernos de Organização e Gestão Curricular*. Instituto de Inovação Educacional.

CNE. (2005). *Avaliação das Escolas: Fundamentar Modelos e Operacionalizar Processos*. Conselho Nacional de Educação.

Decreto-Lei 144/2008. Diário da República, 1.ª série, n.º 144 (28-07-2008).

Decreto-Lei n.º 137/2012. Diário da República, 1.ª série, n.º 126 (2-07-2012).

Decreto-Lei n.º 15/2007. Diário da República, 1.ª série, n.º 14 (19-01-2007).

Decreto-Lei n.º 1/98. Diário da República, 1.ª série -A, n.º1 (02-01-1998).

Decreto-Lei n.º 22/2014. Diário da República, 1.ª série, n.º 29 (11-02-2014).

Decreto-Lei n.º 270/2009. Diário da República, 1.ª série, n.º 190 (30-09-2009).

Decreto-Lei n.º 299/2007. Diário da República, 1.ª série, n.º 161 (22-08-2007).

Decreto-Lei n.º 314/97. Diário da República, 1.ª série -A, n.º265 (15-11-1997).

Decreto-Lei n.º 75/2008. Diário da República, 1.ª série, n.º 79 (22-04-2008).

Despacho Conjunto n.º 258/97. Diário da República, 2.ª série, n.º 192 (21-08-1997).

Despacho Conjunto n.º 268/97. Diário da República, 2.ª série, n.º 195 (25-08-1997).

Despacho n.º 13096/2008. Diário da República, 2.ª série, n.º 90 (09-05-2008).

Despacho normativo n.º 10-A/2015. Diário da República, 2.ª série, n.º 118 (19-06-2015).

Despacho normativo n.º 7-B/2015. Diário da República, 2.ª série, n.º 88 (7-05-2015).

DGEEC. (2016). *Perfil do docente 2014/2015*. Lisbon: DGEEC. Retrieved from http://www.dgeec.mec.pt/np4/98/%7B$clientServletPath%7D/?newsId=148&fileName=DGEEC_DSEE_DEEBS_DEES_2016_PerfilDocente.pdf

DGEEC, DSEE, & DEEBS. (2016). *Educação em números Portugal 2015*. Lisboa: DGEEC. Retrieved from http://www.dgeec.mec.pt/np4/96/%7B$clientServletPath%7D/?newsId=145&fileName=Educac_a_o_em_nu_meros_2015.pdf

Du Gay, P. (1996). *Consumption and Identity at Work*. Sage.

Dussault, G., & Souza, L. (1999). *Gestão de Recursos Humanos em Saúde*. University of Montreal.

Dyer, L. (1983). Bringing Human resource into the strategy formulation process. *Human Resource Management*, *22*(3), 257–271. doi:10.1002/hrm.3930220306

Estevão, C. (1998). Gestão estratégica nas escolas. In *Cadernos de Organização e Gestão Curricular*. Instituto de Inovação Educacional.

Eurydice. (2004). *A avaliação dos estabelecimentos de ensino à lupa*. Retrieved from https://eacea.ec.europa.eu/ressources/eurydice/pdf/0_integral/060PT.pdf

Fialho, I. (2009). A qualidade de ensino e a avaliação das escolas em Portugal. Contributos para a sua história recente. *Educação Temas e problemas – Avaliação, qualidade e formação*, *7*(4), 99-116.

Formosinho, J., Ferreira, F. I., & Machado, J. (2000). *Políticas Educativas e Autonomia das Escolas*. Edições ASA.

Gerhart, B. (2005). Human resource and business performance: Findings, unanswered questions and an alternative approach. *Management Review*, *16*(2), 174–185.

Lawler, E. III, & Boudreau, J. (2009). What makes HR a strategic partner? *People and Strategy*, *14*(1), 14–22.

Lengnick-Hall, C., & Lengnick-Hall, M. (1988). Strategic Human Resources Management: A Review of the Literature and a Proposed Typology. *Academy of Management Review*, *13*(3), 454–470. doi:10.5465/amr.1988.4306978

Levin, B. (1998). An Epidemic of Education Policy: What can we learn for each other? *Comparative Education*, *34*(2), 131–142. doi:10.1080/03050069828234

Lima, L. (1992). *A Escola Como Organização e a Participação na Organização Escolar. Um Estudo à Escola Secundária em Portugal (1974-1988)*. I.E.P.

Magalhães, J. (2010). *Da Cadeira ao Banco - Escola e modernização (séculos XVIII-XX)*. EducaUnidade de I&D de Ciências da Educação.

Mendes, I. M. (2012). *Gestão estratégica de recursos humanos na Administração Pública: mito ou realidade? Perceção dos dirigentes do I.G.F.S.S. I.P. sobre a gestão estratégica de recursos humanos e os possíveis impactos na performance*. Master Thesis in Strategic Human Resources Management, Escola Superior de Ciências Empresariais, Instituto Politécnico de Setúbal, Setúbal.

Nóvoa, A. (1992). Para uma análise das instituições escolares. *As organizações escolares em análise*, *3*, 13-43.

OECD. (2014 a). *TALIS 2013: Results An International Perspective on Teaching and Learning. Teaching and Learning International Survey*. OECD Publishing. doi:10.1787/9789264196261-

OECD. (2014 b). *PISA 2012 Results in Focus What 15-year-olds know and what they can do with what they know*. Paris: OECD Publishing. Retrieved from https://www.oecd.org/pisa/keyfindings/pisa-2012-results-overview.pdf

OECD. (2016). *PISA 2015 PISA results in focus*. Paris: OECD Publishing. Retrieved from https://www.oecd.org/pisa/pisa-2015-results-in-focus.pdf

Peretti, J.-M. (1998). Recursos Humanos (2nd ed.). Lisbon: Sílabo.

Pfeffer, J. (1994). *Competitve Advantage through people*. Harvard Business School Press.

Pordata. (2016). *Portugal. População. Nascimentos e Fecundidade. Nados-vivos de mães residentes em Portugal: total e fora do casamento – Portugal Atualização 2016-04-28*. Retrieved from https://www.pordata.pt/Portugal/Nados+vivos+de+m%C3%A3es+residentes+em+Portugal+total+e+fora+do+casamento-14

Resolução do Conselho de Ministros 95/2003. Diário da República, 1.ª série-B, n.º 174 (30-07-2003).

Santos, M. E. B. (1997). Qualidade das escolas. *Inovação*, *10*(2-3), 161–170.

Stewart, J., & Ranson, S. (1994). Management in the public domain. In D. McKevitt & A. Lawton (Eds.), *Public Sector Management. Theory, Critique & Practice*. Sage.

Tavares, S., & Caetano, A. (1999). A emergência da Gestão de Recursos Humanos Estratégica. In A. Caetano (Ed.), *Coord), Mudança Organizacional e Gestão de Recursos Humanos*. Observatório do Emprego e Formação Profissional.

Ulrich, D. (1998). A New mandate for Human Resources. *Harvard Business Review*, (January-February), 124–135. PMID:10176915

Vinzant, J. C., & Vinzant, D. H. (1996). Strategic Management and total Quality Management: Challenges and choices. *Public Administration Quarterly*, *20*(2), 201–219.

West-Burnham, J. (1994). Strategy, policy and planning. In T. Bush & J. West-Burnham (Eds.), *The Principles of Educational Management*. Longman.

Wright, P., & McMahan, G. (1992). Theoretical perspectives for strategic human resource Management. *Journal of Management*, *18*(2), 295–320. doi:10.1177/014920639201800205

Wright, P., & McMahan, G. (2011). Exploring Human Capital: Putting Human back into Strategic Human Resource Management. *Human Resource Management Journal*, *21*(2), 93–104. doi:10.1111/j.1748-8583.2010.00165.x

ADDITIONAL READING

Adomßent, M., Fischer, D., Godemann, J., Herzig, C., Otte, I., Rieckmann, M., & Timm, J. (2014). Emerging areas in research on higher education for sustainable development–management education, sustainable consumption and perspectives from Central and Eastern Europe. *Journal of Cleaner Production*, *62*, 1–7. doi:10.1016/j.jclepro.2013.09.045

Cebrián, G., & Junyent, M. (2015). Competencies in education for sustainable development: Exploring the student teachers' views. *Sustainability*, *7*(3), 2768–2786. doi:10.3390u7032768

Disinger, J. F. (1990). Environmental education for sustainable development? *The Journal of Environmental Education*, *21*(4), 3–6. doi:10.1080/00958964.1990.9941931

Hailey, J. (1998). Management education for sustainable development. *Sustainable Development*, *6*(1), 40–48. doi:10.1002/(SICI)1099-1719(199803)6:1<40::AID-SD80>3.0.CO;2-Q

Holm, T., Vuorisalo, T., & Sammalisto, K. (2015). Integrated management systems for enhancing education for sustainable development in universities: A memetic approach. *Journal of Cleaner Production*, *106*, 155–163. doi:10.1016/j.jclepro.2014.03.048

Laurie, R., Nonoyama-Tarumi, Y., Mckeown, R., & Hopkins, C. (2016). Contributions of education for sustainable development (ESD) to quality education: A synthesis of research. *Journal of Education for Sustainable Development*, *10*(2), 226–242. doi:10.1177/0973408216661442

Nasibulina, A. (2015). Education for sustainable development and environmental ethics. *Procedia: Social and Behavioral Sciences*, *214*, 1077–1082. doi:10.1016/j.sbspro.2015.11.708

O'Flaherty, J., & Liddy, M. (2018). The impact of development education and education for sustainable development interventions: A synthesis of the research. *Environmental Education Research*, *24*(7), 1031–1049. doi:10.1080/13504622.2017.1392484

Wesselink, R., Blok, V., van Leur, S., Lans, T., & Dentoni, D. (2015). Individual competencies for managers engaged in corporate sustainable management practices. *Journal of Cleaner Production*, *106*, 497–506. doi:10.1016/j.jclepro.2014.10.093

APPENDIX

Figure 1. Model
Source: *Adapted from Barroso (1996), Estevão (1998), and Nóvoa (1992).*

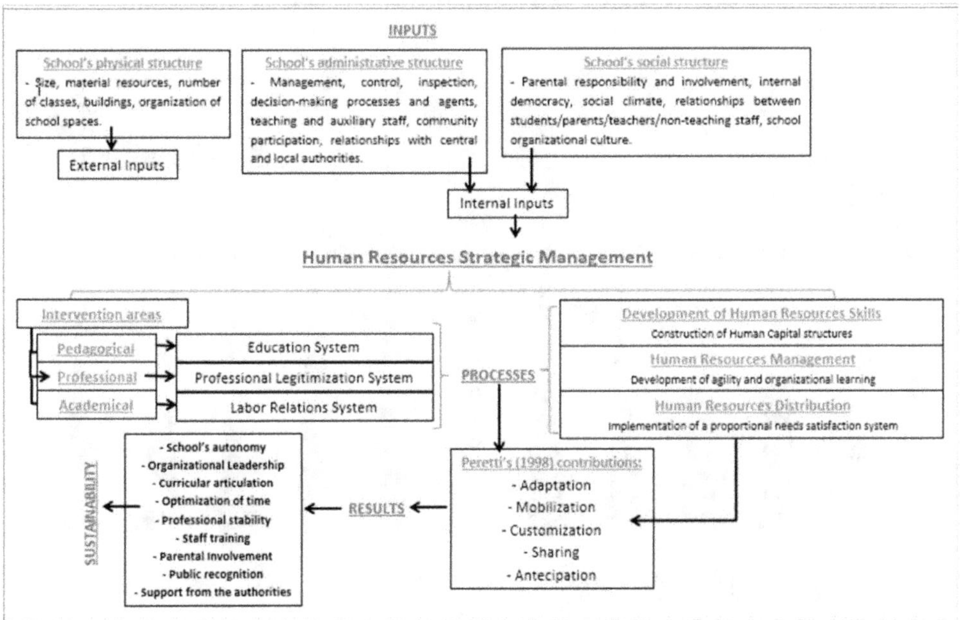

Figure 2. Evolution of the total number of students enrolled in the Portuguese education system, by type of education
Source: *Adapted from DGEEC, DSEE and DEEBS (2016).*

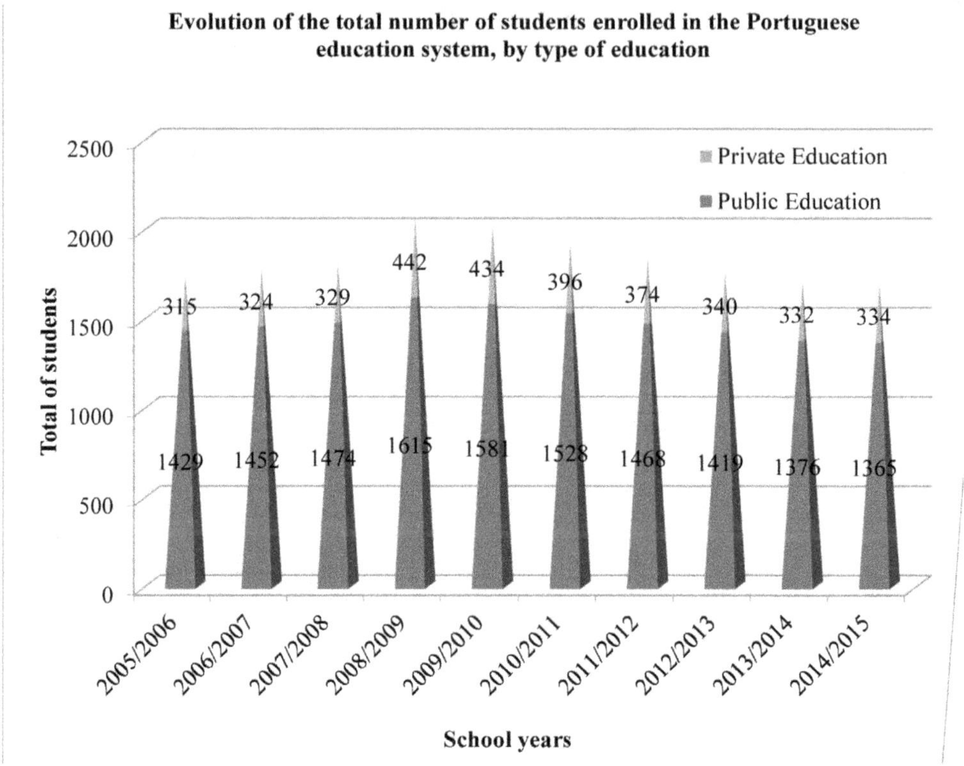

Figure 3. Evolution of the total number of students in the Portuguese education system, by level of education and study cycle
Source: *Adapted from DGEEC, DSEE and DEEBS (2016).*

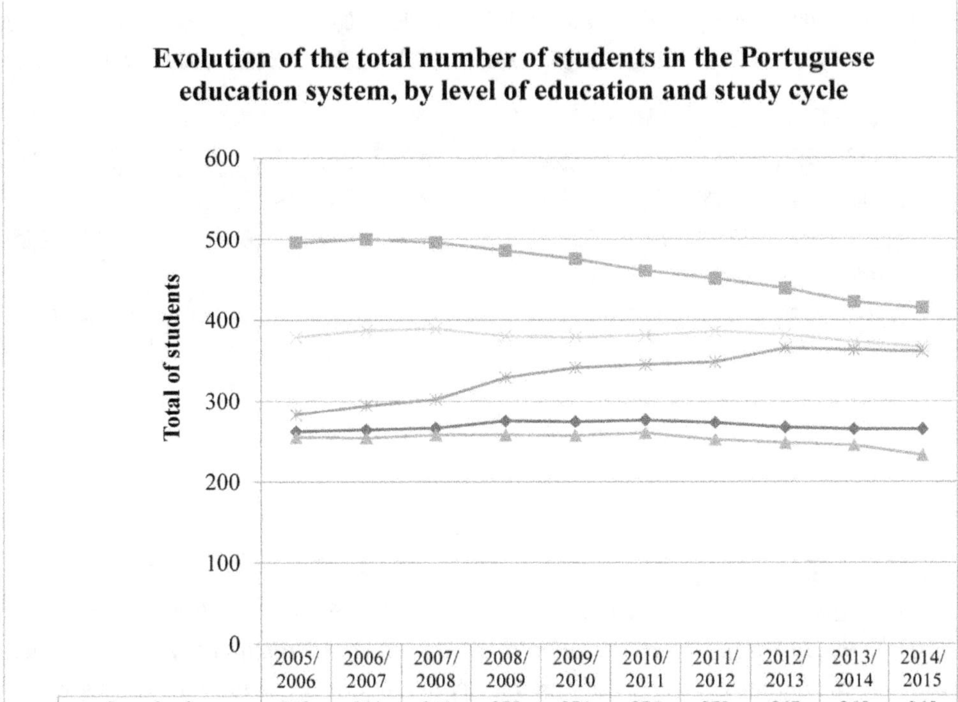

	2005/ 2006	2006/ 2007	2007/ 2008	2008/ 2009	2009/ 2010	2010/ 2011	2011/ 2012	2012/ 2013	2013/ 2014	2014/ 2015
Pre-school	262	264	266	275	274	276	273	267	265	265
1st CEB	496	500	496	486	476	461	451	439	422	415
2nd CEB	255	254	258	258	257	260	252	248	245	233
3rd CEB	379	387	389	380	379	381	386	382	373	367
Ensino Secundário	283	294	302	329	341	345	348	365	363	361

Figure 4. Retention and dropout rate (%) of students enrolled in the Portuguese education system, in Portugal, by level of education and study cycle
Source: *Adapted from DGEEC, DSEE and DEEBS (2016).*

School Year / Grade of Education	2000/2001	2001/2002	2002/2003	2003/2004	2004/2005	2005/2006	2006/2007	2007/2008	2008/2009	2009/2010	2010/2011
1st CEB	8,8%	8,5%	7,6%	6,7%	5,5%	4,4%	4,0%	3,7%	3,6%	3,7%	3,3%
2nd CEB	12,7%	15,6%	14,8%	13,9%	13,0%	10,7%	10,6%	8,0%	7,6%	7,7%	7,4%
3rd CEB	18,2%	19,2%	19,1%	17,8%	19,7%	19,2%	18,4%	14,0%	14,0%	13,8%	13,3%
Ensino Secundário	39,4%	37,4%	33,7%	33,8%	32,1%	31,1%	24,8%	21,0%	19,1%	19,3%	20,8%
Total	91,8%	94,3%	88,2%	84,2%	82,1%	76,1%	67,9%	54,6%	52,1%	52,4%	52,3%

Figure 5. Evolution in the availability of Human Resources for Education (teaching staff) in the Portuguese education system, in mainland Portugal, regarding their contractual link, education's subsystem, academic qualifications and age group
Source: *Adapted from DGEEC (2016).*

Teachers	Categories under analysis	School Years / Education Resources	2005/2006	2006/2007	2007/2008	2008/2009	2009/2010	2010/2011	2011/2012	2012/2013	2013/2014	2014/2015
Childhood Educators	Contractual Link	Permanent contract	8122	9050	8499	8164	7479	7660	7489	7306	7232	7274
		Fixed-term contract	1064	477	400	851	1351	1227	922	881	655	621
	Education Subsystem	Public	10757	11007	10319	10459	10368	10303	9765	9545	9006	8987
		Private	7456	7345	7363	7763	8012	7981	7863	7594	7137	7092
	Academic qualification	Bachelor / Others	3517	3195	2754	2555	1887	1836	2080	1920	2015	7274
		Graduation / equivalent	12912	13305	12991	13671	14292	14341	13371	12972	12118	12074
		Master or PhD	173	207	227	261	302	318	420	538	694	760.
	Age group	< 30 years old	2447	2100	1765	1885	1842	1563	1142	937	879	715
		30 – 39 years old	5015	4794	4350	4487	4364	4308	4041	3672	3220	3144
		40 – 49 years old	7329	7620	7432	7240	7168	6821	6285	5777	5139	4759
		≥ 50 years old	1811	2193	2425	2875	3107	3803	4408	5044	5589	6145
1st CEB Teachers	Contractual Link	Permanent contract	29036	26276	26635	24731	23806	22914	22232	21727	20899	20123
		Fixed-term contract	4451	2370	2746	3827	4433	4033	2516	2608	1934	2550
	Education Subsystem	Public	36449	31543	32105	31094	31293	29604	27264	26789	25201	25084
		Private	2947	2956	3123	3267	3279	3440	3428	3411	3013	3011
	Academic qualification	Bachelor / Others	7854	6149	5306	4340	3514	3414	3214	3001	2722	2521
		Graduation / equivalent	28075	24839	26492	26682	27064	25925	23715	23234	21467	21362
		Master or PhD	315	383	488	566	694	792	1004	1277	1431	1588
	Age group	< 30 years old	7165	4541	4031	3824	3177	2322	1167	781	490	350
		30 – 39 years old	9855	9186	10617	11261	11550	11370	10212	9737	8186	7732
		40 – 49 years old	11409	9819	9597	8778	8695	8317	8220	8570	8681	8968
		≥ 50 years old	7815	7825	8041	7725	7850	8122	8334	8424	8263	8421

Teachers	Categories under analysis	School Years / Education Resources	2005/2006	2006/2007	2007/2008	2008/2009	2009/2010	2010/2011	2011/2012	2012/2013	2013/2014	2014/2015
2nd CEB Teachers	Contractual Link	Permanent Contract	24246	22807	23959	21849	21657	20141	19211	18138	16548	13391
		Fixed-term contract	5263	4968	5105	6658	8030	8617	7006	3769	2975	2550
	Education Subsystem	Public	31707	30067	31327	30944	32285	31062	28419	24149	21503	20927
		Private	3047	2804	2730	3125	3344	3024	2911	2722	2881	2820
	Academic qualification	Bachelor / Others	4641	4336	4291	3700	3054	2893	2492	2254	1851	1929
		Graduation / equivalent	27927	25521	26716	27034	28946	27700	25178	20801	18900	18218
		Master or PhD	707	740	879	1003	1131	1265	1525	1648	1711	1845
	Age group	< 30 years old	4223	3306	2590	2931	2706	2148	1047	437	270	204
		30 – 39 years old	8570	8369	8692	8913	9847	9814	8551	5371	4330	3880
		40 – 49 years old	9385	8635	9175	9218	9785	9433	9243	8481	8099	7739
		≥ 50 years old	10467	10287	11429	10675	10793	10463	10354	10414	9763	9964
3rd CEB and Ensino Secundário Teachers	Contractual Link	Permanent Contract	63351	60491	59326	56592	54775	53146	51847	50962	49711	49769
		Fixed-term contract	12372	13498	15348	20142	21578	22099	18266	12253	10070	2530
	Education Subsystem	Public	80914	79988	80168	82564	82582	80786	75453	68448	65074	65949
		Private	8156	8292	8784	8761	8793	8753	8072	7653	7435	7404
	Academic qualification	Bachelor / Others	5582	3246	4983	4494	3876	3770	3405	3030	2801	2738
		Graduation / equivalent	74525	72959	73994	76194	76224	74324	67866	60274	56415	56661
		Master or PhD	3995	4210	4817	5175	5374	6164	7113	7654	8242	1845
	Age group	< 30 years old	8864	8215	7582	8312	7573	6039	2910	1181	746	595
		30 – 39 years old	28819	27815	27566	28537	27824	27201	23301	17162	13705	12925
		40 – 49 years old	28254	27826	28470	29199	29268	29326	29315	28401	27873	27795
		≥ 50 years old	18165	18559	20176	19815	20809	21692	22858	24214	25134	26937

Figure 6. Evolution in the academic qualifications of non-teaching staff in the Portuguese education system: comparison between the school year 2000/2001 and the school year 2013/2014
Source: *Adapted from DGEEC, DSEE and DEEBS (2016).*

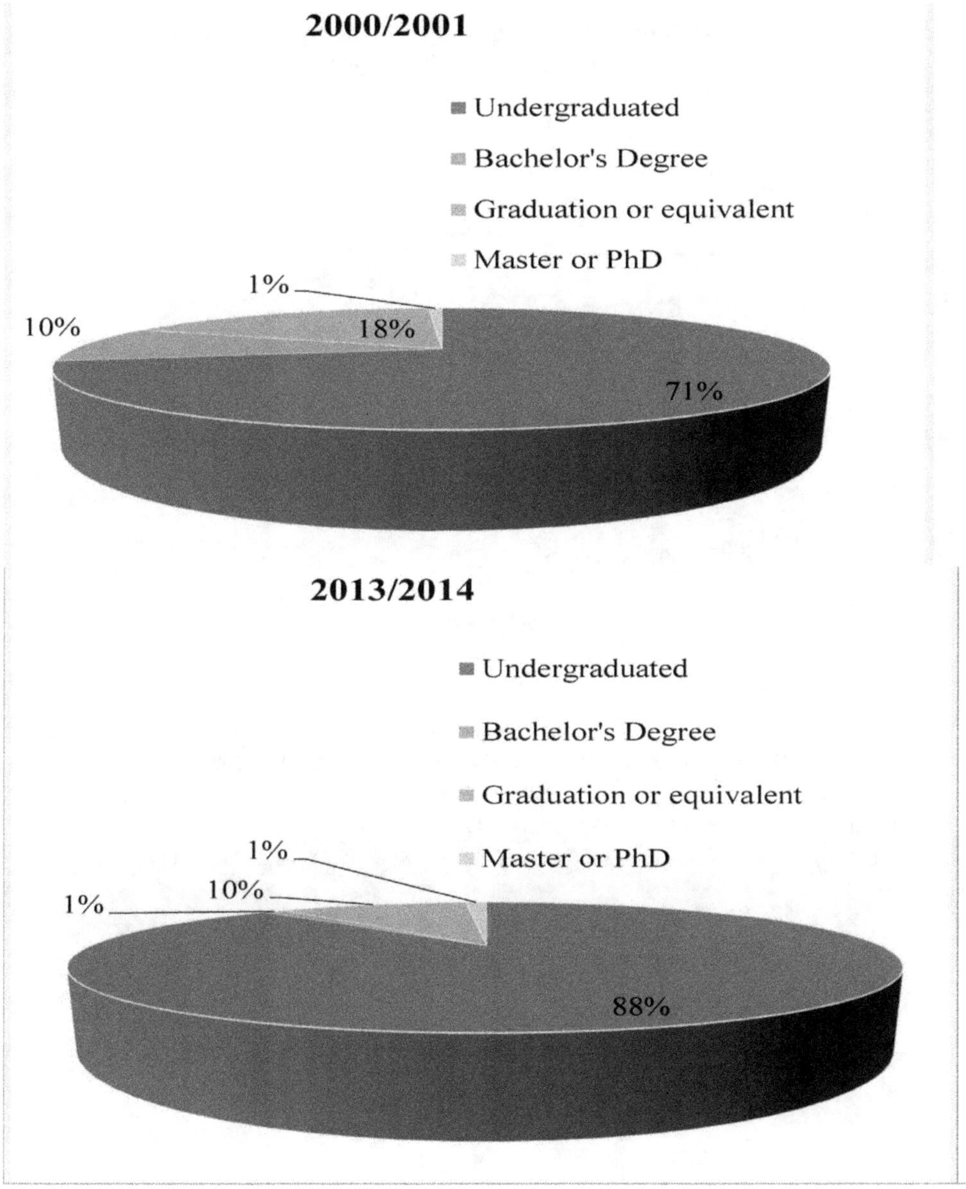

Figure 7. Evolution of the Portuguese State 's expenditure in education, in Portugal, according to a comparison of data presented by various national and international sources
Source: *Adapted from CNE (2015).*

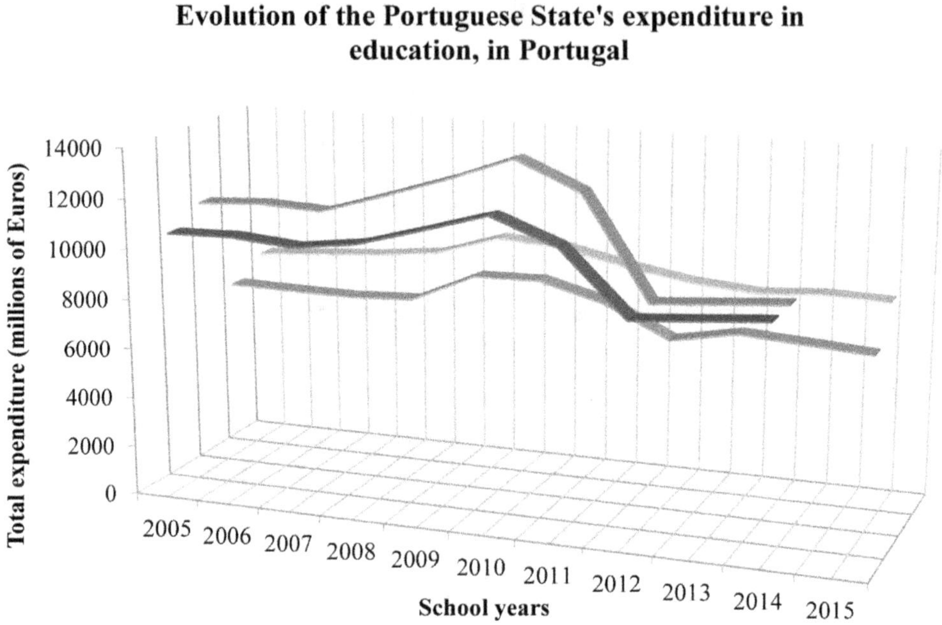

	2005	2006	2007	2008	2009	2010	2011	2012	2013	2014	2015
OCDE/EUROSTAT	10558	10623	10406	10743	11464	12228	11240	8709	8917	9130	
INE	11271	11425	11300	12107	12941	13909	12777	8709	8917	9130	
PORDATA	7316	7263	7232	7349	8507	8559	7879	6622	7108	6945	6756
MEC	8043	8238	8412	8702	9518	9339	8763	8341	8112	8291	8206

Figure 8. Evolution of the Portuguese State's expenditure with education as a percentage of GDP, in Portugal, according to a comparison of data presented by several national and international sources
Source: *Adapted from CNE (2015).*

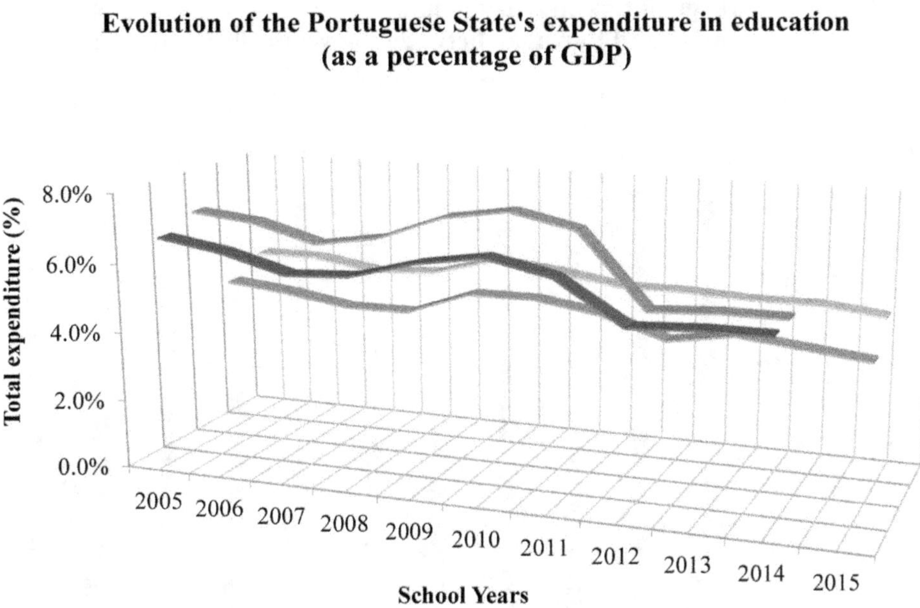

	2005	2006	2007	2008	2009	2010	2011	2012	2013	2014	2015
OCDE/EUROSTAT	6.7%	6.4%	5.9%	6.0%	6.5%	6.8%	6.4%	5.2%	5.3%	5.3%	
INE	7.1%	6.9%	6.4%	6.7%	7.4%	7.7%	7.3%	5.2%	5.3%	5.3%	
PORDATA	4.6%	4.4%	4.1%	4.1%	4.8%	4.8%	4.5%	3.9%	4.2%	4.0%	3.8%
MEC	5.1%	5.1%	4.8%	4.9%	5.4%	5.2%	4.9%	4.9%	4.8%	4.8%	4.6%

Figure 9. Evolution of the Ministry of Education and Science's expenditure in education, by category of expenses, in mainland Portugal, between the years 2005 and 2015
Source: *Adapted from CNE (2015).*

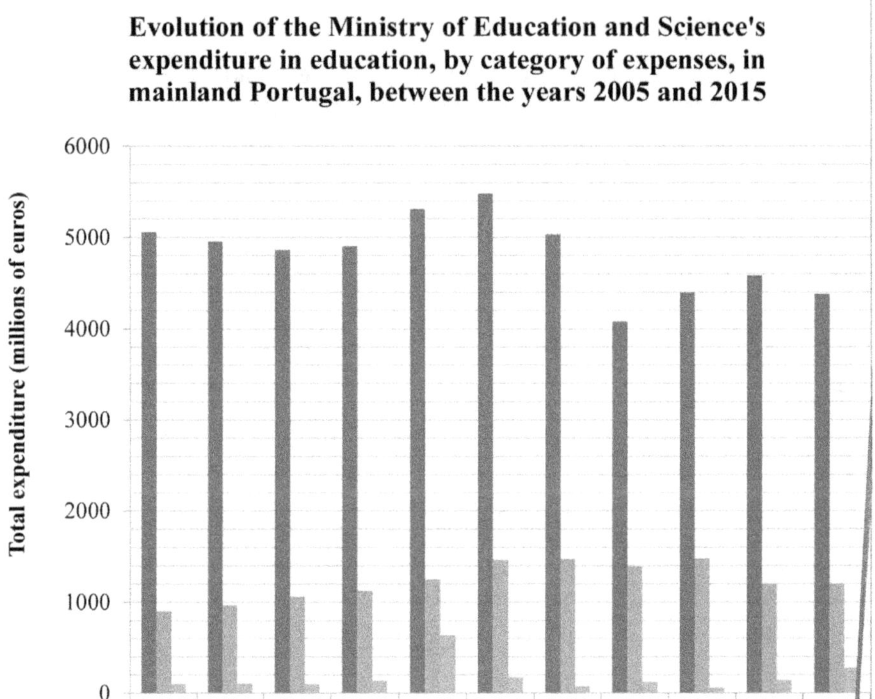

	2005	2006	2007	2008	2009	2010	2011	2012	2013	2014	201
Labor costs	5059	4952	4860	4904	5314	5483	5031	4077	4395	4580	438
Current expenses	902	963	1060	1123	1249	1462	1472	1389	1476	1197	120
Capital expenses	102	109	102	135	634	176	78	123	62	143	27

Figure 10. Teachers' distribution in the Portuguese education system, according to their age group and by nature of the education institution, in mainland Portugal in the school year 2014/2015
Source: *Adapted from DGEEC (2016).*

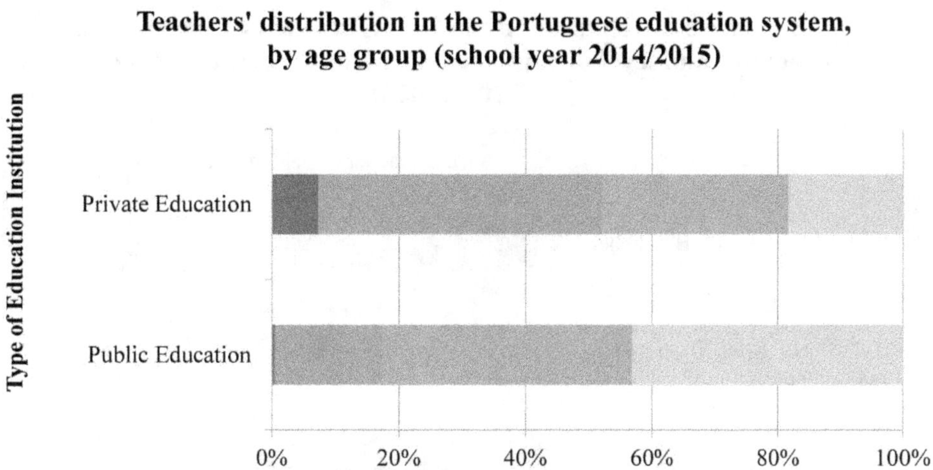

	Public Education	Private Education
< 30 years old	0.40%	7.30%
30-39 years old	17.10%	44.70%
40-49 years old	39.30%	29.60%
≥ 50 years old	43.20%	18.40%

Figure 11. Teacher's distribution in the Portuguese education system, according to their gender and level of completed academical degree, in mainland Portugal in the school year 2014/2015
Source: *Adapted from DGEEC (2016).*

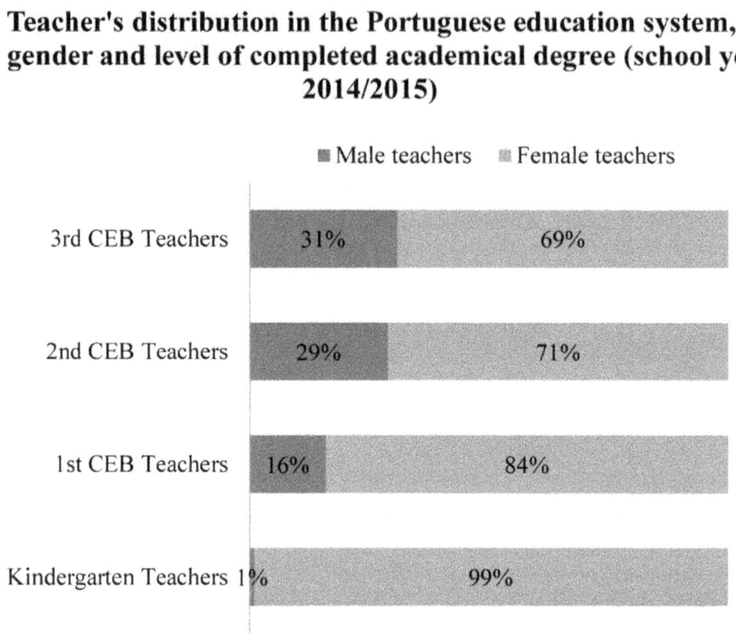

Figure 12. Evolution of the availability of resources (physical capital – establishments, and human capital – Human Resources for Education) in the Portuguese education system, according to their level of education, in mainland Portugal
Source: *Adapted from DGEEC, DSEE and DEEBS (2016).*

School Year / *Education Resources*	2000/ 2001	2001/ 2002	2002/ 2003	2003/ 2004	2004/ 2005	2005/ 2006	2006/ 2007	2007/ 2008	2008/ 2009	2009/ 2010	2010 2011
Pre-school teachers	14 704	14 777	15 414	15 394	16 267	16 602	16 707	15 972	16 487	16 481	16 495
1st CEB teachers	36 722	37 918	37 214	37 251	37 506	36 244	31 371	32 286	31 588	31 272	30 131
2nd CEB teachers	33 222	34 616	34 095	34 754	35 059	32 645	30 597	31 886	31 737	33 131	31 858
3rd CEB and Ensino Secundário teachers	81 724	82 867	81 626	82 099	84 404	84 102	82 415	83 794	85 863	85 474	84 258
Non-teaching staff	85 204	85 540	84 116	83 509	85 273	81 186	75 966	76 009	79 057	77 764	76 932
Education establishments	16 454	16 353	15 782	15 099	14 312	14 073	12 510	11 828	11 514	11 235	10 517

Figure 13. Geographical distribution of Portugal's resident population (number of people by NUTS III) aged between 0 and 24 years, in the year 2015
Source: *PORDATA, 06-06-2016 update apud CNE (2015).*

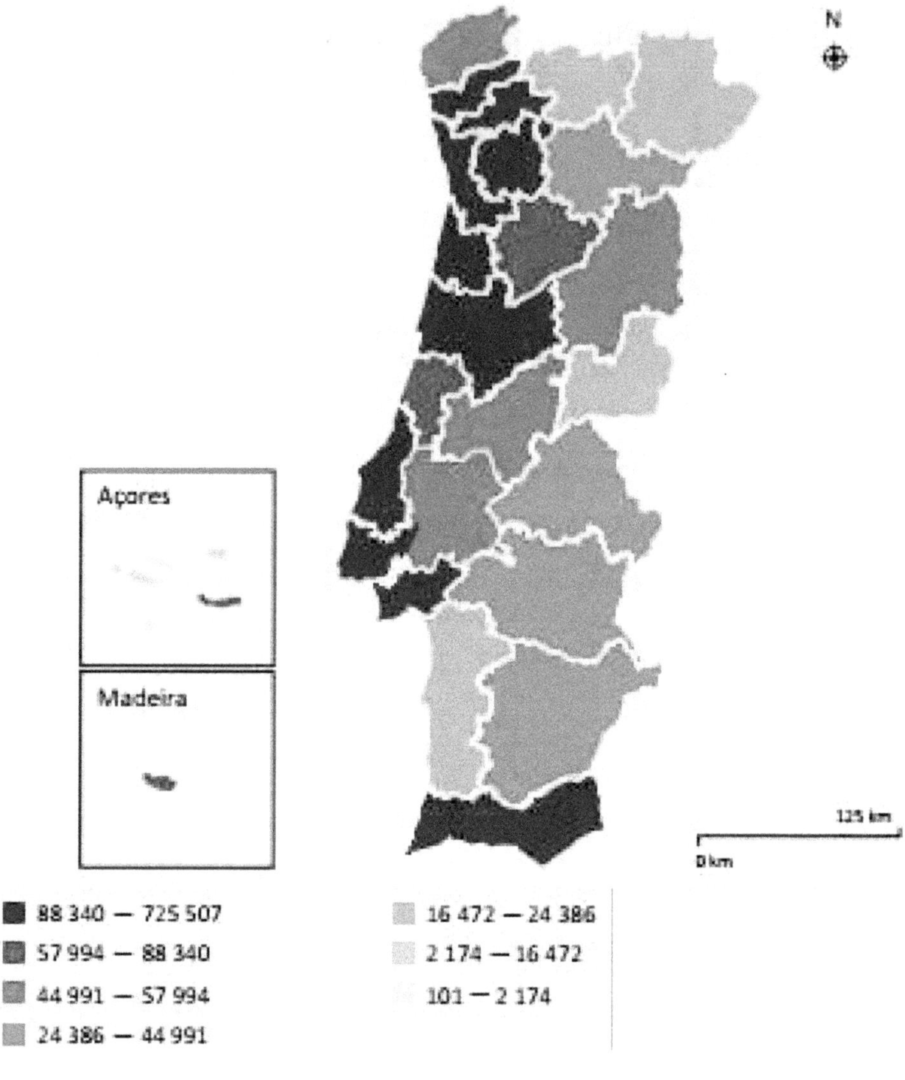

Figure 14. Availability of education establishments in the Portuguese compulsory public education system, in Portugal, by establishment typology and geographical concentration (NUTS II)
Source: *Adapted from DGEEC, DSEE and DEEBS (2016).*

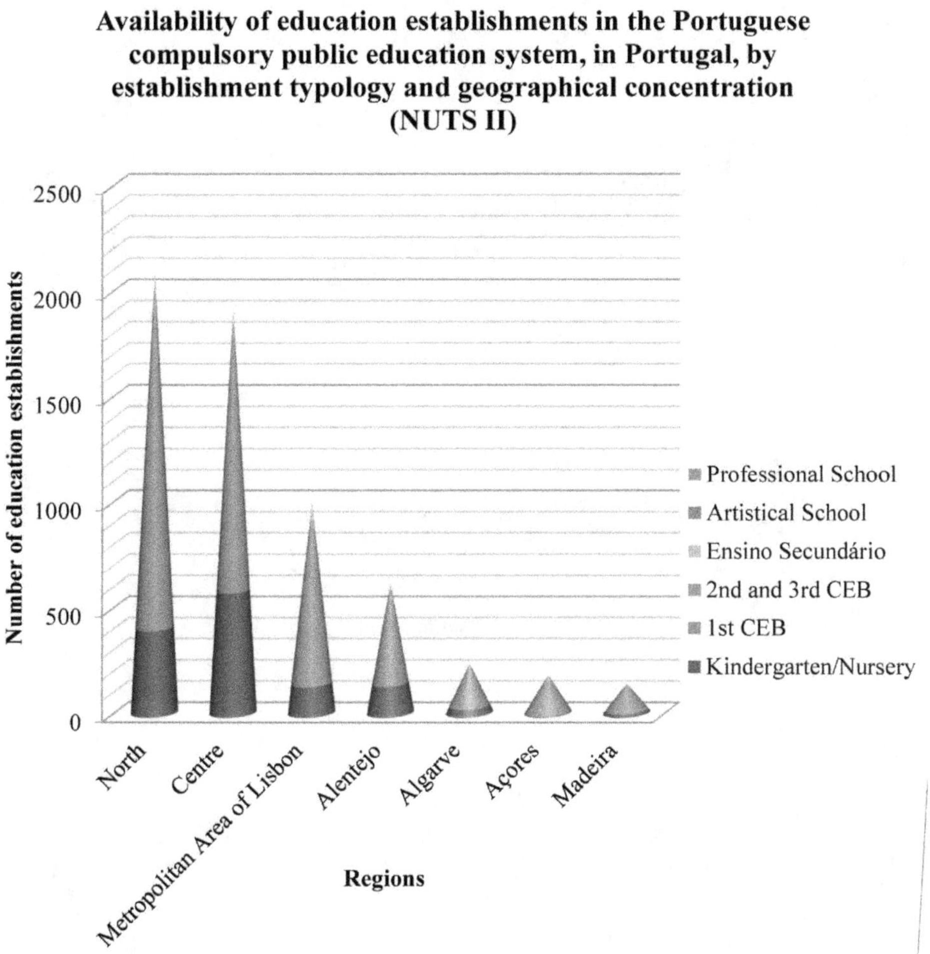

	North	Centre	Metropolitan Area of Lisbon	Alentejo	Algarve	Açores	Madeira
Professional School	8	4	2	3	0	1	3
Artistical School	3	2	4	0	0	1	0
Ensino Secundário	84	82	78	32	16	8	2
2nd and 3rd CEB	100	51	36	21	5	13	12
1st CEB	1489	1202	730	417	177	150	107
Kindergarten/Nursery	401	579	139	140	33	4	12

Figure 15. Availability of education establishments in Portugal, by its institutional nature
Source: *Adapted from DGEEC, DSEE and DEEBS (2016).*

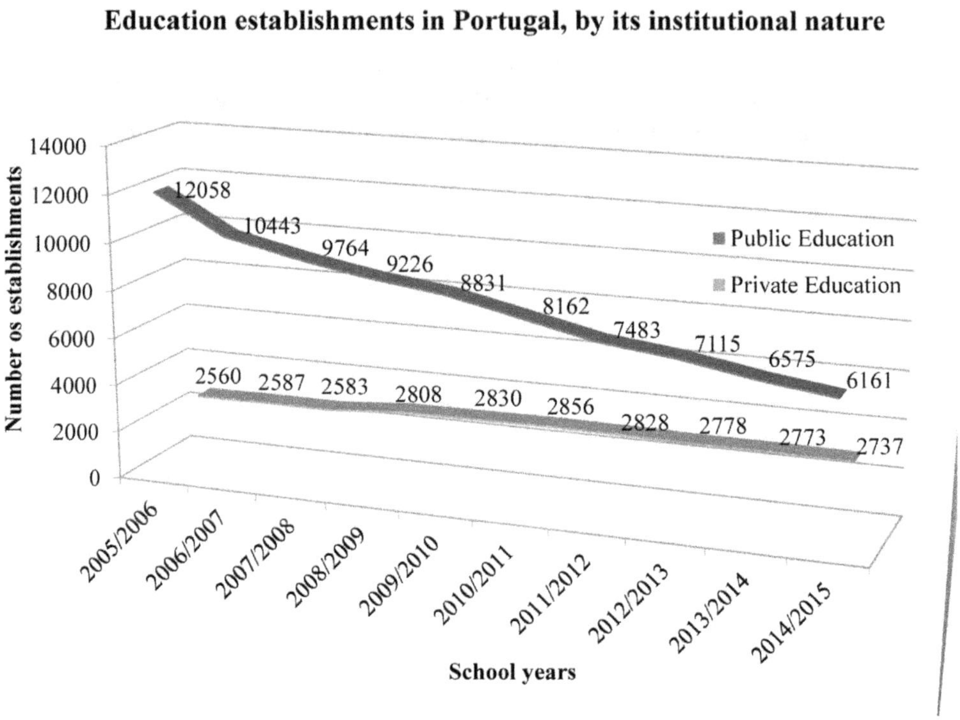

Chapter 7
Assessing the Soft Skills of IT Professionals:
A Proposal for the Information and Communication Technology Sector

Sílvia Diana Moreira de Araújo
Polytechnic Institute of Cávado and Ave, Portugal

Pedro Novo Melo
https://orcid.org/0000-0002-9293-8869
Polytechnic Institute of Cávado and Ave, Portugal & CICS.NOVA, Univeristy of Minho, Portugal

ABSTRACT

This research aims to assess the soft skills of information technology professionals in the information and communication technologies sector. The research method used was a combination of qualitative techniques to survey data about the most valued skills. As major results of this research, the following skills stand out: teamwork, problem solving, organization, leadership, communication skills, dynamism, initiative, autonomy, motivation, creativity, time management, attitude, responsibility, and proactivity. Based on collected data, a cross-cutting skills framework was constructed for five functional areas: trainee, junior consultant, consultant, senior consultant, and business development consultant. It was found that skills are transversal to all functions, and only the indicators vary in relation to the categories.

DOI: 10.4018/978-1-7998-5768-6.ch007

INTRODUCTION

The growing globalization and competitiveness of the business context exerts pressure for there to be a continuous improvement of the processes and services in the market. For this, it is necessary that it aligned all sectors of the company with strategic planning for only then becomes possible to compete in a global market. In this sense, the current context changes require a rapid transformation and adaptation of Human Resources Management (HRM) practices to their surroundings. Thus, the Human Resources manager assumes a more active role, he ceases to perform a purely mechanistic and traditionally administrative activity, to become an agent of change within the organization. Human capital (HC) also assumes itself as the most asset of an organization, since factors such as intellectual capital, technical and behavioural skills, actively contribute to the competitiveness and companies' success. That said, the HRM faces two major challenges: directing the skills and quality of human capital to its own interests through the various processes carried out by Human Resources (HR); and, finally, to outline an organizational structure that allows to maximize the performance of the organization.

The Recruitment and Selection process plays a fundamental role within the organization, if it allows the entry of new employees and may comprise several selection stages to assess the candidates' technical and behavioural skills. Nowadays, a curriculum vitae and a highly qualified candidate, are not the only determining factors in the selection process because, increasingly, technical skills must be accompanied by a set of behavioural or soft skills, since the companies also appreciate the intellectual and emotional abilities of the candidates.

The world is facing a process of constant change, in which companies are experiencing a real revolution, which is mainly due to two major factors with implications for the profile of skills: technological innovation and economic globalization. These factors have been causing profound changes in society and organizations, highlighting the need to gain new skills.

The identification of the transversal skills required for a function is a complex task, especially if the manager only focuses on the technical skills that are important for performing a function. However, it is no longer enough for the employee to master and perform a certain task. Knowledge and / or skills are needed to 'know-how', but also to 'know-to-be' in diverse situations and contexts. According to Ceitil (2016), skills are increasingly an enhancement factor of human capital and a guarantee of greater competitiveness of organizations.

LITERATURE REVIEW

The Recruitment and Selection Process

According to Fernandes (2017), currently, when we refer to the recruitment and selection process, we are talking about one of the richest tools of people management that organizations have. It is a transversal process to any organization and influences its performance because, following the logic that it composed organizations of people, their success ultimately depends on a recruitment and efficient selection process. Recruitment and selection comprise a process that allows the company to provide itself with competent professionals to perform a certain function, it aims to attract candidates more aligned to their culture and objectives. This process is so structured and allows comparison of the powers of each candidate, their personal and/or professional profile, as well as the alignment of the functional profile to the profile of the candidate (Camara et al., 2016).

Currently, the demand for effective people, capable of contributing to the success of organizations, is increasing. In this sense, the aim of the recruitment and selection process is to attract the best candidates, since the organization's goals can only be achieved with the contribution of people. According to Guimarães et al. (2005) the current business world is characterized by a fast pace and strong and agile competition, in the end the success of companies will depend on the talent of their human resources and where those resources are, they need to be found. Companies that are devoid of talents will certainly have difficulties in competing and may, at the limit, put their existence at risk.

Nowadays, organizations adopt a more humanistic perspective, considering employees as strategic partners and an integral part of the company, who have an influential role in the company's performance. In fact, Baylão and Rocha (2014) argue that the human side is valued in these processes that seeks to find people who will be framed in the company's vision. According to Chiavenato (1988) the adoption of this perspective by the organizations, impelled that the process of recruitment and selection was more and more valued and planned to minimize the errors in the choice of candidates. Hence the growing need to improve and adjust this complete process in a long-term perspective.

The recruitment and selection process are no longer just based on the candidates' experience and technical skills. Organizations are increasingly valuing the human being, they are also evaluating their emotional and intellectual potential (Rosa, 1999). It gave the employee a decisive role, and, in this sense, it is not enough to have the technical skills because it is expected that they also have more and better relational skills. If, in the past, we were dealing with an approach in which technical skills and qualifications defined the entire process of recruiting and selecting an individual in a company today, a set of social and emotional skills must also accompany them. In this light, more and more companies are differentiating themselves in the labour market as they increasingly value human capital, seeking to attract the right people for each role and keep them satisfied to add value to the organization. It is essential to align the recruitment and selection process strategically (Guimarães et al., 2005). We are facing the real challenge of the recruitment and selection process: creating an efficient and effective strategy for the development of this process, so that it allows finding the right person, with a whole set of technical and social skills for the job open, with a reduced cost and in a timely manner.

It is also important to clarify the need for the recruitment and selection process to go through the various elements of a company, as it will be through a fluid and effective communication of the company's needs, that the process will run in a more viable way. If the recruitment and selection process results in failures it is essential to understand the reasons for the failure to align a new strategy because it is a complete process that causes losses in terms of time and costs (direct and indirect) to the organization. Therefore, it is essential to adapt the strategy to the objectives. Also, the process of recruitment and selection has changed during the changes in the economic and social fabric. This process is something transversal to any company, and the importance of adapting an effective process is global, aiming at the success of companies. Although its implementation must be worked on within the organization itself, the recruitment and selection process usually follows a sort of standardized path, starting with the disclosure of the vacancy to be filled and ending with the decision-making about admission.

Although complex, the recruitment and selection process are a very important management tool that allows identifying human capital adjusted to the needs, values, and culture of an organization and, enabling it to respond to the challenges of the current market.

This way, the transformation at the level of HRM driven by competitiveness, globalization, continues to change the technology markets, among other factors, also goes through the recruitment and selection processes (Shameem et al., 2012). Although the recruitment and selection process are transversal to any company, the importance of adapting an effective process is global for all organizations, with the aim of their success.

Skills

The different skills seek to differentiate the various natures that jurisdiction may take. According to Ceitil (2016), in an organizational context, it is possible to verify two levels of skills explored by different authors: hard (specific) and soft (transversal), being essential to establish a distinction between both. The transversal skills are characterized by their transversally. They are transferable where they are gained within the exercise of an activity and can occur spontaneously in another context, and they are transversal because they are common skills to several independent contexts, it does not associate them with a specific function. It essentially related them to know-how-to-be that is adjustable to a certain function, and that are expressed in behaviours arising from personal or personality characteristics. In addition, it divided them into personal behavioural skills and interpersonal skills. Whereas personal skills are intrinsic characteristics to the individual (proactivity, ambition, integrity, etc.), it related interpersonal skills to the way the individual interacts and relates (communication skills, teamwork, empathy, interpersonal relationship, etc.) (Câmara, 2017). Successively, technical, or specific skills are directly related to professional specificities that allow the individual to develop their role in a particular sector of activity. Called hard skills, they are technical skills and can be easily measured, unlike soft skills that relate to transversal, human skills and are not directly related to academic or professional training. According to Cardoso (2010), technical skills are understood as the skills recognized in a domain, combining the knowledge gained through training, embodied in a specific diploma, and the know-how, withdraw essentially from experience professional, and that are performing the function. According to Szilárd, Benedek and Ionel-Cioca (2018), communication skills, team spirit, creativity, problem-solving skills, leadership, and innovation, among other characteristics, are some of the necessary soft-skills in the current job market. However, for Swiatkiewicz (2014), opinions are many and differ in relation to the different transversal skills, based on empirical research and literature review. Because of the theme, there is no single definition. Currently, revolutions in the labour market, technologies and communication, cause changes in the entire organizational context. In this sense, the companies' own requirements regarding the resources also suffer some changes, an aspect that ends up influencing the recruitment and selection processes of an organization.

The technology market, mostly, has been generating a great impact on the skills that companies expect to find in candidates. According to Robles (2012), the changes in the current global market, including the fact that we are facing a society of additional information and communication technologies, are factors that enhance the need for many jobs to require behavioural skills combined with techniques. That said, there is a need to re-examine the type of characteristics that organizations seek to integrate into the organizational bosom, as technical skills are no longer sufficient. According to Patacsil and Tablatin (2017) soft skills have an important role for IT students, therefore, they need to be present in the student's curriculum. Soft skills like leadership, communication and management skills are important to the IT industry.

Technical skills (hard-skills), although important for performing the function, become insufficient because of the lack of transversal skills (soft-skills). Nowadays, soft skills assume a crucial role in the labour market, as they are essential to enter and maintain the market because they allow the individual to reach their potential and, bring about results in the organization's efficiency. Increasingly, there is an awareness of the importance of soft skills, which are already critical factors for the success of organizations, especially in the areas of ICT. However, and according to Ribeiro (2017), there is still a wide gap between the skills that employers need and the skills that young graduates have. Currently, recruiters are looking for candidates who add value and make a difference in the workplace using soft skills as it allows organizations to maintain or gain competitive advantage in the market. In addition, the lack of soft skills can compromise the career of an individual who only has technical skills (Robles, 2012).

With the global modernization of the labour market and the sharp development of ICT, the demand for professionals with this type of skills combined with technical knowledge is increasing. Thus, the success of these professionals is based on the combination of hard and soft skills.

Hard skills emphasize 'know how' and are directly related to academic training, complementary training, professional experience, among other variables that are related to the individual's technical know-how. They are essentially all those skills that a manager can analyse in a curriculum vitae. These are measurable and tangible skills (Ribeiro, 2017). According to Klaus, 75% of long-term success in a job results of the presence of soft skills. Consecutively, only 25% of success depends on technical skills and/or knowledge (Robles, 2012). According to Cunha (2016), the transversal skills, combined with the technical skills, promote the potential of each member of the organization and their effective performance of the function. Ribeiro (2017) also emphasizes the need to know which specific or technical and transversal skills allow defining the concept of human capital. With this, the combination and balance of these two types of skills is fundamental because, although the individual is technically fit for the exercise of a function, he may not get the intended results because of the adopted behaviour (Costa, 2015). Thus, the existence of a technical profile, no longer ensures the permanence in the work environment, as both skills are complementary. Therefore, it is essential that soft-skills training initiatives are evaluated, as well as hard-skills training. Litecky & Prabhakar (2004) refer that some studies present contradictory results on the importance of soft and hard skills. Some research has indicated more importance for social skills; others also strongly believe that greater importance should be given to technical skills.

Although soft skills are current and important tools to deal with issues such as the advancement of technology and competitiveness, it remains to take care of this aspect of transversal skills and to develop behavioural skills within the organizational bosom, since these excel in scarcity. In addition, it is important to mention the challenge of finding methods to assess this type of skills, since they are intangible and difficult to define.

Skills Profile Analysis

Today, with increasing market competitiveness and more unstable organizational environments, it has forced organizations to change their management practices and policies. It is in this sense, that most organizations abandon their traditional approaches to more strategic approaches that imply a greater focus on human capital and, mainly, on the major strengths necessary for change. This aspect gives rise to a central management tool in HRM practices: The Analysis and Description of Functions (ADF).

The functions are grouped with a certain content and complexity, and for them to be successfully exercised, individuals must have technical knowledge, but also attitudes and behaviours that allow them to achieve success in performing a certain function. It is necessary to have a specific portfolio or profile of skills aligned with each function. ADF has proved increasingly important for organizational sustainability, as well as for the development of human capital, as the individual skills of employees (technical and behavioural) influence the overall performance of the organization (Couto, 2018). In addition, it is considered an essential tool in the various practices or subsystems of HRM and, in the management of an organization's employees.

According to Chiavenato (1988) it is necessary to describe and analyse the positions to know their content and specifications to manage the human resources applied in them. However, while the Analysis concerns the moment of collection and identification of the activities, conditions and information of a certain function, the Description makes up a summary of the tasks, objectives and responsibilities of the information collected, resulting in a profile of skills associated with a function. That said, the ADF tool is a structured process that allows the collection, analysis, and systematization of information regarding the profile of a function. This allows a distinction to be made between the various functions existing in an organization, to identify the skills necessary for its performance. For Chiavenato (1988), the analysis and job description "(...) is basically a written survey of the main significant aspects of the position and the duties and responsibilities involved" (p.173). Therefore, to develop a skills profile, it is necessary to carry out an ADF, which forms the basis of many HRM policies and practices because it allows the planning and alignment of the function's objectives to the organization's objectives.

With Recruitment and Selection, the ADF is a fundamental tool because it helps to attract and keep candidates with the most appropriate profiles for the organization's objectives and culture. According to Paiva (2013), this management tool is valued in the recruitment and selection process, allowing to observe a complete set of skills that aims to choose the ideal candidate. For Quintela (2016), the ADF refers to the first moment before the recruitment and selection process because it is from there that it provided us with the information to proceed with the respective process. This profiles the function's profile, both at a technical and behavioural level, through the collection of data that, subsequently, allows the HR manager to know how to identify the candidate with the most appropriate profile for the function in question. Also, Chiavenato (1988) highlights the importance of ADF in the recruitment and selection processes, to determine the ideal profile of the candidate.

The definition of the skills profile of a function is helpful for an organization because it ensures the results from the definition of the skills necessary for the effective performance of the tasks. This factor not only contributes to the achievement of the organization's mission, but it is helpful for employees as it allows them to know what the organization expects of them, both at a technical (knowledge) and behavioural level (attitudes and behaviours) and even gives them the possibility of developing and increasing their potential.

According to Cardoso (2010), the definition of the skills profiles is a procedure inherent to the selection and recruitment process, both internally and externally, since it allows the identification of the skills intended for a function, whether technical and/or behavioural. It allows you to reduce undesirable costs by hiring candidates who have undesirable skills or who are not under the profile required for the function and, therefore, the best candidates must be those who come closest to the defined profile.

Although, to define a skills profile, it is always essential to know what the organization expects from individuals, and it is necessary to develop a process for selecting the skills that are associated with the organization's success and effective performance. Only then will it be possible to build in an aim and effective way, a set or profile of skills necessary for each function, adjusted to the objectives of the organization. It is also important to mention the dynamic character of the analysis and description of functions, since it only represents the content of the function at a moment, which can be reviewed and adjusted periodically.

A company's Competency Model comprises a set of generic and specific competencies that have been identified as essential for the company to achieve its business objectives (Câmara et al., 2016). Although the skills management model is a very useful tool within an organization for it to be efficient and effective it is necessary to adjust to the current challenges of the market and the skills needs foreseen in the organization, both in the present and in the future. The HRM has been using skills management models in its practices, but sometimes the difficulty in implementing these models is largely because of the organization's lack of understanding of their benefits. In short, the analysis and description of a skills profile aligned to a certain function, represents an added value to the capacity of organizations, allowing to know the characteristics, skills, aptitudes, and knowledge necessary to occupy the position. At the organizational level, skills profiles are necessary to ensure that it achieved the desired results, thus representing the basis of any HRM work.

RESEARCH METHOD

Considering that each paradigm comprises a certain set of methods, theories, and ways of defining data that are suited to its assumptions, the research method used fell on a combination of qualitative and quantitative techniques so that it can achieve a broader vision. Reality that we intend to study (Freixo, 2011). As previously mentioned, the research aims to identify profiles of transversal skills in information and communication technologies through a case study.

This research focuses on unique people, in a certain context that is very specific to them. Therefore, it should find a social representation, based on the principles of saturation and diversity of the data collected.

The study is exploratory and descriptive therefore this type of research allows us to understand the object in question on the ground, through contact with real and current situations, using techniques that allow the interaction with the object studied. It allows, therefore, to know the dimension of the phenomenon, how it manifests itself and the facts by which it relates, to analyse the information collected from the collaborators and direction to identify the profile of transversal skills that a computer consultant should have.

In a first phase, the technique used was Document Analysis. Through access to specific internal documents with useful information to study the object in question, (goals, values, mission, organizational and descriptive structure functions), it was possible to better match the objectives and requirements of the research, which would not otherwise would be.

Using the Interview as a method of collecting information, in a second phase, was linked to some advantages that this technique presents, for example: flexibility, opportunity to deepen the elements of analysis collected, rich interpretation of details, diversity of data and the importance of results (Quivy and Campenhoudt, 1995). The managers directed at all elements of the Direction and Management to collect information about the HR function and the applied practices, as well as the behavioural factors considered most important in the functions, the interview technique.

Regarding Direct Observation, in a third phase, it proves to be an effective and differentiating method, because it allows to capture behaviours at the exact moment and in themselves. Here, the researcher himself directly collects the information, directly appealing to his sense of observation. For the exposed work, participant observation is also considerable, since there was an interaction with the most diverse actors because it allowed for integration and participation in the various processes. Although, it was possible to assist on-the-spot tasks performed by employees, because there was a certain openness in this aspect. It was possible for the researcher to be attentive to the appearance or transformation of behaviours and to the effects they produce in the observed contexts.

ANALYSIS AND DISCUSSION OF RESULTS

The IT area is an area that lacks highly specialized resources, in a context where opportunities are diverse. Thus, the role of an IT recruiter and its practices throughout the selection process are fundamental and must be under the strategic role of the organization. The lack of criteria for the evaluation of profiles can lead to errors, such as bad hiring, therefore, the instruments for an efficient recruitment and selection process are indispensable.

For decades, it made the potential candidate evaluation based on his academic profile, analysis of diplomas and/or qualifications gained. However, there was an authentic break with this traditional thinking, verifying that this type of skills is independent of an individual's performance (Ceitil, 2016). The importance that transversal skills assume in companies, today, has become clear. People are no longer seen as a resource endowed only with knowledge and/or technical skills, and treated as beings endowed with attitudes, behaviours, emotions, knowledge, and aspirations. There are some authors (Mitchell, Skinner and White, 2010) who support the view that transversal skills are equally important and of great relevance within the organization. According to them, these are essential attributes and tools that allow people to achieve concrete and positive results from their performance.

This research aims to identify the profile of transversal skills in information and communication technologies for the further development of instruments to be applied in recruitment and selection. Therefore, it is important to define what skills are necessary in view of the new challenges that companies face and to adapt a skills profile for the subsequent creation of instruments that allow the improvement of human resources intervention. It should train companies with agile instruments for the successful success of their recruitment and selection process, as well as to understand and identify which transversal skills are most valued by IT companies. The case study was developed in a small and medium reference company in the Information Systems area, having already been awarded some awards and certifications as SME Leader and SME Excellence. The choice for this company was mainly because it is a company with technological bases and with a greater incidence in Computer Engineering profiles and because it is a small and medium-sized company.

With the analysis of the data collected through the interviews carried out with the managers and considering the objectives proposed for the research, it identified five thematic areas below:

- Thematic area I - Cross-cutting skills most valued in IT profiles.
- Thematic area II - Variation of transversal skills according to the different functions.
- Thematic area III - Importance of integrating a profile of transversal skills in the Recruitment and Selection processes.
- Thematic area IV - Benefits and challenges of a skills-based HRM.

Transversal Skills Most Valued in IT Profiles

Characterizing the transversal skills of a computer consultant is not a simple task, since there is no single descriptive and consensual of the transversal skills necessary for a particular role of a consultant. By analysing table 1, it is possible to see that there are several aspects that the interviewees consider important to highlight, even using several examples. Therefore, the data resulting from the interviews conducted resulted in the identification of sixteen transversal skills: Responsibility, Dynamism, Autonomy, Involvement, Initiative, Coherence, Communication, Resilience, Emotional Intelligence, Proactivity, Teamwork, Attitude, Leadership, Problem Solving, Organization and Humility.

In a first analysis, two of the interviewees identify **Responsibility** as an active element, as can be seen through the analysis of the speeches of (interviewee 1, 2, 3 – I1, I2, I3):

I1 (…) that what we value most is the sense of responsibility (…) I2 (…) must have relatively prime responsibilities, because this is the company's culture, pass responsibility (…)

According to I1 (…) *dynamism, that it is not closed in its world, although this does not exclude people, but brings them added value* (…). The intention of the current bosses is to be different, to stand out, which happens in the interviewee's perspective for knowing how to be dynamic in an organizational context.

The three interviewees mentioned the subcategory Autonomy, as can be confirmed by the speeches of I1 (…) *autonomy is, for more senior profiles, completely mandatory (…); (…) And autonomy, are aspects we try to work on, although this is not always the case* (…) I2 (…) *have to work in a team, autonomy, because the company's culture is that,* (…) I3 (…) *must be able to present own ideas, look for solutions, be independent and not always driven* (…).

Involvement also presents itself as one of the least representative subcategories. This characteristic is important in that, according to the I1 (…) *the involvement that the person has with the company and shows* (…).

From interpreting the speeches, the notion of **Initiative** also stands out to make contributions in other functions, as well as participation in the search and presentation of suggestions that contribute positively to the work context, as can be confirmed by the speeches of the I1 (…) *most appreciated, is the capacity for initiative that people have. Leaving your duties, suggesting things that have little to do with your work* (…) I3 (…) *I think it is an important characteristic/behaviour, as well as the search for solutions* (…).

The subcategories **Consistency** I1 (…) *and in the technical interview subsequently conducted, we also assessed the coherence of the people,* (…), **Resilience** I1 (…) *a highlight for a last skill that I haven't said yet, which for me is very relevant, resilience* (…), **Emotional intelligence** I1 (…) *how the person can manage the relationship with the client, the part of emotional intelligence* (…) and **Proactivity** I1 (…) And the proactivity towards the client, giving different solutions (let's apply this to the technical part), (…) were only highlighted by one of the interviewees.

Through the analysis of the interviews, the subcategory **Communication** was highlighted by three interviewees, since it is an active element in the role and functions that the consultants play, as can be seen in the speeches of I1 (…) *the whole set of soft-skills is relevant, from the part of how to communicate with the client,* (…) I2 (…) *but also, the ability to communicate and speak to an audience, sometimes demanding, and be comfortable in that role* (…) I3 (…) *consultants in their areas make a presentation to colleagues, giving them a set of skills and the ability to make and prepare presentations, to debate what they do* (…).

Teamwork is another one among the several subcategories identified in terms of the transversal skills most valued by managers I2 (…) *teamwork will always be something important, so I would say yes, we must pay attention to these aspects* (…) I3 (…) *whether the requirements are understood in terms of soft skills, be it presentation of ideas, driving, group participation* (…).

It also identified leadership as a transversal skill, albeit with little representation I2 (…) *That is, if I am hiring a team leader, I will privilege some aspects* (…). Also, the **Attitude** subcategory I2 (…) *what we value most, besides the academic part, which also has an important component, is the attitude of the consultant, the presence* (…), **Problem solving** I2 (…) *your ability in solving situations that are sometimes not technical, but are everyday situations* (…), **Organization** I2 (…) *we organize workshops that stimulate, on the one hand, the organization of consultants' knowledge* (…) e **Humility** I3 (…) *humility in the sense of listening to ideas and* *being able to debate them, I think it is an important characteristic / behaviour* (…) stand out for their lower representativeness.

That said, the data resulting from the interviews conducted resulted in the identification of 16 transversal skills (table 1).

Table 1. Transversal skills identified by the interviewees

Transversal Skills	Total number of evidence
Responsibility	6
Dynamism	2
Autonomy	10
Involvement	1
Initiative	5
Consistency	1
Communication	2
Resilience	1
Emotional Intelligence	1
Proactivity	1
Teamwork	2
Attitude	1
Leadership	1
Problem Solving	1
Organization	1
Humility	1
Total Skills / Evidence	**37**

Following the data presented in the table, Responsibility, Autonomy, Initiative, Communication and Teamwork are highlighted as the characteristics of IT profiles. These are the subcategories that have the greatest representativeness. The recognition of the interviewees for each of these subcategories is clear, and it is possible to verify that these are differentiating characteristics at the time of recruitment.

It is difficult to reach a consensus on the characteristics that a computer engineer must have, however, the opinion of the interviewees is not very distant from each other. In an increasingly competitive market and where growth and technological innovation prevail, the ability to be autonomous in the development of knowledge, the capacity for initiative in the search for new learning, effective communication in a social and / or professional environment, work in team and the ability to be responsible, are essential characteristics for a prosperous and sustained growth in the company.

Variation of Transversal Skills According to the Different Functions

This thematic area corresponds to the answer to one aim of this research, to understand if the transversal skills identified vary according to the different functions assumed in the company. According to all respondents, transversal skills are modelled according to the profile that is at stake and, therefore, the level of demand will be different depending on it, as can be seen through the analysis of the speeches of:

I1 (…) autonomy is, for more senior profiles, completely mandatory. The sense of responsibility, of course, is also required, requiring less in the more junior profiles, and we hope to develop this type of skills, internally (…)

I2 (…) obviously with variations depending on the function for which it is being contracted. (…) Therefore, despite being transversal, they are modelled according to the profile in question (…)

I3 (…), have a slightly different applicability, but yes, I think it is transversal from the Project Manager to the Junior Programmer (…).

Therefore, the transversal skills identified, and the level of demand differ according to the profile that is intended, within the areas in which the company develops.

The Importance of Integrating a Profile of Transversal Skills in the Recruitment and Selection Processes

Within the third thematic area, it should understand the importance of integrating a profile of transversal skills in the respective recruitment processes, from the perspective of the interviewees. Although the organization does not have a formalized descriptive of transversal skills for each function, the three interviewees consider it relevant to establish and formalize a descriptive that meets the transversal skills necessary for the exercise of a certain function. I1 (…) *The whole set of soft skills is relevant (…) and this is, in a non-formal way, very relevant in decision-making for hiring* (…) I2 (…) *And we even have an interview process in which the first one intends to be a non-technical interview, already a little to assess this profile as well (…) I3 (…) they are more difficult to evaluate, however, they have a quite large weight* (…).

Benefits and Challenges of a Skills-Based HRM

The **development of the business area** is only identified by one interviewee. He considers transversal skills are fundamental in some business areas because they stimulate their development, as can be seen through the analysis of the discourse of the I1 (…) *considering some areas of our business, it is essential to have some soft skills developed* (…).

The **quality improvement in the integration process** is identified as the most representative subcategory. Through the analysis of the discourse of I1 (…) *brings the possibility of integrating people who would fit even in our profile* (…) I2 (…) *in the medium term it will bring us employees with better quality and suited to the company's culture* (…).

Understanding the requirements is only identified by the I3 (…) *If I can identify in a more correct way if they understand the requirements in terms of soft skills, (…), if you can identify that, which is increasingly important, if you can identify a priori, better* (…). According to the interviewee, understanding the requirements of a profile at the level of transversal skills will allow to streamline the entire process, from recruitment to selection.

The interviewees also identified some challenges of a skills-based HRM: the integration of tests, the relevance in the recruitment process, the job market and the benefits of integrating these processes.

In a first analysis, two of the interviewees identified the Integration of tests as a challenge, as can be seen through the analysis of the speeches of I1 (…) *I would say that it is to extend too much a recruitment process, or to bring tests to which people, at the limit, can feel invaded in their space* (…) I3 (…) How to maintain quality and implement better processes, without creating so much entropy in the process itself (…). The mindset of the respondents follows the difficulty in integrating tests to identify transversal skills without hinder or create entropy in the recruitment process, making it by itself invasive and / or time-consuming.

Also, the subcategory **Relevance in the recruitment process** is identified by two of the interviewees, I1 (…) *convince the relevance of integrating these processes in the recruitment process* (…) I2 (…) if we don't take care of it and if we don't do that planning, and cut short on that path, eventually there's also going to be something going wrong here (…). In the opinion of the interviewees, it is a hard task to convince them of the importance of integrating means that allow the identification of these skills in the recruitment and selection process. Interviewee 2 also highlights the fact that if there is no planning in this direction, the results may not be very positive at the level of the organization.

The labour market subcategory is also identified as a challenge, as can be seen from the speeches of I1 (…) there is another major challenge, particularly related to our labour market, which is the lack of resources (...) I3 (...) *professionals have so much to look for and, therefore, they may not be as available for another type of more time-consuming process* (...).

The analysis of the interviews also allows to verify that the subcategory **Benefits of integrating these processes** was the one that got less representativeness, having only been identified by one interviewee, as can be seen through the analysis of the speech I1 (...) *to show that the eventual integration of these processes really brought benefits, therefore, from the part of the integration to the part of the evaluation of the results of the determined action* (...). For the interviewee, the challenge is to show the benefit of implementing processes that allow the identification of skills in individuals.

Knowledge of the technical profiles, but also behavioural ones, is fundamental for an organization. In constant change, not only in the foreign market but also in the relations between organizations and employees, it is imperative to have tools that allow working on attracting and keeping capital and/or human talent.

In addition, the data got showed that all participants in the study consider the divergence of transversal skills according to the profile that is analyzed within the areas in which the company develops.

Based on the analysis of the interviews, it was concluded that there is a list of transversal skills most valued and evidenced by the interviewees, with emphasis on responsibility, autonomy, initiative, communication skills and teamwork. With a more residual value, it highlighted 11 skills. It identified 16 transversal skills.

In terms of benefits highlighted by the interviewees, and regarding the definition of a transversal skills description, it identified the development of the business area; improving quality in the integration process and understanding the requirements.

However, the informality identified in the company under study does not favour the existence of structured action plans aimed at the development and continuous improvement of HR processes and human capital. The lack of a structured relationship between skills and behavioural indicators does not benefit the sustained development of the HRM model.

The needs that are being felt in companies, many of them driven by the constant development and change of the labour market, reinforce the perspective that it is increasingly essential and necessary to provide companies with instruments that serve their needs, boost their growth and that fit their specific values, culture, and aim, aiming at the development and continuous improvement of human capital (Spencer & Spencer, 1994; Ceitil, 2016).

In this sense, and in order to provide companies with instruments that allow their internal development, promoting a more strategic, efficient and oriented HRM for the present and future needs of HR, as well as monitoring at the level of the external market, we proceeded with the initially proposed objective: the analysis and identification of a description of transversal skills associated to each function and category, in the company in question, so that it is possible to establish a relationship between the skills and the respective behavioural indicators.

It is proposed, therefore, a set of guidelines showed by Câmara (2016), which allows the recruiter to conduct a process with the least possible probability of error to get a match between the candidate and the desired profile.

Although these guidelines have been proposed by the author described in the sense of added ability to make certain task, it is proposed to apply five levels of proficiency and complexity, smaller or larger, the defined skills framework:

1st Level- *Trainee*: shows that the individual has a knowledge base about the skills. Knows the concept and applies the skills in a contextualized way, albeit with some difficulties, since he has no experience. Acts through rules, requiring superior support to guide and monitor the performance and development of skills. This is a skills profile that corresponds to a young graduate, since he still does not have enough experience to pass to the next level, however, he already has a well-supported theoretical knowledge.

2nd Level – *Junior Consultant*: The individual already has some contextualization concepts in the skills performance needing, occasionally, guidelines and monitoring for a better development of the skill.

3rd Level – Medior *Consultant*: The individual has already accumulated experiences that allow him to apply the contextualized skill in different situations. The individual has a holistic view of the application of skills, which allows him to take actions with a well-defined eventual goal.

4th Level – *Senior Consultant*: In this state, the individual no longer needs support, monitoring, or guidelines. Intuitively apply skills through diverse experiences that guarantee a high level of performance.

5th Level – *Business Development Consultant*: This is the level that represents the best performances, those that stand out in relation to the others. The individual can excel in a very positive and intuitive way.

In this regard, the behavioural skills profile was elaborated based on the information collected in the interviews, document analysis and direct observation and which can serve as a reference for a transversal skills profile in IT companies (table 2):

Table 2. Behavioral skills profile

Skills Level	C1	C2	C3	C4	C5
Category	**Trainee**	**Junior Consultant**	**Medior Consultant**	**Senior Consultant**	**Business Development Consultant**
Professional experience	**0-1 years**	**1-2 years**	**3-5 years**	**5-10 years**	**<10 years**
Teamwork	Promotes dialogue with colleagues, accepting contributions and discussing ideas.	Keeps colleagues abreast of useful or relevant information. Occasionally encourages the participation of colleagues. Ensures the transmission of information.	Keeps colleagues informed about what is useful or relevant. Encourages the participation of team members and ensures clarity of information. Is flexible and oriented towards the goals of the team, sharing them. Respects the opinions and work of others. Always available to help.	Encourages the participation of team members. Creates empowerment. Can share successes and failures. Uses the best people to carry out the team's tasks, always. Always available to help.	Encourages the participation of team members. Creates empowerment and enthusiasm on others. Always shares successes and failures. Helps colleagues. Evaluate the results and learn from them. Creates a positive vision of the future.
Problem Solving	Can identify simple problems and always try to implement some resolutions.	Sometimes adopts a reactive attitude to the problems that occur, introducing or implementing mechanisms that prevent unknown occurrences of the identified problems.	It can identify and solve simple problems and some complex ones that occur. It often adopts the attitude of implementing forms of resolution that prevent an unknown occurrence of the problems that have arisen.	Shows ability to identify and expect problems of different levels of complexity, being able to solve them in most cases. Reveals a proactive attitude in implementing mechanisms that prevent the occurrence of expected problems, often avoiding breaks in productivity and performance.	Ability to identify, predict and solve problems of any level of complexity. A very keen sense of proactivity in the development and implementation of mechanisms that almost completely prevent the occurrence of problems with the greatest impact. Reveals great commitment and availability in the study and development of mechanisms for continuous improvement and elimination of problems and inefficiencies.
Organization	Collects and structures the information required to carry out the assigned tasks. Prioritizes and organizes the accomplishment of the assigned tasks, although with some difficulties.	Collects and structures the information required to carry out the assigned tasks. Prioritizes and organizes the accomplishment of the assigned tasks and can sometimes exceed the agreed deadlines.	Shows concern in preparing to carry out the tasks assigned, seeking to collect and structure the information required for this purpose. Seeks to prioritize and organize the performance of the various tasks assigned, although it can sometimes compromise the last quality of the work to meet the agreed deadlines.	prepares with great efficiency the accomplishment of the assigned tasks by collecting and structuring the information required for this purpose. shows high capacity of time management, efficiently prioritizing the accomplishment of the various tasks under its responsibility. Systematically complies with the agreed deadlines, being able to expect them, always ensuring a high level of last quality. Reveals a considerable capacity to accommodate new unplanned tasks, always trying not to jeopardize the quality of work already in progress.	Shows total efficiency in the collection and structuring of all the information required to carry out the tasks, often expecting the preparation activities. Shows total ability to prioritize and organize the performance of multiple tasks under his responsibility, always optimizing the time available. Never exceeds the deadlines established for the completion of the assigned tasks, frequently expecting them without loss of quality in the eventual result. Can accommodate new, unplanned tasks without noticeable damage to the timely completion and last quality of work already in progress.

continued on following page

Table 2. Continued

Skills Level	C1	C2	C3	C4	C5
Leadership	Identifies the goals of the working group, even if it does not transmit and delegates the responsibilities.	Identifies and communicates the goals of the working group, although with some difficulty. Is occasionally, concerned with giving some guidance, support, and feedback to colleagues.	Knows and clearly communicate the goals of the working group. Involves employees in setting goals, almost always. Usually delegates many responsibilities, assigns tasks, guides, provides support, and gives some feedback to colleagues.	Knows and clearly communicate the goals of the working group. Involves employees in setting goals, almost always. Usually delegates many responsibilities, assigns tasks, guides, provides support, and gives some feedback to colleagues. Gives feedback to colleagues very often.	Ability to clearly communicate the goals of the working group. Always involves employees in setting goals. Delegates many responsibilities, challenging tasks, guides and supports colleagues in carrying out their work. Always gives feedback to colleagues.
Communication	Can express himself, add some value and have an active listening, even if he does not have a totally effective communication.	Can provide and ask for simple clarifications, clearly and logically. Has a good communication capacity even with some communication failures, which he is not aware of.	Can transmit information, ideas, and opinions orally and in writing in a clear, logical, timely and synthetic way. Pays attention to the problems that may arise in the design process. Is aware of the reactions and behaviours of its interlocutors, to correct communication failures. When communicating, add value and help to solve problems.	Can transmit information, ideas, and opinions orally and in writing in a clear, logical, timely and synthetic way, captivating the audience. Is very attentive to the reactions and behaviours of his interlocutors, always bridging any communication failures. Its intervention is looking forward to unlocking complex problems.	Is a born communicator, transmitting information in an interesting and persuasive way and showing great confidence, even in hostile environments and difficult situations. Shows great helicopter vision, always realizing the consequences of what it proposes and expecting solutions. Can communicate with many audiences, adapting their communication style to each situation and avoiding communication failures. Has the final say in solving problems, and its communication is the one having the best solution.
Dynamism	Even though having little knowledge of the target audience and the brand image, can sometimes present new ideas and concepts.	Occasionally the ideas and concepts presented are original. Almost always pay attention to what is going on in the market. Does not like to take risks. Reasonable knowledge of the target audience and brand image.	Presents potential ideas, original concepts, and improvements in projects. Is well informed and aware of what is going on in the market. Has an observant and curious spirit. Able to take a chance on new ideas. Knows the target audience and the brand image well. Is concerned with testing the adequacy of what it has designed.	Presents originality in the ideas, concepts, and improvements in the projects. Is well informed and aware of what is going on in the market. Very observant and curious. Great ability to take risks with new ideas. Knows the target audience and the brand image very well. Is concerned with testing the adequacy of what it has designed.	Presents originality in the ideas, concepts, and improvements in the projects. Is well informed and aware of what is going on in the market. Very observant and curious. Does not hesitate to take risks with new ideas that believes in. Fully identifies with the target audience. The conceived ideas are invariably adequate.

continued on following page

Table 2. Continued

Skills Level	C1	C2	C3	C4	C5
Initiative	Can expect and detect opportunities. Sometimes has some capacity for initiative in approaching new ideas and / or concepts and in the support of teammates.	Sometimes suggests new opportunities in performing its functions. Occasionally identifies windows of opportunities, although its intervention is limited.	Systematically seeks new challenges and opportunities. Identifies situations, opportunities, and obstacles in proper time. Acts on the opportunities and obstacles detected. Is decided in crisis situations. Usually originates the action, even if it does not have complete information.	Is innovative in its function, seeking to capitalize on the opportunities that arise. Expects situations, opportunities, and obstacles. Acts on the opportunities and obstacles detected quickly. Is always decisive in crisis situations. Always originates the action, even if he does not have complete information.	Reveals a great vision of the future in its activity, incorporating the most recent ideas and innovations. Expects situations, opportunities and obstacles that are often not obvious to others. Detects opportunities and acts on them quickly and effectively. Is decisive in crisis situations, always revealing clarity of thought. Takes the first step and originate the action, even if you do not have complete information. Involves others and encourage innovative ideas.
Autonomy	Has some autonomy in situations of minimal impact during work.	Has some autonomy in situations of minimal impact during work, sometimes requiring specific support.	Rarely hesitates to decide, often dispensing specific support and trying to avoid delays in the processes under his responsibility.	Never hesitates to decide, always dispensing any specific support and not allowing delays in the processes under his responsibility.	Systematically decides in due time and with total autonomy, always considering the timetable and time requirements within the process to take place in the follow-up.
Motivation	Shows reasonable determination to achieve a goal, not considering himself defeated at the first attempt.	Shows determination to achieve a goal. Continues to pursue an aim, even if defeated at first. Can distinguish achievable and unattainable goals.	Is persevering in the quest to achieve a goal. Does not give up, despite being rejected one or more times. Identifies with the company's values and "wears the shirt"	Always show great determination to fully achieve the objectives. Keeps the composure and cool heartedness in the face of rejections and continues his action without fainting. Is decisive in the search for success. Embodies the company's values and live them day by day.	Is an example to be followed by perseverance in pursuing exceeding goals. Is not satisfied with less than total success in the missions he has. Has great dignity and calmness when rejected, and proceeds with the same determination in the search for the desired result. Is considered a paradigm of the company's values and an example to follow.
Creativity	Even though having little knowledge of the target audience and the brand image, can sometimes present new ideas and concepts.	Occasionally the ideas and concepts presented are original. Pay attention, almost always, to what is happening in the market. Does not like to take risks. Reasonable knowledge of the target audience and brand image.	Presents potential ideas, original concepts, and improvements in projects. Is well informed and aware of what is going on in the market. Has an observant and curious spirit. Able to take a chance on new ideas. Knows the target audience and the brand image well. Is concerned with testing the adequacy of what it has designed.	Presents originality in the ideas, concepts, and improvements in the projects. Is well informed and aware of what is going on in the market. Very observant and curious. Great ability to take risks with new ideas. Knows the target audience and the brand image very well. Is concerned with testing the adequacy of what it has designed.	Presents originality in the ideas, concepts, and improvements in the projects. Is well informed and aware of what is going on in the market. Very observant and curious. Does not hesitate to take risks with new ideas that believes in. Fully identifies with the target audience. The conceived ideas are invariably adequate.

continued on following page

Table 2. Continued

Skills Level	C1	C2	C3	C4	C5
Time Management	Prioritizes and organizes the accomplishment of some simple tasks.	Prioritizes and organizes the accomplishment of tasks or various tasks, rarely presenting some difficulties.	Seeks to prioritize and organize the performance of the various tasks assigned, occasionally meeting the agreed deadlines.	Shows a high capacity of time management, efficiently prioritizing the accomplishment of the various tasks under its responsibility. Systematically complies with the agreed deadlines, being able to expect them, always ensuring a high level of last quality. Shows a considerable capacity to accommodate new unplanned tasks, always trying not to impair the quality of work already in progress.	Shows a total ability to prioritize and organize the performance of multiple tasks under his responsibility, managing in the best way and always optimizing the time available. Never exceeds the deadlines established for the completion of the assigned tasks, frequently expecting them without loss of quality in the result. Can accommodate new, unplanned tasks without noticeable damage to the timely completion and last quality of work already in progress.
Attitude	Shows reasonable determination to achieve a goal, not considering defeated at the first attempt.	Shows determination to achieve a goal. Continues to pursue an aim, even if defeated at first.	Is persevering in seeking achieving a goal. Does not give up, even if it is rejected one or more times. Keeps a positive behaviour in changes in the search for results. Identifies with the company's values and "wears the shirt".	Always show great determination to fully achieve the objectives. Keeps the composure and cool heartedness in the face of rejections and continues his action without fainting. Is decisive in the search for success. Embodies the company's values and live them day by day.	Is an example to be followed by perseverance in pursuing exceeding goals. Is not satisfied with less than total success in the missions he has. Has great dignity and calmness when rejected, and proceeds with the same determination in the search for the desired result. Reconciles the interpersonal relationship perfectly with seeking goals. Is considered a paradigm of the company's values and an example to follow.
Responsibility	Can decide sometimes and may require some specific support in the processes under his responsibility.	Rarely hesitates to decide, although he may require some specific support in the processes under his responsibility.	Does not hesitate to decide in most cases, often dispensing specific support and trying to avoid delays in the processes under his responsibility.	Never hesitates to decide, always dispensing any specific support and not allowing delays in the processes under his responsibility.	Systematically decides, in due time and with total autonomy, always considering the timetable and time requirements within the process to take place in the follow-up.
Proactivity	Can expect situations and detect opportunities occasionally. Has initiative, although sometimes hesitating to take the risk.	Aware of change. Sometimes detects opportunities and obstacles. Occasionally identifies windows of opportunity, but its intervention is slightly reduced.	Identifies situations, opportunities, and obstacles in proper time. Acts on the opportunities and obstacles detected; In crisis situations, it reveals prudence. Rarely gives rise to action if it does not have complex information.	Expects situations, opportunities, and obstacles. Acts on the opportunities and obstacles detected quickly. Is decisive in crisis situations. Usually originates the action, although it does not have complete information.	Expects situations, opportunities and obstacles that are often not obvious to others. Detects opportunities and acts on them quickly and effectively. Is decisive in crisis situations, always revealing clarity of thought. Takes the first step and originate the action, even if he does not have complete information. Involves others and encourage innovative ideas.

According to the guidelines outlined in the table above, behaviours differ according to the category of the company's functions. With a Junior Consultant, is unlikely to have such an intuitive skill in solving problems as the one that a Senior Consultant has. The same can be seen in the communication ability of each level of skill because a profile with a level of greater complexity (C5) already has a very elaborate discourse and effective communication in contrast to a newly licensed profile (C1) that, although having a well-supported bag of theory, will only develop this skill when his interpersonal development in work and/or social.

CONCLUSIONS

One of the current challenges facing organizations is to conduct the recruitment and selection process, which, in most cases, is not successfully accomplished because of the lack of adequate criteria and instruments, which can generate large losses in terms of time and costs. The structuring of the recruitment and selection process allows for greater assertiveness when hiring new employees, this when aligned with the description of technical and behavioural skills to the strategic planning of the organization.

The general goal of this research was to propose the analysis and identification of a profile of transversal skills, aiming at a final list of skills inherent to the functions and professional categories. In addition, the elaboration of this research made it possible to get to know employees better, their potentials and limitations and, even, to analyse and reveal an improvement in failures in the HR processes, this aspect being of paramount importance.

Regarding transversal skills, there was a growing interest and importance that they gained in a professional context, both by the individuals involved in the study and by several authors in the literature review. Of the transversal skills valued in the case study, the following stand out: teamwork, problem solving, organization, leadership, communication skills, dynamism, initiative, autonomy, motivation, creativity, time management, attitude, responsibility, and proactivity.

The major limitation during the development of this research is the limited number of interviews and the difficulty in collecting documentary information in the studied company.

It is worth mentioning the importance of carrying out, in the future, a study that aimed to answer the same objectives of this research but using the construction of an instrument that would enhance a qualitative and quantitative analysis, generalizable to the population of Portuguese technological companies.

REFERENCES

Baylão, A., & Rocha, A. (2014). *A Importância do Processo de Recrutamento e Seleção de Pessoal na Organização Empresarial*. XI Simpósio de Excelência em Gestão e Tecnologia.

Camara, P. (2017). *Dicionário de Competências* (1st ed.). Editora RH, Lda.

Camara, P., Guerra, P., & Rodrigues, J. (2016). HUMANATOR XXI - Recursos Humanos e Sucesso Empresarial (7ª ed.). Alfragide: Publicações Dom Quixote.

Cardoso, A. (2010). Recrutamento & Seleção de Pessoal (7ª ed.). Lisboa, Porto: Lidel – edições técnicas, Lda.

Ceitil, M. (2016). *Gestão e Desenvolvimento de Competências* (2nd ed.). Edições Sílabo.

Chiavenato, I. (1988). *Recursos Humanos*. Editora Atlas.

Costa, N. (2015). *A importância das competências transversais (soft skills) na formação do Engenheiro*. Masters Dissertation, Lorena Engineering School, University of São Paulo, Brazil. Available in: https://sistemas.eel.usp.br/bibliotecas/monografias/2015/MIQ15031.pdf

Couto, V. (2018). *Análise e Reformulação de um Descritivo de Funções*. Masters Project, School of Economics and Management, University of Minho, Portugal. Available in: http://hdl.handle.net/1822/56970

Cunha, I. (2016). *The efficacy of 1-day soft skills training on master students' performance*. Masters Dissertation, ISCTE Business School, University Institute of Lisbon, Portugal. Available in: http://hdl.handle.net/10071/13243

Fernandes, V. (2017). *Recrutamento e Seleção: as práticas de Recrutamento e Seleção nas Startups*. Masters Dissertation, Higher Institute of Accounting and Administration, Polytechnic Institute of Porto, Portugal. Available in: http://hdl.handle.net/10400.22/11283

Freixo, M. (2011). *Metodologia Científica: Fundamentos Métodos e Práticas* (3rd ed.). Piaget Institute.

Guimarães, M., & Arieira, J. (2005). O Processo de Recrutamento e Seleção como uma ferramenta de Gestão, *Rev. Ciências Empresariais da UNIPAR*, *6*(2), 203–214.

Litecky, C. R., Arnett, K. P., & Prabhakar, B. (2004). The paradox of soft skills versus technical skills in IS hiring. *Journal of Computer Information Systems*, *45*(1), 69–76.

Mitchell, G., Skinner, L., & White, B. (2010). Essential Soft Skills for success in the twenty-first century workforce as perceived by business educators. *Delta Pi Epsilon Journal*, *52*, 43–53.

Paiva, C. (2013). *As redes sociais online – A perspetiva do potencial candidato*. Masters Dissertation, ISCTE Business School, University Institute of Lisbon, Portugal. Available in: http://hdl.handle.net/10071/6658

Patacsil, F. F., & Tablatin, C. L. S. (2017). Exploring the importance of soft and hard skills as perceived by IT internship students and industry: A gap analysis. *Journal of Technology and Science Education*, *7*(3), 347–368. doi:10.3926/jotse.271

Quintela, D. (2016). *Recrutamento e Seleção de Pessoas: A perceção dos avaliadores*. Masters Dissertation, Higher School of Industrial and Management Studies, Polytechnic Institute of Porto, Portugal. Available in: http://hdl.handle.net/10400.22/7946

Quivy, R.; Campenhoudt, L. (1995). *Manual de investigação em ciências sociais*. Lisboa: Gradiva – Publicações, S. A.

Ribeiro, E. (2017). *Soft Skills no mundo laboral atual. A criação de uma nova empresa*. Masters Project, Department of Languages and Cultures, Aveiro University, Portugal. Available in: http://hdl.handle.net/10773/18347

Robles, M. (2012). Executive Perceptions of the Top 10 Soft Skills Needed in Today's Workplace. *Business Communication Quarterly*, *75*(4), 453–465. doi:10.1177/1080569912460400

Rosa, E. (1999). *Recrutamento e Seleção de Pessoal. Quais as vantagens deste processo para uma organização?* Recuperado em 3 setembro, 2018, available in: https://www.oswaldocruz.br/download/artigos/social22.pdf

Swiatkiewicz, O. (2014). Competências transversais, técnicas ou morais: Um estudo exploratório sobre as competências dos trabalhadores que as organizações em Portugal mais valorizam. *Cadernos EBAPE.BR*, *12*(3), 663–687. doi:10.1590/1679-395112337

Szilárd, S., Benedek, A., & Ionel-Cioca, L. (2018). Soft Skills Development Needs and Methods in Micro-Companies of ICT Sector. *Procedia: Social and Behavioral Sciences*, *238*, 94–103. doi:10.1016/j.sbspro.2018.03.012

Chapter 8
Leadership:
A Fundamental Tool for Dynamic and Intelligent Management – An Application

Maria Heliodora Matos
https://orcid.org/0000-0002-2036-3947
University of Minho, Portugal

Carolina Feliciana Machado
https://orcid.org/0000-0002-9685-1576
Interdisciplinary Centre of Social Sciences (CICS.NOVA.UMinho), University of Minho, Portugal

ABSTRACT

The main aim of this chapter is to examine the role played by leaders as determinant agents of a special type of banking institution such as the mutual agricultural credit banks (or Caixas de Crédito Agrícola Mútuo [CCAM]). In other words, it looks to study and understand to what extent employees with leadership roles use their position to maximize CCAM performance through their direct intervention on the teams. Considering that both leaders and subordinates have their own aims (organizational/professional and personal, respectively), and in order to find the necessary equilibrium between them, it also seems to be relevant to better understand the way leaders can mobilize and support organizational subordinates. In sum, issues like leadership, motivation, satisfaction, and personal development are studied in this chapter.

DOI: 10.4018/978-1-7998-5768-6.ch008

INTRODUCTION

Conscious of their individual limitations, people usually joint to organizations to achieve common goals that would be impossible to obtain individually. Indeed, people need the organization in the same way as organization needs people. Organizations recruit and select employees to themselves and through them look to reach organizational objectives. Employees have individual goals that are struggling to obtain individually and often serve the organizations to reach them in the most appropriate way.

The interdependence of needs between individuals and organizations is huge since both their lives as well as their objectives are inseparably linked. While employees bring skills, knowledge, abilities, capacity to learn and undefined level of performance, the organization imposes responsibilities (defined and/or undefined), some within or below their current capabilities, other demanding a long-term learning. It can, therefore, appear conflicts between the goals that organizations seek to achieve and the goals that, individually, each employee seeks to obtain.

To better be aware of the change in human resource management paradigm, we must reflect the role played by leaders within organizations. Leadership is a major condition for the success of any organization.

There are numerous agents in organizational leadership, not only in the top of the hierarchy, but also in all hierarchical levels, both vertical and horizontal. Leaders are, essentially, just that: conductors of people, elements that can, through a variety of attributes, guarantee that the other stakeholders act in harmony to what has been required. Leadership assumes an important role in the management of organizational human resources. To succeed, they need to identify with organizational policies otherwise, it is only a matter of time until these policies become unsuccessful.

The aim of this chapter is to study the nature of the task played by those who hold formal leadership positions within the mutual agricultural credit banks, or *Caixas de Crédito Agrícola Mútuo* (CCAM). This is of particular interest given that CCAM have a configuration that is different to that of other credit institutions, being more decentralized and having ways of implementing organizational policies and practices that are applied in a different way when compared with the Banking Industry in general.

Considering the highly significant role these figures play in organizational activity and the increasing importance that the issues related with the behavioral side of human resource management are coming to have, in terms of employee motivation, satisfaction and development, it is vital to understand the way in which institutional leaders can act as facilitators or hinderers of these characteristics.

CRÉDITO AGRÍCOLA (CA) IN PORTUGAL

Even though its roots date back to the late 15th century, the true CA (Crédito Agrícola / Agricultural Credit) was formed in 1911 with the Decree-Law of that date which regulated the establishment and functioning of the *Caixas Agrícolas*, or Mutual Agricultural Credit Banks. CA is one of a small number of private financial institutions with entirely Portuguese capital.

It is a cooperative-based banking institution within a model that is well represented in the European Union financial systems, especially in Germany, France, The Netherlands, and Italy. In Portugal, the only cooperative financial group is *Crédito Agrícola*, which includes *Caixas Agrícolas* as organizations rooted in local communities, which becomes a vital element for the development of the regions where they operate and constitutes an example of company that simultaneously pursues social and business goals.

The existence, in the financial system, of entities such as *Caixas Agrícolas*, with marked regional trend allows compensating the centripetal trend of banking business developed in large and globalized markets. Their presence in rural environment limits the imbalances between the countryside and the city, while, at the same time, opens the access of rural populations to the banking system.

Caixas Agrícolas are agricultural organizations inspired by democratic ideals and the social element, and its leaders take into account not only economic performance but also their social vocation, differentiating factor that includes them in the social economy (Pina, 2014).

In accordance with CCAM *Serras de Ansião* Directors' Report for 2010 (2011, pp. 13-14): "At European level, cooperative banks have a market share of over 20%, provide a service for 170 million clients from 4.100 local banks with 65.000 branches, and employ hundreds of thousands of people. Some of the cooperative banks are among the leading institutions in Europe". In harmony with their cooperative nature, CCAM act as boosters of regional development.

At the end of 2010 financial year, CA's liquid assets stood in more than 13 billion euro, representing around 8% of the Portuguese Gross Domestic Product (CCAM *Serras de Ansião*, 2011). According to data provided by the Human Resources Department Director of *Caixa Central*, the 2012 CA's liquid assets have also stood at around 14 million euro. This testifies to its national significance and its importance in the country's financial system.

As stated by the above-mentioned Report, in some districts of the Portuguese mainland CA network represents over 30% of the existing banking network, has 10% of the national ATM (Automated Teller Machine) network - 400 of these in places where no other type of banking service exists – and almost 15.700 on-site EFTPOS

(Electronic Funds Transfer at Point of Sale) machines, as well as having a 7% share in the cards market, in function of its service provision as acquirer.

According to the Legal Framework of Mutual Agricultural Credit and Mutual Agricultural Credit Banks (RJCAM)[1], CCAM are credit institutions, in the form of cooperative associations, whose object is to provide agricultural credit to their associates as well as performing other acts inherent to the banking activity. Regarding territorial scope, the Legal Framework states that their scope is local; no mutual agricultural credit banks may be incorporated if they propose to operate outside of the municipality where their head office is based, except in cases where there is no such institution operating in adjoining municipalities or if the spill over results from merging of existing CCAM. Likewise, authorization will not be granted to new *Caixas* proposing to operate in a municipality where another is already based or has opened an agency.

It must be borne in mind that the organizational structure of CCAM is, as already mentioned, highly decentralized, with a central body – *Caixa Central* – and its ramifications (686 branches belonging to 85 CCAM and to *Caixa Central* at the close of 2011) dotted throughout mainland Portugal and Azores, which has a considerable degree of autonomy.

Figure 1. Evolution of the number of branches of SICAM (includes the branches of the Caixa Central)
Source: CA Crédito Agrícola (2011, 2012)

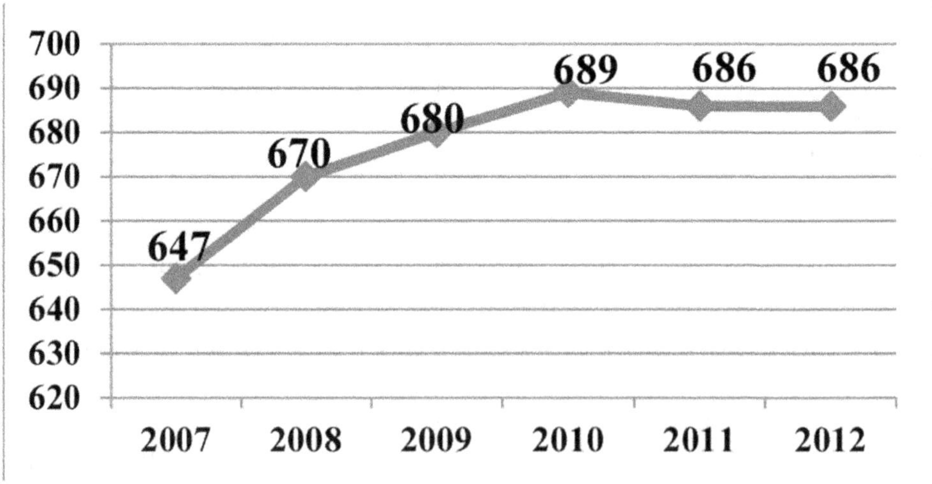

CA has about 4,300 employees, being 3,845 affected to different work spaces (branches and seats of CCAM and *Caixa Central*) and the remaining to companies of the Group (CCCAM, 2011). Being, in 1994, about 236 CCAM, the decrease is not due to restriction of activity or reducing the number of branches, but, fundamentally, to bank concentration procedures (mergers) with the consequent creation of synergies.

In 1991, with the new legal regime for CA, *Caixa Central* came to have functions and powers for directing, supervising and with financial representation of SICAM, the Integrated Mutual Agricultural Credit Scheme, with a regime of co-responsibility being established between it and its affiliates such that monitoring of liquidity and solvency is based on consolidated accounts.

In addition to these *Caixas*, there are another 45 branches belonging to five CCAM[2] that are not included in SICAM and that function as completely independent credit institutions.

There is a continuing decrease in the number of employees, maintaining a pace that accompanies the contraction of the number of CCAM (92 in 2008; 88 in 2009; 85 in 2010, and 84 in 2011 and 2012). The number of branches of total SICAM suffers variations in growth direction, from 2008 to 2010 (670 to 689), to stay stable in 2011, and 2012 (686). The explanations for the option by the growth and subsequent stability of the number of branches can be presented in the Sustainability Report 2010 (2011, p. 4): "The Agricultural Credit, as a representative in Portugal of Cooperative Banking, has preserved a business model that allows a greater resistance to the crisis, through the high capitalization and liquidity. The "recipe" is the best interpretation of the concept of proximity banking, supported by an extensive network of branches and points of contact with the respective associates and customers. A model that many, before the crisis, called as "obsolete"..." and: "the cooperative matrix gives the CA a unique nature in the Portuguese financial system, because it enables a close and solid relationship with local communities"(2011, p. 11).

According to SICAM Consolidated Balance Sheet for 2011, staff breakdown is 55% men and 45% women. Geographical distribution of employees is as follows: Northern Zone (16 CCAM) – 20%; Central Zone I (22 CCAM) – 17%; Central Zone II (25 CCAM and *Caixa Central*) – 36%; Southern Zone (21 CCAM) – 24.2% and Azores (1 CCAM) – 2.8%. The mean age is 42.8 years, and Secondary and Higher Education are the most common qualification levels, representing 85.9%. In terms of functional distribution, the highest percentage – 69.6% of employees – is concentrated in the categories Highly Qualified and Qualified Professionals.

The big difference between SICAM and the rest of the banking industry lies in the fact that the latter has a strong predominance of central and back-office services, which are less important in CA.

It is appealing to note that each CCAM client can also be an associate[3], and that there is a formal leadership body – the Board of Directors of CCAM – whose members are elected by these associates for periods of three years, with re-election being allowed. It is not an essential requirement that all the members in these positions have specific management skills, and this gives rise, on various occasions, to a number of ambiguities and disconnections since these people occupy positions that are higher up in the hierarchy than the managers (professionals who are expected to have management knowledge and skills).

TYPES OF LEADERSHIP

For numerous decades, the study of leadership has been a significant part of the literature on organization and management behavior. The leadership is central to the functioning of organizations, constituting one of the most important determinants of their performance and survival. Palma (2012) exposes that "To the leadership we can attribute the ability for all members of the same organization unite for a common goal. This feat leads leadership to be understood as primarily responsible for the existence of organizations, in the truest sense of the word – while organized systems coordinated and directed towards the same purpose". However, there is still a controversy about what leadership is or what characteristics that organizational actors must possess to be considered leaders. Stogdill (1974, p. 259) settled that "there are almost as many definitions of leadership as there are persons who have attempted to define the concept." According to Bass (1990a), there are numerous definitions of the concept present in the literature, almost as many people who dedicated themselves to study the topic.

A brief definition of leadership can be given by Buelens, Kreitner and Kinicki (2002, p. 450), according to which leadership is nothing more than "A process of social influence, where the leader seeks the voluntary participation of subordinates in an effort to achieve the organizational objectives". It is possible to find another definition of organizational leadership in the Globe Study (Global Leadership and Organizational Behavior Effectiveness) of 62 Societies, where its organizers define this concept as "an individual's capacity to influence, motivate, and enable others to contribute to the effectiveness and success of the organization of which they are members" (Dorfman & House, 2004, p. 56).

Preston (2001, p. 62) defines leadership as "the process of influencing others to act in such a way as to be achieved the goals of the leader and the organization. A successful leader establishes a goal, reaches it, and during the proceedings, satisfies those whose opinion must be respected ". For Machado (1994, p. 37), "Leadership is the ability to read, interpret and make operative the organizational culture, leading the corporate collective to know and to incorporate this force". In the opinion of Castro (1994, p. 50), leadership "assumes certain requirements, including of physical and psychological nature, and to be true requires that it is accepted freely by the others". Henderson, Liden, Glibkowski and Chaudhry (2009) expose that the leader is perceived as a role model and influence decisively the performance of his employees. Refer to the leader as a behavioral, emotional and cognitive level model, which infects the followers in a positive or negative manner.

For Yukl (1989), many definitions involve an influence process but appear to have little else in common and reflect deep disagreement about identification to leaders and leadership processes. In the opinion of the author (1989, p. 253), "it is better to use the various conceptions of leadership as a source of different perspectives on a complex, multifaceted phenomenon."

Mayo and Nohria (2005) stated that the word "leader" emerged to define the executives in general. But, when we seek to understand the different types of leadership, it becomes clear that the same word can be applied to one of the three essential archetypes of executives: Entrepreneurs, Managers and Leaders. Certain periods may seem, at first glance, almost ideally suitable for the emergence and predominance of a particular kind of executive. It turns out, however, that the three types coexisted and coexist and are common in each period. In fact, there are the three archetypes that were and are vital to maintain the vitality of the capitalist system. Entrepreneurs create new businesses; managers make them grow and optimize them, and leaders transform them into critical inflection points.

Pelletier, as referenced in Barracho (2012), points out that when a boss manipulates or coerces his subordinates there is no leadership. Leadership is the ability to raise the voluntary participation of persons or groups in the pursuit of the goals set. In order that leadership can be exerted the group must allow the leader influence. For Barracho (2012), exert influence is to use authority, influence through implicit acceptance on the part of the group members. As stated by the author (2012, p. 65), "the opponents claim that this thesis is restrictive, because it deletes important influence processes. For example, the same attempt to influence can lead to the commitment of some people and in some situations however raise the resistance in other people or in other situations".

The leadership always implies a process of control and influence over others. The different definitions of leadership diverge as to the type or style: the use of authority, with the distribution of rewards and punishments as a way to exert influence over the followers, opposes non coercion. Leadership is expressed or displayed through the interaction between people and implies, necessarily, to be followed. For a person to be able to influence, other(s) should be swayed. Moreover, the leader and the follower(s) shall be minimally organized about some common purpose or agreement, than its scope depends at least in part, of the relationship between the leader and the follower(s).

Given the importance of this concept, there will be presented several theories on organizational leadership, as well as the links between leadership and organizational change.

Theoretical Perspectives

Following the foregoing, the theme of leadership within the framework of organizational behavior has been object of great attention by researchers. Even in the forties, in the twentieth century, prevailed the concern about identifying the physical attributes, skills and personality traits of leaders. It was presumed the existence of universal traits that empower individuals with certain characteristics to take as effective leaders in any situation.

Checking the existence of contradictory results Stogdill (1948), while reviewing 124 studies, concluded that individuals do not become inevitably leaders and are not necessarily effective due to the possession of a certain combination of traits that researchers focused on behaviors resulting in behavioral approaches: task orientation and guidance to people / relationship. It has been argued that, if these behaviors were practiced at high level, would generate effectiveness in any situation (Janusian leadership). An effective leader would be one that was able to integrate the needs of working with the human ambitions.

According to Cunha and Rego (2009) the universal logic would, however, lead to concessions contingent because the thesis that there would be traces for effectiveness in any situation has been replaced by the idea of the necessity of "situational experts", i.e., each situation requires different personality traits, so that an effective profile in a situation may not be effective in another. Afonso (2009) stressed that for the first time the search of the secret of effective leadership considers external factors to the leader.

For Hersey and Blanchard (1993), the effective leader is the one who adapts his behavior to situational characteristics, denoting the ability to act as a "chameleon" to adjust the behavior to the requirements of the situation, i.e., every situation requires different leadership behaviors. To assume that any style is appropriate, since it fits the situation and produces results, the chameleonic leadership suggests, to Cunha and Rego (2009, p. 31), "a markedly instrumental conduct: the use of strategies that work, even if those collide with the respect and dignity of the people".

Cunha and Rego (*op. cit.*) argue that the dichotomy universality-contingency came, in a way, to be challenged by the logic of leadership based on values, resulting in two major approaches as the focus falls on the personality (integrity leader) or about the behaviors (coach). Afonso (2009) highlights that this orientation is imbued with aspects related to the leader's personality and behaviors. Thus, when the highlight is the personality, the approach is the integrity leader; when the stress is more on behaviors, the approach is that of the coach.

Northcraft and Neale (1994) believe that the traditional theories of leadership can be divided into two categories. The first focuses on the leadership aspect: personality or behavioral traits. The second focuses on the type of influence the leader has: either universal or contingent on the situation or context.

The universal theories assume that leadership behavior or the leader's traits are always the same, regardless the followers or situation. These theories presuppose that the leader responds in a consistent or universal manner to many different situations. The contingency theories assume the opposite. Leadership and leader behavior are dependent upon the followers and situations that occur. Hence, the leader can adjust the manifestation of personality or behavioral traits to respond to the demands of the situation.

According to Jesuino (2005) and Barracho (2012) the functionalist paradigm (the "why" to be leader) aims to expand the route of gradual accumulation that ends in the Contingency Theories but less impregnated by ideological belief of individual effectiveness. On the other hand, the genetic paradigm ("how" do we lead?) is heavily tributary of cognitive and social psychology and situates its concern in symbolic interactions underlying inter individual level leadership processes and in an inter or intra group.

The concept of organizational leadership was, and still is, object of study by various theoretical currents, where some favor the personality traits and other favor the leader's behavior, adopting the proposed taxonomic schema by Jago (1982), according to which it is possible to distinguish two major criteria: the criterion of universality versus contingency and the criterion of the traits versus behavior of leaders. According to the typology of Jago (Barracho, 2012), the four types of theories are summed up thus having underlain the idea of organizational effectiveness:

- Type I theories – correspond to theories about the personality traits of the leader.
- Type II theories – focuses on the behavior of the leader and his leadership styles.
- Type III theories – address the leadership phenomena in a contingency perspective which consists of equate the effects of situational variables on the personality of the leader.
- Type IV theories – while maintaining the contingency perspective, equate the effect interactions about situational variables and characteristics of the leader's personality.

We will then analyze the theoretical perspectives: universal (types I and II) and contingencies (types III and IV).

Type I Theories

In Type I theories leadership is seen as a trait or a characteristic that the leader should possess to exercise an effective leadership within a group or organizational context. These theories are an attempt to define the psychological traits of subjects able to lead effectively in any situation (Jesuino, 2005). At the time of its formulation (beginning of the twentieth century) it was thought that leaders were born leaders, that they could not learn the issues of leadership (Teixeira, 1998), that existed innate characteristics or personality traits, social, physical or intellectual, which differed leaders from the followers (Caldeira & Correia, 1996). According to Cunha, Rego, Cunha and Cabral-Cardoso (2007) existed the idea of great men with innate features that led them, of course, to leadership positions, enabling them to effectively exercise this role.

Management research, focused especially on leadership, dates to the 1950s, having been initiated, essentially, by Robert Bales and Douglas McGregor. Bales (1950) emphasized the importance of groups and leadership, which he classified into two types: task leadership and social leadership. In the former, the leader carries out his/her activity by focusing on group objectives in terms of productivity. In the latter he looks to meet objectives, working on the development of cohesion and encouraging collaboration between group members.

McGregor (1960) states that there are two ways of seeing the human nature: a negative one – Theory X – in which, basically, people see work as a sacrifice to be avoided and hence need and prefer to be directed and controlled; and a positive one – Theory Y – which believes that work can be as natural as rest or leisure, that workers appreciate shouldering responsibilities and prefer self-control. According to the author (1960) managers will have different behaviors towards their subordinates in terms of styles of direction and control, in accordance with the extent to which they accept the assumptions of Theory X or Theory Y.

The Theories of Leadership Traits postulate that there are innate characteristics or personality, social, physical, and intellectual traits that differentiate leaders from non-leaders or followers (Caldeira & Correia, 1996). According to Teixeira (1998) this approach is based on the acceptance of the idea of the born leader.

Ghiselli (1971) identified thirteen components of leader traits characteristics, of which the six most relevant would be: supervisory ability, need for occupational achievement, intelligence, decisiveness, self-assurance, and initiative. House (1977) put forward the theory of the charismatic leader, suggesting that the great leaders basically employ four personal characteristics: dominance, self-confidence, a need to influence others and a strong sense of their own moral values. Leaders with these "traits" are more charismatic than those who have them in a lesser extent.

Silva (1994, p. 41) states that "… reading and studying the more recent theories on leadership help to contextualize and consolidate some of the acquired experience. However, I do not believe it is possible to "manufacture" leaders, since I believe that leadership is an innate ability that can be developed and perfected".

According to Cattell and Stice (1954) the personality traits determine the emergence of leaders and may vary the form of election of them – by pairs or by identifying external observers. These authors (1954) indicated the following personality traits as the most important for the exercise of leadership: emotional stability, dominance, conscientiousness, boldness, shrewdness, absence of anxiety, deliberate control of will and absence of nervous tension.

However, several problems emerged that led researchers to discard the universal traits of leadership, among which are: too many exceptions to the general rule that effective leaders possess a superior intelligence than subordinates (Jesuíno, 2005); the acceptance of the fact that those who are effective leaders in some situations, may not be in all of them. It would be necessary to consider the structure and organizational history, the nature of the objectives and goals of the group, task characteristics and the followers' traits and characteristics (Gibb, 1969; Stogdill, 1974).

Still, Calder (1977) has outlined a theory that returned to give some interest to the leadership traits theory. For him, the attribute of leadership is a trait that needs to exist only in the perceptions of the others, especially the followers. People "label" the others as owners (or not) of leadership qualities and, in this perspective (Assignment Theory of Leadership), are leading watchers who are in the limelight and not the leader (or potential leader). In this theory, the role of leadership begins with the observation of the subjects' behavior and their consequences, or with "inferred observations", like simple observations, that may imply the existence of behaviors which, otherwise, would not be observed (Calder, 1977).

There are three criteria for the acceptance / rejection of these observations as evidence of leadership (Calder, 1977): distinction and kind of behavior, temporal and spatial consistency of behavior and possibility of rejection of other traits that cause the behavior.

Essential to the allocation of leadership and for proper compatibility with these criteria is the "implicit theory of leadership" by the observer, which specifies the existence of perceived differences between leaders and followers.

There are other authors whose studies also assign some relevance to the theories of the traits, with a particular focus on House (1977) who proposed the theory of charismatic leadership. The charisma can be seen as a form or special type of leadership and, for this author, the great leaders that create charismatic effect on the followers have four personal characteristics: dominance, self-confidence, need to influence and strong conviction in the moral correctness of his ideas. Charismatic leaders have an enormous need for power, a high self-confidence and a great conviction in their own beliefs and ideas.

In this process of charismatic leadership, Klein and House (1995) created an affinity with the fire triangle. In this sense, the charisma is the result of three factors:

- Leader – spark or ignition temperature
- Followers - Fuel or flammable matter
- Enabling environment – oxygen

Like the fire, the charismatic leadership only happens in the simultaneous presence of the described elements.

According to Yukl (1999) charisma contains a positive side, when it is used in satisfying the followers needs as well as a more negative one when points to the leader self' aggrandizement. The downside of charismatic leadership is called destructive leadership (Padilla, Hogan & Kaiser, 2007). For these authors, the most destructive leaders are charismatic and develop their leadership based on a toxic triangle.

There are five critical factors listed to the figure of the Destructive Leader: charisma, power personalization, narcissism, negative experiences of life and hate ideology. As regards to Susceptible Followers Padilla et al. (2007, p. 180) divide them into Conformists and Conspirators. The former are characterized by having unmet needs, low self-esteem and low maturity; the seconds are characterized by ambition, vision and perfidious values. Conducive Environments present instability, perceived threats, cultural' values, lack of control and balance and ineffective institutions.

Padilla et al. (2007) referred to the existence of five destructive leadership structures:

1. Destructive leadership is rarely absolute or entirely destructive, since there are good and bad results in most situations of leadership.
2. The process of destructive leadership involves domain, coercion, and manipulation, rather than influence, persuasion and compromise.
3. The process of destructive leadership has a selfish orientation, which concentrates more on the needs of the leader than the needs of the larger social group.
4. The destructive effects of leadership are results that compromise the quality of the followers' life and undermine the main organization objectives.
5. Destructive organizational results are not exclusively the product of destructive leaders but are also a consequence of susceptible followers and conducive environments.

In the opinion of Barracho and Martins (2010) the traits ideas have been progressively losing consistency, in part due to the revision of studies (124 in 1948 and 163 between 1949 and 1970) by Stogdill. His preliminary findings show that a leader with certain traits could be effective in a situation but ineffective in another, as well as two leaders with different traits could be successful in the same situation. The other conclusions did not allow to maintain the principle that certain traits are essential for effective leadership, allowing only note that an individual with certain traits would have a higher possibility of becoming an effective leader and the relative importance of these would depend on the situation. Began to be recognized the grounds that leaders' personal characteristics cannot be separated from the context where they operate.

Type II Theories

Since the traits or profile theories did not prove to be effective in explaining the leadership phenomenon, researchers began to analyze the specific behaviors that differentiated leaders from followers. Organizational leadership theories inserted in this typology emphasizes the specific behaviors that distinguish leaders from followers. Born during World War II, the first behavioral leadership approaches supported the idea that the leaders' behavior directly affects the working group effectiveness (Buelens et al., 2002).

Hence, within the Behavioral Theories of Leadership there arose the studies by the University of Ohio, the University of Michigan, the Likert's classification and the managerial grid of Blake and Mouton. Beginning in 1945, a group of researchers at the University of Ohio identified two dimensions in leader behavior which they called initiating structure (to what extent the leader sets objectives and structures his/ her own tasks and those of the subordinates to better meet objectives) and consideration (to what extent the relations between leader and subordinates are characterized by mutual trust and respect for subordinates' ideas and feelings). They classified the various forms of leadership into four types, according to the level of consideration and initiating structure manifest in the behavior of a given leader (Teixeira, 1998; Caldeira & Correia, 1996).

For Teixeira (1998), although we cannot conclude about what would be the best leadership style, there is however suggestions that the combination of those dimensions to obtain the most suitable style would depend on the circumstances of the situation.

Studies were carried out at University of Michigan to research leader behavior characteristics linked with effective performance. Two aspects of leadership behavior were identified in these studies, designated task-orientation, and relationship-orientation. According to Barracho (2012), in these studies emerged three dimensions of leadership behavior that were designated for orientation for tasks (level where the leader sets and structure his own work and that of subordinates with a view to the objectives achievement), orientation for people (greater importance afforded to interpersonal relationships, in which the leader cares about their subordinates and with their well-being) and participatory leadership (understood as the degree to which the leader involves the subordinates in decision making).

Subsequent investigations suggest that effective leaders have a simultaneous moderate orientation for people and for tasks. In these studies, conducted in Michigan by Likert, it was suggested the existence of two opposite poles in the states of leadership: automatic and democratic directivism. Likert (1967) determined the leadership styles (management), the quality of relationships at work and the labor climate in the optics that individuals are actors (resources at stake).

Likert (1967) believed that effective management is strongly geared to subordinates, being based on communication to generate a greater concentration of ideas and objectives. He posited four styles of leadership: exploitive authoritative, benevolent authoritative, consultative, and participative. For the author (1967), the faster organizations abandoned the first two styles and adopted the latter two, the more efficient they would become, seeing their productivity increase as well as the satisfaction of their human resources.

Likert's thesis was that organizational effectiveness – productivity and satisfaction of individuals – increases as organizations change their autocratic systems to the participatory systems.

In the same way as the Michigan group, studies from Ohio University refused to identify the behavioral styles of effective leadership from direct observation. However, there are distinctions from the methodological point of view because the Ohio guided studies to the factorial analysis of questionnaires to the purpose of identifying the main behavioral dimensions of leadership: *Behavior Descriptive Questionnaire* with 30 items (1957), the Stogdill's version of this scale with 20 items (1963), *Supervisory Behavior Description Questionnaire* with 40 items (1957) and *Leadership Opinion Questionnaire* with 40 items (1957). The Ohio studies acknowledged some relevant roles to the leader: membership, representation, organization, domain, communication, recognition, production, and action. The ideal style is that in which there is a high guidance for both the task and people (Barracho, 2012). The two factors – consideration and structuring – were not considered, as in studies at the University of Michigan, as opposite ends of the same continuous, but as separate dimensions and conceptually independent. A leader can have high or low scores on the two dimensions as a high and low on another.

Building on previous studies, especially those of the University of Ohio, Blake and Mouton presented a two-dimensional matrix, named The Managerial Grid, to demonstrate the existence of a better style of universal leadership (Blake & Mouton, 1964).

For Blake and Mouton (1964), the levels of effectiveness vary between managerial styles. For them, managers with the best performance are those who utilize the style 9.9 in the grid, providing an excellent organizational (or group) climate, while, at the same time, the concern for production is of clear importance in their actions and behavior.

Despite the valid contribution and the great receptiveness of the Blake and Mouton model, it began to be challenged in terms of its applicability to the training and selection for managerial positions. Also, because the concerns were centered on the analysis of the context in which it is practiced leadership, situational theories will emerge.

Type III Theories

The Type III theories resulted from the idea that relates the leaders' personal characteristics with the situations in which they are involved. This typology enlists in the contingency approach of leadership and one of the most representative models of this is the Contingency Model, elaborated by Fiedler. Fiedler´s (1967, 1978) contingency theory of leadership posits that effective group performance depends on a good combination between leadership style in interacting with subordinates and the level to which the situation is controlled by the leader and gives him leverage. In the opinion of Buelens et al. (2002, p. 459), Fiedler's model is based on the following proposition: "The performance of a leader depends on two interrelated factors: the degree to which the situation gives control and influence to the leader — that is, the possibility that (the leader) has to carry out successfully the task; and the basic motivation of the leader – i.e. if the self-esteem (of the leader) depends, mainly, of the fulfillment of the task or in having good relations with others". To Fiedler (as referenced in Buelens et al., 2002), leaders assume two types of guidance: task-oriented and people-oriented. He argues that whatever the leadership' style it can't be deemed as the most effective in any situation; that is to say, nobody becomes a leader merely due to personal attributes, but thanks to the coexistence of various situational factors and the interaction between the leader and subordinates. Therefore, the best leading performance does not depend only on his personal attributes, but also on his interaction with the subject and the situation-specific factors. Thus, there are three key dimensions that determine the actions of the leader (Buelens et al., 2002): relationship leader-subordinate, the task structure and position of power.

In cases where a leadership style is not suitable for a particular situation, the leader should be replaced or situational conditions should be modified to fit the leader - Situational Engineering (Cunha, Rego, Cunha & Cabral-Cardoso, 2005).

For Teixeira (1998), the work developed by Fiedler allowed concluding that in very favorable or unfavorable cases, the more effective leadership is task-oriented, while the people-oriented leadership will be more suitable for intermediate situations. Afonso (2009) considers that the merits of Fiedler's work have to do with the fact that, in the leadership area he had been one of the first academics to formulate and test empirically a contingency approach about the effectiveness of leadership and demonstrate that it depends not only to the leader style (or personality), but also to the characteristics of the situation in which the leader and the team are acting, i.e. the external factors to the leader.

Fiedler´s model helped to destroy the idea that leadership ability is innate and that there can be a best leadership style regardless the circumstances. As Jesuíno (2005, p. 117) states, "the paradigm of contingency will however remain as epistemological framework, in fact adopted fairly widespread in the field of organizational behavior."

Type IV Theories

The inability of the above models and theories to produce consistent results led to the approaches focused on contextual influences. In other words, and in consonance with the taxonomy of Jago, there appear the contingency theories of second generation or of type IV.

The theoretical perspectives of Type IV disclaim, as Fiedler's theory, which the effective leadership depends on the specific situation. However, the attention of the apologists of these new theories focused on the behaviors of leaders, in the detriment of their personality traits.

The "Path-Goal" theory developed by House and Mitchell (1974), and House (1996) is closely linked with Vroom's expectancy theory on the motivation of people. It is, according to Barracho (2012), probably the most comprehensive of all contingent models and is based on the valence motivation model – Vroom expectancy theory.

According to this theory subordinates' performance is more effective if the leader clearly defines the task, provides training, helps them to work more effectively and establishes rewards directly related to performance level. Leadership is seen in terms of impact on motivation and meeting the needs of subordinates. According to this model, the leader's behavior is acceptable when employees perceive it as a source of present or future satisfaction (Buelens et al., 2002). The task of the leader is to help subordinates to achieve their goals and objectives, as well as ensure that these are compatible with the mission of the group or the organization. There are four types of leadership style within this model: directive, supportive, participatory and results oriented (Luthans, 2002; Buelens et al., 2002).

House (1971) identified four types of leadership behavior: directive, supportive, participative and achievement oriented. The nature of the situation faced by the leader depends on two groups of contingency factors: environment (of the task) characteristics and subordinates' characteristics. Adopting the most appropriate leadership style according to his/her interpretation of these contingency variables, the leader can increase motivation and work satisfaction, clarifying the goals and the paths to achieve them.

The approach proposed by Tannenbaum and Schmidt (1973) is the view of leadership as something that involves a variety of styles, going from being highly centralized around the manager (the leader enjoys full authority while subordinates have no influence over decisions), to being totally decentralized with subordinates having the maximum of freedom, within the limits defined by the leader (subordinates can have full participation and influence in taking decisions). They posit a continuum of leadership behaviors based on the premise that the choice of an effective style depends fundamentally on three groups of factors: leader characteristics, subordinate characteristics, and the situation requirements. The authors argue that the democratic leader tends to be more relationship-focused while the authoritarian leader is more task-oriented. Essentially, leaders must choose the leadership style that will maximize the chances for effectiveness.

Although it has evolved since then, another contingency model of leadership is that of "Leader-Participation" developed by Vroom, Yetton and Jago, which relates leadership behavior with participation in decision making. "Vroom and his collaborators sought to determine the most desirable form of collaborator participation in decision making, taking into account the different types of task structures, i.e., whether routine or non-routine" (Caldeira & Correia, 1996, p. 6-12). The model is based on a complex decision tree that guides one towards the most appropriate leadership style(s) for the situation. Vroom's model confirms the trend for research on leadership to focus more on the situation and less on the person.

To Luthans (2002), leaders will be more appreciated by subordinates and will exert greater influence on them if their behavior fit their needs and values, as well as the requirements of the specific situation of work. For its part, the main premise of the model designed by Vroom and Yetton and Jago's is that the leader has five forms of decision that can be used in different circumstances (Clarke et al. 2005, p. 293).

These decisions, to evaluate its effectiveness, are subject to four criteria: decision quality, decision acceptance, decision costs and the subordinates' development (Cunha et al., 2005). What these authors seek, above all, is the determination of the most desirable type of employees' participation in decision-making, considering the different types of routine and non-routine tasks structures (Caldeira & Correia, 1996).

For all this, it is possible to understand that the model proposed by Vroom, Yetton and Jago confirms the trend towards research on organizational leadership to focus more on the concerned situation and least in the leader.

Hersey and Blanchard's Contingency Approach to Leadership (also called the "Tridimensional Leader Effectiveness Model" – 1993), assumes that managers should use different leadership styles according to the situation, i.e., the leadership style to be utilized depends on subordinates' level of maturity / readiness. To be effective, it is important that leaders "... have not only the ability to diagnose the most appropriate leadership style for the situation, but also that they can correctly apply that style" (Hersey & Blanchard, 1993, p. 112).

To the two original strands of situational leadership – *task orientation* and *relationship orientation* – a third was added, *effectiveness*, which the authors refer to as being the Environment. They deem this pertinent as they believe leader effectiveness depends on the way leadership style fits the specific situation (Hersey & Blanchard, 1993). The leadership style varies according to the levels of the follower maturity/readiness. Maturity should not be understood as a personal feature or personality trait, but as something that has to do with a specific task, namely, maturity represents the ability and willingness of individuals taking responsibility for their own behavior and orientation has two components: the maturity at work and the psychological maturity (Barracho, 2012).

As with leadership styles, the levels of subordinate development are situational: a person or group can be specialists, safe and motivated for performing a given task, but could be less competent, motivated, or committed for another.

For Hersey and Blanchard (1993) it is the subordinate who determines the leader behavior, i.e., it is the leader who must adapt to the subordinate's level of development. This latter premise opposes the Situational Engineering advocated by Fiedler and is called *Behavioral Flexibility* (Cunha, Rego, Cunha & Cabral-Cardoso, 2005).

New Theories of Leadership

Almost all leadership theories presented behind relate to transactional leadership, that is, "leaders who guide or motivate his followers towards the goals established through prizes and rewards for their behavior" (Caldeira & Correia, 1996, p. 6.15). The new theories of leadership present different types of leaders who, through their personal vision and energy, inspire followers and have a significant impact on their organizations. The buzzwords are transformational or change leadership, servant leadership, teaching organizations as opposed to learning organizations, and substitutes for leadership.

Maillet (1988) proposed to define leadership relying on four components: influence, voluntary aspect, notion of excess and relationship. The leader exercises influence on his team to propose a vision or directions that promote the enthusiasm and the incitement and confidence as a mean of coordination. The team participates in a voluntary mode in the project initiated by the leader. This conception is near the proposed by Bass (1990). In Bass definition of transformational leadership as opposed to transactional leadership is a relationship based on punishment and retribution. Transformational leadership comprises four dimensions: charisma or idealized influence, motivational inspiration, rational and individualized stimulation consideration.

Messeghem and Pierson (2002) state that leaders present different approaches depending on the focus in one or another of these dimensions. They assume, also, different roles which are sometimes ambivalent. This ambivalence is necessary for the leader to ensure the proper functioning of the group.

Quinn (1984) mentioned two pairs of internal / external opposition, stability / change that allow him to put in evidence the eight papers that the leader is expected to play. According to the figure below, the circular representation which he proposes shows that certain roles are contradictory. Perret (as referenced in Messeghem & Pierson, 2002), from the frame of analysis, highlights the paradoxical side of change by showing that the leader takes on roles that may prove to be contradictory. In the conduct of change, the leader's competencies are expressed in his breadth of reconciling these different roles.

Figure 2. The different roles of leadership
Source: Adapted from Messeghem and Pierson (2002, p. 56).

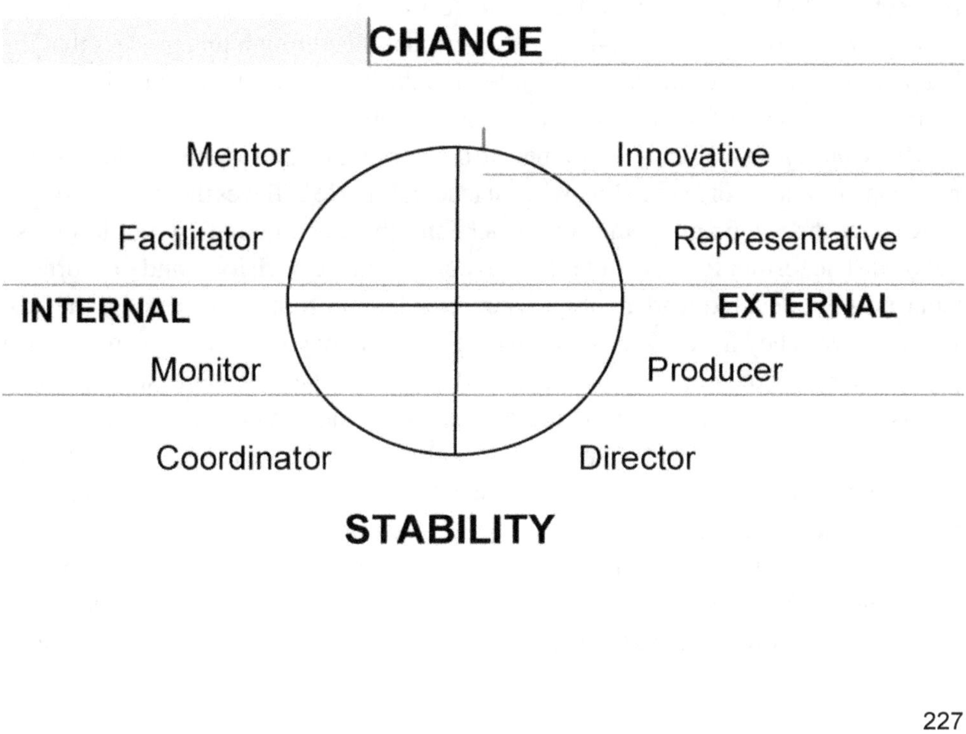

Messeghem and Pierson (2002) explain that the top of the schema brings up a whole series of papers that reference to change and innovation. Upon taking a groundbreaking paper, the leader must be able to print a direction or a vision that corresponds to a representation of the company and the environment in the future. The leader assumes the role of representative that is enhanced by the elimination of the borders of the company. In the internal quadrant /change, the leader takes a function of human resources by encouraging initiatives and to give the possibility to subordinates to participate in defining the direction. As regards the role of facilitator, the leader must help others in their tasks with the ability to delete himself. In the external quadrant / stability, the leader image corresponds to the transformational leadership model, and must be able to translate its vision in the form of plan. He must define the objectives and the rules and define a system of incentives. The pair internal / stability refer to the control and coordination. The leader must avoid the destruction of the organization. Horizontal coordination contributes to reduce this risk. The leader's effectiveness lies in its ability to perform these multiple and contradictory roles.

According to Yukl (1989, p. 269), "Transformational leadership refers to the process of influencing major changes in the attitudes and assumptions of organization members and building commitment for the organization's mission, objectives and strategies." Transformational leaders motivate their followers to do more than they intended by transforming group expectations. The effect of the influence is to empower subordinates to participate in the process of transforming the organization. For Rojot (2000, p. 19), "Overall, the transformational leader is one who is able to transcend or make in-depth changes to situations. It is through this process that the biggest changes are introduced in attitudes and the basic underlying hypotheses that sustain the activity of members of the organization".

Blanchard (2000), already mentioned in the overview of the contingency leadership, presents a new form of leadership: servant leadership. He believes there are two types of leaders: those who are primarily leaders and those who are, first and foremost, devoted. The former type of leader tries to control, make decisions, and give orders. Contrariwise, devoted leaders only assume leadership if they feel it to be the best way to serve. They have no sense of owning the post they hold; they see it as an act of service rather than of ownership. They delegate or share power easily, and their aim is to help workers to become more free, more autonomous, more capable and more effective. According to Blanchard (2000), this type of leader helps to produce good results, modeling and encouraging behaviors and values in line with a shared vision. With them, both, workers and organization are winners.

Tichy and Cohen (1998) believe that organizations can no longer be learning organizations but must become teaching organizations if they wish to survive in today's highly competitive global market. Teaching organizations are more able to achieve success and maintain it thanks to the constant focus on leader development. The actual leaders of these organizations are creating the next generation of leaders, teaching the critical issues facing the business and how to deal with them. Consequently, these organizations have an intellectual capital to continue ahead.

Leadership substitution theory tries to identify situations where the influence of the leader behavior is null or of very little significance for subordinate performance. Zenger, Musselwhite, Hurson and Perrin (1991, p. 50) consider the existence of what they term the "two sides of power" in a leader: "The first is based on the submission of others. The second is based on the concern for group objectives, finding objectives that drive the group, on helping the group to formulate them, on providing means to attain them and giving group members the sense of competence they need to work". It is the latter aspect of power that belongs to a leader who learns how to lead within a team environment.

Zenger et al. (1991) advocate the need for a new team, a new form of leadership. Instead of cutting the work into chunks, effective leaders are conducting the reintegration of work processes, through the delegation of many of their traditional duties in small groups of frontline employees (high performance teams), in autonomous teams, teams semi directed, in self-directed teams, in work teams or only in teams of plan.

Zenger et al. (1991) consider that, depending on the type of organization, teams can be permanent or temporary, functional, or cross functional, conventionally supervised or, to various degrees, autonomous. These organizations want a new type of leader – one who is more strategic and reactive to clients, workers, and the organizational imperatives. This new type of leader reacts to the changing nature of the workforce and the pressures of competition. This kind of leader appears because teams of collaborators are acquiring the experience to take charge of the duties which were the responsibility of the managers, in addition to the fact that the management of organizational change is being one of the fundamental concerns of the managers / leaders.

The need of the managers to have political skills to implement decisions was defended by Pfeffer (1992). In his view, both hierarchical via as the strategy to develop a strong culture or a strongly shared vision, are not sufficient for such implementation reason why it is necessary to resort to political action.

Goffee and Jones (2005) stated that employees do not follow a leader who invests little of himself in his leadership behavior. People want to be led by someone "real". This is due, in part, as a reaction to the turbulent times in which we live, and a response to the widespread disenchantment with politicians or businessmen. Both the leaders and the followers associate authenticity with sincerity, honesty, and integrity. Authenticity cannot be the product of sheer manipulation, but accurately reflect aspects of the true interior of the leader and therefore cannot be a mere act of representation. For the authors, the great leaders seem to know what the personality traits that should reveal when and to whom. Like chameleons, they can adapt to situations they face and the people who lead without, however, losing their identity in the process.

Mayo and Nohria (2005) argue that the long-term success of a leader is not coming from the sheer force of his personality or the breadth and depth of his skills. Without the ability to anticipate and to adapt to changes in business conditions, the personality and skills are strengths but merely temporal.

The long-established practice of coaching is a new current in the field of leadership. According to Cunha and Rego (2005), it corresponds to leader actions geared to a higher value: helping others to find their own path of self-development. Hence, it is down to the leader to help each worker to find the best way to express his/her talents. For Urrutikoetxea (2003, p. 174), coaching can be described as a process that helps "a person to develop their skills so as to know how to do four things well: have a good knowledge of their strong points, be aware of their weaknesses, know how to capitalize on their strengths and compensate for their weaknesses".

For Afonso (2009), a coach provides workers with a variety of actions: developing a training process based on teaching learning rather than learning from teaching; helping them to discover their development potential; helping them to establish goals based on self-assessment; recognizing merit and stimulating their creative potential instead of controlling it.

Coaching is another element of the great change in the organizational world which sees knowledge as a first resort: complex organizations with simple people are giving way to simple organizations with complex people able to invest in their own development, with the cooperation of the organization where they work, for the good of their current work and future employability (Afonso, 2009).

The brief outline presented demonstrates that the concepts and meaning of leadership have changed over the time. Because people, organizations and societies are constantly changing, it is to be imagined that they will assume new meanings and will have new approaches in the future.

METHODOLOGY

This work falls within the field of Management and Psychology and entails the use of quantitative research methods. Quantitative method presents distinct characteristics of the qualitative method mainly in what concerns the process of data collection and the way how they are analyzed (Bryman, 1988; Bullock, Little & Milham, 1992).

Quantitative method implies quantification, both in data collection and its statistical processing, whether simple (mean, median and standard deviation) or more complex, such as measures of association, regression analysis or factor analysis (Richardson et al., 1985). This method represents, at first, the desire to ensure precision of results and avoid bias in data interpretation and analysis while seeking to attribute a safety margin when making statistical inferences (Richardson, Peres, Correia, Peres & Wanderley, 1985).

This method is often applied in descriptive research that seeks to discover and classify the relation between variables in addition to ascertaining the causal link that might exist between phenomena (Richardson et al., 1985). According to Gil (2008), the quantitative method has advantages but also some disadvantages associated with the survey, such as the difficulties in terms of design and the lack of guarantee for the understanding of the sense of the issues, as well as problems linked to the rate of return.

For this study, a quantitative research instrument was utilized, in the form of a questionnaire put to a group of CA workers with leadership functions. This instrument is explained below, as is the application methodology. Some of the difficulties that were faced at the time of the respective application within CA institution are highlighted. The limitations of the study will also be presented in this chapter.

Leaders' Survey

The survey of leaders by questionnaire involved the staff who had positions of responsibility over other workers, such as Managers and Under-managers, Commercial Directors, Coordinators and Sub-coordinators, and Heads of Service, *inter alia*.

Questionnaires have frequently been used for assessing and monitoring economic and social phenomena. As well as allowing for the synthesis and monitoring of organizational changes over time, questionnaires also make it possible to compare between different elements of a set or between dimensions of the same element. For Parasuraman, Zeithaml and Berry (1988), the fact that there is an increasing amount, in the services sector, of research oriented to the workers satisfaction and motivation, demonstrates a growing use of questionnaires as support methodology of management and organizational practices.

The questionnaire administered to CA leaders was an instrument that had as purpose to enable the researcher to measure, compare and monitor the visions of these professionals on different items related to their practices of leadership, and the way they interacted with their subordinates. The choice of this instrument was based on its broad character, once studied various aspects related to the practice of leadership, as well as other cross-cutting dimensions such as organizational climate. This is a research tool which could allow CA decision-makers systematically seize the perceptions and opinions of the organization's human resources, about a number of dimensions that are part of the organizational environment, being able to capture changes and continuities in an organization, if it is applied periodically.

The units of analysis targeted in the questionnaire were all CA branches in mainland Portugal and the Azores. In what concerns the search universe, according to the data obtained in the Consolidated Balance Sheet of SICAM 2010 (2011, p. 5), it consisted of 3,875 employees. From these, the ones that have interest for the leaders' study were those designated in the 2010 Company Consolidated Balance Sheet as "Senior Management" (228 people), "Middle Management" (81 people) and "First Line Management" (583 people). In total, it yielded 892 professionals.

For this study it was performed the calculation of the sample size, considering the universe concerned (892<100,000), which was finite in nature, in which the confidence level established was 2 sigma (95.4) with an allowed estimation error of 5%. Thus, taking into consideration these factors, it was used the following formula to calculate the sample:

$$n = \frac{4pqN}{e^2(N-1) + pq4}$$

where:

n = sample size

N = size of the universe

4 = confidence level chosen in terms of deviations– sigma (2 sigma squared)

P = proportion of characteristics researched at the universe, calculated as a percentage (according to Richardson et al., (1985), 50% is the most common in social research)

q = proportion of the universe not having the researched characteristic (100 – p)

e = allowed estimation error

From the universe of 892 people emerged through this formula, it will be required a sample of 276 elements needed to the development of the research. However, despite several contacts developed by the researcher, it was not possible to achieve this number of respondents. Indeed, it was possible to have the collection of information only from 85 elements (30.8%).

The sampling techniques used in this study, were non-probabilistic - quota sampling and purposive sampling. For Ghiglione and Matalon (2001), quota sampling seeks to reproduce in the sample how certain variables are distributed in the population being studied. As for purposive sampling, for Richardson et al. (1985), the sample units (staff) are chosen because they have certain characteristics established in the research objectives. In this case, resided the fact that these professionals are employed by CA and belong to the required professional categories. Here the objective was to select these sample units until the sample dimension was completed. However, this aim was not possible to obtain.

The period for applying the questionnaire was from December 2010 to March 2011, with the delivery deadline being extended until May 2011 to complete the sample.

The study was done using a questionnaire made available via an online platform, with a direct link being sent by email straight to the diverse *Caixas*, with a view to the professional groups of that institution signing up. In terms of the variables studied, the aim was to associate the opinions and perceptions of CA leaders regarding their relationship with subordinates and the relationship between them, as well as the working atmosphere in their work team. In addition, other variables focused on dealt with workers' satisfaction and professional training. Therefore, those filling out the questionnaire were asked to identify the Professional Group of the Collective Bargaining Agreement (CBA) that they belonged to, seeing as this was decisive for the analysis in question.

Those filling out the questionnaires were guaranteed anonymity, and that the answers were to be treated on an aggregate basis, to protect the identity of the professionals who had agreed to take part.

Concerning the distribution of the questionnaires, as previously mentioned, it was decided to put them on an online platform to allow them to be filled out quickly and be more convenient for the sample subjects, providing a larger guarantee of confidentiality and anonymity, as well as being less costly in terms of time and money. There are many other reasons in opting for this research instrument to the work in question over and above the advantages associated with the dimensions it has (Motivation, Satisfaction, Organizational' climate). According to Richardson et al. (1985), with a questionnaire one can obtain information from many people simultaneously or within a reasonably short time span, the data processing can be done faster and easily than if having recourse to other techniques (such as interviews, for example) and, since this questionnaire was anonymous, respondents could feel more at ease expressing their opinions.

Limitations of the Study

Regarding the questionnaire, the first limitation that can be found is in the use of a sample for the study as an alternative of applying it to the entire population of staff with leadership functions within the institution. However, one can easily recognize the decision to use a sample, not only due to the size of the universe in question, but also because of the time available for carrying out the study and the number of leaders involved.

An additional limitation has to do with the type of sample chosen. As it is not probabilistic, it is not possible to make inferences for the leader population of CA in its entirety, and its results will have to be considered only in the scope of the sample analyzed.

The third limitation relates to the non-fulfillment of the quotas for all staff with leadership functions – the different categories represented in the sample are not in the same scopes as in the population (Senior Management, Middle Management and First Line Management).

RESULTS ANALYSIS AND DISCUSSION

The statistical analysis of the results is presented in Table 1, showing the descriptive and inferential values for CA staff in leadership positions.

Initializing with gender, we find that the large majority (87.1%) is male, with no more than 12.9% of the sample being female. The data that emerged from the application of the quantitative methodology indicates a disproportion in the representativeness of the two genders in the studied population. Results for total employees in CA allude to a 55% male, 45% female breakdown, but it was not achievable to discriminate by functional distribution (2011 SICAM Consolidated Balance Sheet).

Table 1. Gender of leaders

	N	%
Male	74	87,1
Female	11	12,9
Total	85	100,0

In what concerns the leaders age, the main age groups were from 46 to 55 years (41.2%) and 36 to 45 years (40%). This was found to agree with the 2011 SICAM Consolidated Balance Sheet, even if the age ranges presented in that document were [45 – 49] and [40 – 44].

In Table 3, one can see that the large percentage of respondents has between 10 and 12 schooling years (54.1%), with those having a university degree being in second place (16.5%). Merely 8.2% of respondents have postgraduate and masters' qualifications. The sum of items "Attendance at Higher Education", "*Bacharelato*"[4], "University Degree", "Postgraduate course" and "Masters" comes to 37.7% of the total, it being rare or unusual at the present time to find recruits to the Banking Industry with only Secondary Education, reflecting the high demands and complexity of the

Table 2. Age of leaders

	n	%
26 - 35 years	8	9,4
36 - 45 years	34	40
46 - 55 years	35	41,2
> 55 years	8	9,4
Total	85	100

banking activity. The fact that 54.1% of the leaders have qualifications relating to 10-12 years in the secondary school may be linked with the situation reflected for the age group [46-55 years].

Table 3. Academic qualifications of leaders

	n	%
9th Grade	7	8,2
10th – 12th Grade (GCSE)	46	54,1
Bachelor's degree	1	1,2
Higher Education attendance	10	11,8
University degree	14	16,5
Masters	4	4,7
Post-graduate course	3	3,5
Total	85	100

Analyzing which of CA business units are most represented in the sample, *Central Alentejo* has the largest share (12.9%), closely followed by *Costa Azul* (10.6%), *Noroeste* (10.6%) and *Pombal* (9.4%). Regarding the type of employment contract for CA leaders, the huge majority have a permanent employment contract (94.1%), with the other category having low values, seen in Table 4.

Table 4. Type of employment contract of leaders

	N	%
Term contract	5	5,9
Permanent contract	80	**94,1**
Total	**85**	**100**

As for the Group IV[5], in accordance with the CBA of CA leaders, one can see that the largest segment (69.4%) belongs to Group I, with another 17.6% belonging to Group IV. Group III (8.2%) and Group II (4.7%) are the least representative.

Table 5. CBA group of leaders

	n	%
Group I	59	**69,4**
Group II	4	4,7
Group III	7	8,2
Group IV	15	17,6
Total	**85**	**100,0**

In relation to wage levels, the most representative CBA level is Level 10, with 23.5% of respondents, followed, in second place, by those with Level 11 (11.8%) and, in third place, those with Level 12 (10.6%). There was also another 8.2% who did not answer to this question.

Table 6. Leaders Salary levels

	N	%
7	3	3,5
8	8	9,4
9	7	8,2
10	20	**23,5**
11	10	11,8
12	9	10,6
13	7	8,2
14	3	3,5
15	1	1,2
16	4	4,7
18	5	5,9
n/a	1	1,2
n/a	7	8,2
Total	**85**	**100,0**

Table 7 presents the posts of employees with leadership functions at CA. Most respondents are Branch Coordinators (52.9%), with another 10.6% being Administrators and 8.2% being Commercial Coordinators.

Table 7. Functions performed by the leaders

	N	%
Branch coordinator	45	**52,9**
Manager	3	3,5
Assistant manager	6	7,1
Administrator	9	10,6
Auditor	3	3,5
Area coordinator	1	1,2
Advisor	1	1,2
Commercial coordinator	7	8,2
General coordinator	3	3,5
Advisor coordinator	1	1,2
Client manager	2	2,4
Branch sub-coordinator	1	1,2
Diretor	1	1,2
Head of unit	2	2,4
Total	**85**	**100,0**

The 2011 SICAM Consolidated Balance Sheet gives the subsequent distribution of leadership functions[6]: Senior management – 6.0%, Middle Management – 2% and First Level Management – 15.2%.

Table 8 shows the allocation of functions performed by leaders according to the respective CBA Level, highlighting the most expressive percentages.

With regard to the number of working hours per week, most leaders work more than 40 hours (58.8%), with only 15.3% of the respondents working 35-hour week, as seen in Table 9.

With the purpose of partially solve the problem of paying for extra work, the banking institutions have adopted a system of partial or full exemption from fixed working hours. For this reason, not only managers but also a high percentage of banking professionals presently has no fixed working hours. However, analyzing the 2011 SICAM Consolidated Balance Sheet (2012, p. 31), one can see the evolution of extra working hours' number, by gender, between the years 2009 and 2011, shown in Table 10. With the exclusion of 2011, females have worked the greatest number of overtime hours.

Table 8. Functions performed by leaders according to CBA level

	7	8	9	10	11	12	13	14	15	16	17	18
Branch coordinator	2,4	17,1	9,8	**34,1**	12,2	14,6	4,9	0	0	4,9	0	0
Manager	0	0	0	**33,3**	0	0	0	**33,3**	0	0	0	**33,3**
Assistant manager	0	0	0	16,7	**50,0**	33,3	0	0	0	0	0	0
Administrator	0	0	0	0	0	0	14,3	14,3		14,3		**28,6**
Auditor	0	0	**33,3**	0	**33,3**	0	0	0	**33,3**	0	0	0
Area coordinator	0	0	0	0	**100,0**	0	0	0	0	0	0	0
Advisor	0	0	0	0	0	**100,0**	0	0	0	0	0	0
Commercial coordinator	0	0	16,7	**33,3**	0	0	**33,3**	16,7	0	0	0	0
General coordinator	0	0	0	0	0	0	**66,7**	0	0	0	0	33,3
Advisor coordinator	0	0	0	**100,0**	0	0	0	0	0	0	0	0
Client manager	**100,0**	0	0	0	0	0	0	0	0	0	0	0
Branch sub-coordinator	0	**100,0**	0	0	0	0	0	0	0	0	0	0
Diretor	0	0	0	0	0	0	0	0	0	0	0	**100,0**
Head of unit	0	0	0	**50,0**	0	0	0	0	0	**50,0**	0	0

Table 9. Number of hours worked by leaders

	n	%
35 hours	13	15,3
40 hours	21	24,7
+ 40 hours	50	**58,8**
n/a	1	1,2
Total	**85**	**100,0**

Table 10. Evolution of overtime hours (includes working days and mandatory rest days)

	2009	2010	2011
Female	8.168	7.610	6.525
Male	7.317	6.225	7.406
Total	**15.485**	**13.835**	**13.931**

Leadership

Table 11. Number of years working at CA

	n	%
6 a 10 years	4	4,7
11 a 15 years	15	17,6
16 a 20 years	16	18,8
+ 20 years	50	**58,8**
Total	**85**	**100,0**

Table 12. Leaders' behavior (values in %)

	Never	Rarely	Sometimes	Often	Always	n/a
Clear attitudes towards employees	0	0	7,1	42,4	**49,4**	1,2
Specific tasks for different employees	2,4	2,4	10,6	**57,6**	25,9	1,2
Establishing deadlines for jobs / tasks	1,2	1,2	29,4	**47,1**	18,8	2,4
Defining performance standards for employees	2,4	3,5	24,7	**44,7**	23,5	1,2
Standardization of procedures for employees	1,2	2,4	9,4	27,1	**58,8**	1,2
Compliance with requirements by employees	0	1,2	1,2	23,5	**74,1**	0
Informing employees as to what is expected of their work	1,2	0	11,8	**42,4**	**42,4**	2,4
Deciding on what has to be done and how	1,2	0	4,7	**49,4**	44,7	0
Checking that the role of each employee is understood by all	1,2	0	10,6	**48,2**	40	0
Debating ideas with employees	1,2	1,2	8,2	**55,3**	34,1	0
Making every effort to ensure that belonging to the work team is enjoyable	**47,1**	20	10,6	8,2	14,1	0
Keeping information for the leadership	**52,9**	36,5	5,9	3,5	1,2	0
Refusing to explain decisions	**61,2**	27,1	4,7	4,7	2,4	0
Making decisions without consulting employees	16,5	29,4	**41,2**	10,6	2,4	0
Treating all employees impartially	3,5	4,7	7,1	17,6	**65,9**	1,2
Being open to change	0	1,2	5,9	30,6	**62,4**	0
Showing friendliness and accessibility	0	0	4,7	44,7	**50,6**	0
Ability to put into practice employee suggestions	0	0	15,3	**58,8**	25,9	0
Giving employees forewarning of changes to be made	1,2	0	10,6	43,5	**44,7**	0
Concern with employee well-being	0	0	2,4	21,2	**76,5**	0

Studying Table 11, one can see that the greater percentage of respondents have been with CA for more than 20 years (58.8%), representing substantial levels of longevity at the organization (no more than 4.7% of employees have been with the organization between 6 and 10 years).

In Table 12 are the results for the leaders' self-assessment on their leadership behaviors.

The larger proportion believes that managers always have clear attitudes towards their subordinates (49.4%), frequently allocating explicit tasks to different workers (57.6%). The establishment of deadlines for activities is something that is frequently cited in leaders' performance (47.1%), as is the defining of performance standards for employees (44.7%). There seems to be a concern to standardize organizational practices, since 58.8% of leaders establish the consistency of procedures for their subordinates, and at the same time, 74.1% refer to the meeting of requirements by workers.

The respondents also frequently or systematically take the initiative to notify workers as to what is expected of their work (42.4% for each), and frequently decide on the jobs to be done and the *modus operandi* to be used (49.4%). Leaders' habit of checking that the role of each employee is understood by the work team also occurs regularly (48.2%), and they frequently support debating of ideas with employees (55.3%).

Establishing organizational goals is something that is natural to modern organizations' activity, and, while expecting these to be met (or exceeded), the expectation is also to increase levels of employee motivation and satisfaction. Fundamentally, goals should be SMART (*specific*; *measurable*; *agreed*; *attainable*; *realistic*) (Cunha et al., 2005).

For Locke, as mentioned in Montserrat (2006), setting goals is one of the major determinants of behavior and he insists on the key role of feedback in work. People generally search for feedback, particularly from their line manager, which is felt as an exhibit of consideration by the manager. The less the psychological distance relevance, the higher is the receptiveness to feedback. If the psychological distance influences receptiveness to feedback, sharing in drawing up goals influences acceptance and improves performance.

One of the items that hierarchical superiors appear not to rate as important is a good work environment, since 47.1% of the leaders never make any endeavor to guarantee that belonging to their work team is a satisfying experience and 20% rarely do so. Most of them affirm that they never keep information to themselves (52.9%), just as 61.2% never reject to explain the justification for their decisions, albeit 41.2% occasionally make decisions without consulting subordinates.

Crozier, as cited in Montserrat (2006, p. 93), stresses that power within an organization results from the ability to manage uncertainty. Thus, "… the more that the zone of uncertainty controlled by an individual is crucial to the success of the organization, the more power that individual will have". This attitude explains the command and maintenance of information, the private agreements, the confidentiality and the irrationality of certain behaviors. The power of each depends on the unpredictability of his/ her behavior and the level of control over the organization. The relationship with authority and level of responsibility of agents within the organization establishes the stage of motivation.

For Delaire (1986), at the present time one needs to do more than control; animate and motivate, and the quality of the relation between those giving orders and those who must carry them out is fundamental.

As for the management of employees, the great majority of respondents (65.9%) state that they constantly deal with them objectively, with 50.6% admitting that they constantly behave in a friendly and accessible way. Openness to change is a question that, for 62.4% of the leaders, is a constant part of their nature, and 58.8% of them consider that they habitually can put into practice employee suggestions. A considerable portion of the leaders (44.7%) always inform their team in advance about the changes needed to be implemented, while 76.5% say that they are always concerned with the well-being of their subordinates.

According to Montserrat (2006), the personality and status of those wielding power is indispensable. The capacity to lead is based both on the effectiveness of the

Table 13. Relationship among team members and between them and the leader (values as %)

	Strongly disagree	Disagree	No opinion	Agree	Strongly agree	n/a
Employees have difficulty maintaining a good relationship	**45,9**	43,5	3,5	4,7	1,2	1,2
Employees can be trusted	0	0	3,5	37,6	**57,6**	1,2
There is a friendly atmosphere among employees	0	3,5	5,9	**51,8**	37,6	1,2
Cooperation from employees when solicited	1,2	1,2	2,4	**50,6**	43,5	1,2
Existence of friction between leader and employees	**51,8**	32,9	4,7	7,1	2,4	1,2
Help and support in performing tasks by employees when requested	0	1,2	2,4	**55,3**	40	1,2
Good team work from employees when solicited	0	1,2	4,7	**65,9**	27,1	1,2
Good relationship with employees	0	0	1,2	43,5	**54,1**	1,2

wielders of power and on the acceptance of their subordinates those who recognize that power. The author (2006, p. 92) believes that "An organization is a group of people whose power relations must be regulated. If the sense of importance of each one has to be developed, it is preferable to appeal to their expectations rather than their obligations".

According to most of the leaders, as seen in Table 13, it is not the case that employees have difficulties in maintaining a good relationship (45.9% strongly disagree and 43.5% disagree), with 57.6% believing that the workers can be trusted. A large percentage (51.8%) of the leaders believes that there is a pleasant atmosphere among employees, and 50.6% think that they team up when requested to. An expressive segment of the leaders (51.8%) strongly disagrees with the idea that there is friction between them and their team, and 54.1% rate highly the quality of the relationship they have with them. 55.3% of the managers as well agree that employees help and support them in performing tasks when demanded, with 65.9% of leaders agreeing that there is good teamwork between the members of the group.

Table 14. Work group atmosphere (values as %)

	Never	Rarely	Sometimes	Often	Always	n/a
Co-operative	0,0	1,2	7,1	**52,9**	38,8	0,0
Quarrelsome	38,8	**49,4**	11,8	0,0	0,0	0,0
Welcoming	0,0	0,0	8,2	**52,9**	37,6	1,2
Rejecting	**47,1**	41,2	7,1	3,5	0,0	1,2
Enthusiastic	1,2	0,0	15,3	**62,4**	21,2	0,0
Indifferent	**45,9**	38,8	11,8	1,2	0,0	2,4
Productive	0,0	1,2	7,1	**62,4**	29,4	0,0
Unproductive	**58,8**	32,9	7,1	1,2	0,0	0,0
Warm	0,0	0,0	18,8	**58,8**	22,4	0,0
Cold	**52,9**	34,1	12,9	0,0	0,0	0,0
Gratifying	0,0	0,0	14,1	**57,6**	27,1	1,2
Frustrating	**44,7**	38,8	16,5	0,0	0,0	0,0
Interesting	0,0	0,0	15,3	**60,0**	24,7	0,0
Boring	**42,4**	41,2	12,9	3,5	0,0	0,0
Effective	0,0	0,0	11,8	**64,7**	18,8	4,7
Ineffective	**44,7**	40,0	8,2	3,5	1,2	2,4
Pleasant	0,0	0,0	7,1	**65,9**	25,9	1,2
Unpleasant	**48,2**	43,5	5,9	2,4	0,0	0,0

Turning to the various types of work group atmosphere, exposed in Table 14, for 52.9% of the leaders, this is frequently co-operative, as for 49.4% it is rarely difficult. For 52.9% of managers, their work group is frequently friendly; with 47.1% considering that it is never rejecting for the team members. There is habitually enthusiasm within the leaders' team (62.4%), while 45.9% state that there is not at all indifference.

Productivity is a key issue that is regularly present in work teams. According to 62.4% of managers, and 58.8% of leaders their employees are never unproductive. Equal percentage (58.8%) confirms that there is "human warmth" in their work, while 52.9% have never seen any show of coldness in the workplace. For 57.6% of the leaders, the group atmosphere is gratifying and for 44.7% is never frustrating. For 60% of leadership, the group atmosphere is interesting, while for 42.4% it is never boring. In terms of group effectiveness, for 64.7% of the leaders the group atmosphere is regularly effective and is not at all inefficient for 44.7%. The majority of the leaders states that the group atmosphere is often agreeable for members (65.9%) and is never unpleasant for 48.2% of the leaders.

Guillén-Gestoso and Guil-Bozal, as cited in Gonçalves (2012, p. 80), state that "In the modern systemic view of the organization, the climate is an undeniable reality in business. Arising from the relations between the various subsystems and which constitutes a link between individuals and environment. In concrete terms, the climate is the perception of a group of people who form part of an organization and establish various interactions within a work context". Therefore, the work climate can be dissimilar in function of the department, of the leadership and even of each incident.

Table 15. Openness towards employees (values as %)

	No	No opinion	Yes	n/a
Knowledge of each employee's areas of professional and personal interest	2,4	7,1	**88,2**	2,4
Concern with employees' professional and personal development through ongoing training and regular feedback	0	7,1	**90,6**	2,4
Habit of assessing employee levels of satisfaction	11,8	4,7	**81,2**	2,4
Encouraging employees to give ideas on the strategy and decisions for the work group	1,2	2,4	**94,1**	2,4
Openness towards employees talking about problems of a professional and personal nature	1,2	1,2	**95,3**	2,4
Self- assessment of employees that is analyzed by the leadership and by other members of the work group	27,1	15,3	**55,3**	2,4

Leaders owned that they know of each employee's areas of professional and personal interest (88.2%) and were also concerned with employees' personal and professional progress (90.6%). Assessment of subordinates' performance seems to figure among their activities (81.2%), while self-assessment is permitted, this being analyzed by the leaders and other members of the team (55.3%). Staff can make contributions towards optimizing the strategy and decisions of the work group (94.1%) and may also talk about problems of a personal and professional nature (95.3%), seen in Table 15.

The results of the preceding table comply with the view of Montserrat (2006), who believes it to be one of the leader's major functions to concurrently address the angst of the organization fight stress and the psychological burden of work, and transmit energy, creating a motivating environment. To favor the relational dynamic, the leader may have alternatives to different approaches such as boosting the sense of recognition and belonging, sharing emotions in a social setting or the management of stress situations. In terms of motivation, listening and, above all, empathy, the capacity to put oneself in the shoes of another, understanding their emotions, play an essential role.

Table 16. Training and performance assessment at the branch (values as %)

	Never	Rarely	Sometimes	Often	Always	n/a
How often the leader proposes professional training for subordinates	2,4	3,5	21,2	**52,9**	12,9	7,1
How often performance assessment occurs at the branch	5,9	5,9	**32,9**	30,6	18,8	5,9

According to the leaders' responses, professional training for the members of their work teams is habitually proposed, as seen in Table 16, with a view to perfecting their professional and personal skills. The ever-higher demands of the banking activity, in function of the complexity of their products and services, has resulted in the banking organizations being very concerned to have human resources with up-to-date skills and knowledge. CA strategy has also included improving the standard qualifications of their professionals. According to SICAM Consolidated Balance Sheet (2010, p. 22; 2011, p. 22), the number of training hours – internal and external – achieved a total of 76,758 in 2010 and 128,816 in 2011, an increase of 67.8%. Investment in internal and external training hours in 2010 was € 543,000 and € 627,000 in 2011, representing an increase of 15.5%.

Figure 3. Mandatory promotions CBA

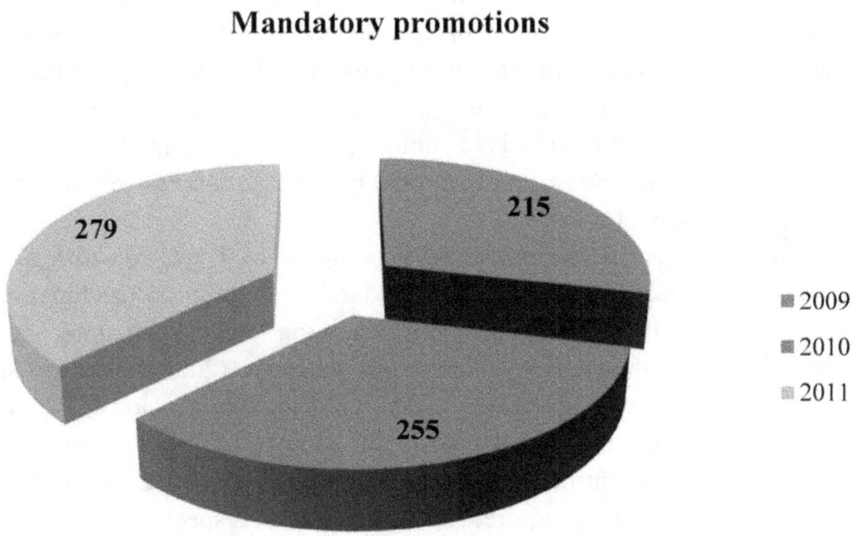

Figure 4. Promotions on merit

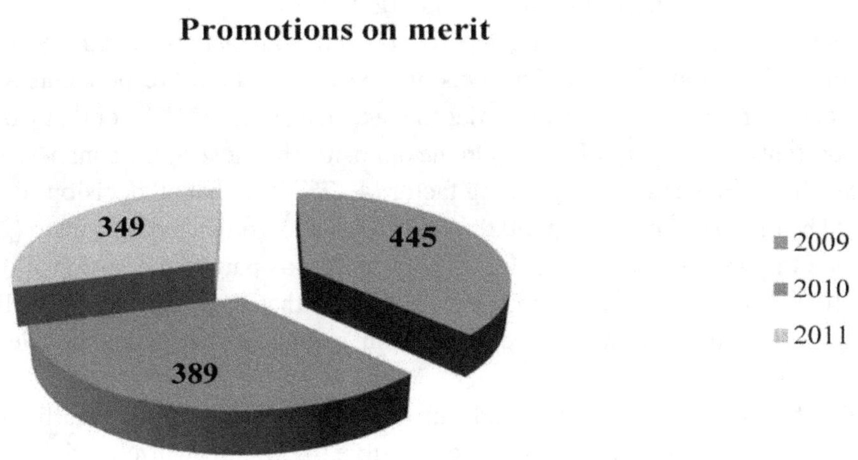

In matters of performance assessment, we notice that it occurs with some regularity (but not much) in the responses from the most significant portion of managers (32.9%), even if, in the teams of 30.6% of them, performance is assessed more frequently. One of the most important effects that performance assessment has on employees' careers is promotion through merit and the consequent increase in salary. Figures 3 and 4 show how obligatory promotions (in function of the CBA) have grown – because of the levels of seniority and longevity at CA – and the use of promotions on merit and as a motivational factor has progressively reduced as a way of falling personnel costs.

Decision making and what this involves is a subject closely related with leadership. The questionnaire aimed to capture CA leaders' perception on the loneliness of power, subordinate to organizational efficiency, on "decision making" and the "role of leaders". In the study by Tomei and Fortunato (2008), there are several types of decision that must be taken in isolation. With the diffusion of the practice of participative management, this type of decision is occasionally criticized as being centralizing and controlling. Most participants in the above-mentioned study emphasized that authority is required, but that one must respect individuality while keeping individualism at bay in the participative practice of opinion forming. As well, the search for consensus is an indication of lack of leadership. However, there are situations where consensus is not possible. It is at such times that solitude sets in: the person is alone and has nobody with whom to discuss things, having previously heard everybody's opinion.

As for the attributes referring to concepts of solitude in leadership held by CA leaders who took part in the study, Table 17 sums up the percentages of the various attributes, with the most important being highlighted.

As seen in Table 17, the larger percentage of managers declared not to have feelings of isolation when making decisions (38.6%), and most respondents believe it is not imperative to be lonely in order to make decisions (51.2%), or that isolation is important or necessary (51.2%). On the opposite, the most significant percentage agrees that solitude is a demotivating factor (41.7%) and that all decisions can and should be discussed in a group (50.0%). According to Tomei and Fortunato (2008), solitude in power is intrinsic to the decision maker as part of the responsibility of directing an organization. All interviewees stated that they experienced solitude, and noted the physical importance of isolation, sometimes necessary for a decision to be made.

The fact that decisions are made in a group does not automatically signify that the right people cannot be held responsible (47.0%), because participation in management modifies the sense of solitude in power, as stated by 61.9% of the leaders questioned. In the words of Tomei and Fortunato (2008: 24), "The company director who calls too much for collective decisions in important matters for the

Table 17. The solitude of the leader (values as %)

	Completely disagree	Disagree	No opinion	Agree	Strongly agree
Sensation of isolation in decision making	12,0	**38,6**	15,7	28,9	4,8
Importance of isolation in decision making	25,0	**51,2**	9,5	14,3	0,0
Solitude of a leader is important and necessary	26,2	**51,2**	11,9	8,3	2,4
Solitude as a demotivating factor	6,0	15,5	13,1	**41,7**	23,8
All decisions can and should be discussed and made as a group	4,8	25,0	8,3	**50,0**	11,9
Group decision making complicates the allocation of responsibilities	12,0	**47,0**	16,9	22,9	1,2
Participative management alters the sense of solitude in power	1,2	10,7	15,5	**61,9**	10,7
There is group participation in decision making on matters that are fundamental to the organization	0,0	10,0	16,3	**58,8**	15,0
Solitude in power affects only organizations whose decisions are very centralized	3,6	24,1	19,3	**45,8**	7,2
Solitude as something positive for the function of leader	18,3	**52,4**	14,6	14,6	0,0
Organizational climate and human behavior can be responsible for the sense of solitude	0,0	8,4	22,9	**59,0**	9,6
Sense of solitude can create stress in those holding power	4,8	7,2	22,9	**55,4**	9,6
Stress causes a sense of solitude in those holding power	6,0	24,1	19,3	**41,0**	9,6

company strategy assumes a position that can easily be mistaken and interpreted as lack of courage and, in the case of a failure, the responsibility comes to be not his own, but that of the group". Thus, in the opinion of the participants of the study, the leader has to know how to give people space but, at the same time, exercise his/her authority. In addition, Boyd and Gumpert, as mentioned in Tomei and Fortunato (2008), advise that sharing apprehensions or doubts with colleagues can give rise to conflicts. Leaders, even though being always surrounded by people, do not feel happy sharing their decisions and worries, because those people around them are seen as competitors.

Most of the leaders agree that there must be a group participation in making decisions on questions that are fundamental to the organization (58.8%) and that solitude in power influences only those organizations where decision making is very centralized (45.8%). Referring to Tomei and Fortunato (2008), most of the study participants confirmed an ability to stand high levels of solitude in the decision-making process, even deeming it to be necessary. The need to have this ability grows in proportion to the increase in responsibilities. "Solitude can also be deemed one of the most important characteristics in a leader's attitudes and behavior as the demands become stronger and more varied" (Tomei & Fortunato, 2008, p. 25).

For 52.4% of the respondents, solitude is not a positive element for leaders to perform their function, and the sense of solitude may cause stress for those in power (55.4%). In addition to this, 41.0% of them agree that stress produces a sense of solitude among power holders and that this feeling of solitude may be fostered by the organizational climate and by human behavior (59.0%). Jackson et al., as mentioned in Tomei and Fortunato (2008), highlight the importance of the task of the leader, who needs to have the stability to harmonize the needs for solo participation with the needs of his/her team members. Balance is only viable through the management of solitude in power.

CONCLUSION

The overall effectiveness of an organization lies in its human efficiency and management. Occupying human resources, a central position in the conduct of business, without them no strategic plan can be executed. In addition to other variables, personal fulfillment of human being goes, too, by the possibility that he will have to develop and evolve as a professional, typically housed in a company or institution. The development of his professional activities depends on what may be characterized as an exchange ratio that he should enable, and the organization.

As a result of several conditions, environmental or individual, organizations must anticipate change and development through a leadership with a strategic vision that goes beyond the context in which they operate. It is the opinion of Correia (1999) that this strategic vision can become the base of technological revolutions, targeting of customers and competitors, of development of new levels of competence and internal motivation and, yet, as a source of guidelines influence of governments and supranational organizations. Herein lies the greatest challenge that the leadership cannot evade, aware of its responsibility towards customers, employees, investors and, ultimately, to the society. Leadership represents the company in all its relations with the outside world being, simultaneously, the real connecting link of the social system that is any organization. Increased responsibilities while agent of change and transformation of companies; as an educator, in the sense that with its direction and guidance it modifies behaviors and attitudes, as well as the organizational culture.

A successful organization requires from the leadership the ability to identify, at the right time, opportunities to respond promptly and adequately. It also requires an improved diagnostic capacity to detect, at the same time, threats and difficulties in order to neutralize or buffering them. According to Cunha (2002), the organizational change should not only address about the planned change, i.e., built according to a pre-established process but also on emerging changes – which draws on a non-envisaged in response to problems and local specificities. Cunha et al. (2005) argue that the conduct of change may pass, in considerable measure, by the creation / modification of organizational contexts, which is a role that leaders can instruct. To leverage the capabilities of organizations renewal and rejuvenation, they create organizational environments that encourage responses adjusted to the environment.

The aggressive environment in which organizations find themselves has created the need to find new administration models capable to lead to success. Modern management trends involve a growing need for leaders and not simple operational, and for diverse hierarchical structures. The new business organization models are moving towards greater flexibility and the extending of responsibilities, fostering the development of motivation and skills within an organization.

From the results observed, some appealing conclusions could be drawn about the leaders for the area of Human Resources Management. First, according to the leaders, subordinates at CA can take part in decision making for matters that influence their daily work and, at the same time, there is a concern to ensure that the organizational climate is constructive and promotes the meeting of set objectives and targets. Linked with this, it was observed that the communication flow at CA appears to be fluid and clear, such that all are conscious of the most relevant information for everyday matters in the organization.

Another significant conclusion in this study was that, according to CA leaders, the atmosphere within the work group they lead is reasonably positive and promotes good general and individual performance. This was to be expected from the results obtained concerning to the work group climate.

In Human Resources Management it is known that followers give large significance to the way that leaders are concerned with their needs and aspirations within the organizations. It appeared that the leaders have adopted a proactive role in the relationship with their subordinates, enabling them to be involved in work group strategies or being interested in what the followers give most significance to in personal and professional terms. This signifies that, for the leaders, their leadership appears to go further than the simple task-based relationship, involving more complex components that are determining factors for good performance by team members.

Another critical aspect dealt with in this research was the observation that the leaders had, as to the way that their position sets them apart from other employees. It could be seen that most of CA leaders like better to apply their leadership not from a posture of isolation / solitude, but on a more pacifying and participative basis, involving more people in decision making and strategy adoption. Another important result was the association that leaders are aware of between the manifestation of stress in their activity and the solitary practice of leadership, which may be one of the causes why they prefer participative management in their teams.

Therefore, leaderships in CA appear to assume an open character with their teams, showing concern with the well-being and the personal and professional progress of their employees. The question of status is still there but, from the results analyzed, we could maybe state that at CA it is less powerful than at other Credit Institutions, something that is possibly not unconnected with the fact that the organizational structure of CA is different from the others.

Modern management trends have a propensity to decrease the importance and interest of supervision in favor of setting up autonomous teams with a good capability to respond fast to all situations that might compromise the success of processes. These teams are given challenging and pragmatic objectives and they are given the opportunity to carry out their own prior self-control at times of institutional control the employees have empowerment. At the same time, the flattening of enormous hierarchical pyramids and the decrease to two or three levels means the end of vertical careers. Promotions have become more horizontal, and talk is more of "practicing management" with no hierarchical relation of the manager-coach. This is a movement towards multi-functioning, the extension of responsibilities to more important processes or the management of change projects that establish careers in modern companies (Ceitil, 1994). This change has led to the decrease in the number of managers and the increase in the leaders in an organization. Ceitil (1994) believes that there being more leaders in an organization is something that cannot be separated from the need to have better leaders who can be leaders of leaders, stressing the fact that the function of a leader within a company is, more and more, to foster employee motivation and skills, in other words, promoting their growth.

Human resources are increasingly seen as the indispensable resource of an organization and the leader, whether present or future, must focus the efforts on taking full advantage of this resource in its most fragile and multifaceted dimension: that of the individual. Leader should be able to talk to everyone in the organization, mobilize that individual's intelligence and energy through inherent motivations and inspire free adhesion to the company project. Organizations leaders should, more than ever, be managers and strategists who are conscientious to the human capital of the organization and give more time and effort to that role. They should be largely concerned in training employees, assessing heads, identifying talents, and conceiving new forms of organization. From this point of view, they should prove themselves to have outstanding skills in the domain of human resources strategic management.

REFERENCES

Afonso, P. L. (2009). *Liderança - Elementos-chave do processo*. Escolar Editora.

Bales, R. F. (1950). *Interaction process analysis: A method for the study of small groups*. University of Chicago Press.

Barracho, C. (2012). *Liderança em Contexto Organizacional*. Escolar Editora.

Barracho, C., & Martins, C. (2010). *Liderança e género*. Edições Sílabo.

Bass, B. M. (1990). From Transactional to Transformational Leadership: Learning to share the vision. *Organizational Dynamics, 18*(3), 19–31. doi:10.1016/0090-2616(90)90061-S

Bass, B. M. (1990a). *Handbook of Leadership: A survey of theory and research*. Free Press.

Blake, R., & Mouton, J. S. (1964). *The managerial grid*. Gulf Publishing Co.

Blanchard, K. (2000). Leadership by the book. *Executive Excellence, 17*(3), 4–5.

Buelens, M., Kreitner, R., & Kinicki, A. (2002). *Organizational behaviour*. McGraw Hill.

Caldeira, P., & Correia, M. F. (1996). *Psicologia das Organizações*. APB-ISGB.

Calder, B. J. (1977). An attribution theory of leadership. In B. M. Staw & G. R. Salancik (Eds.), *New directions in organizational behaviour* (pp. 179–204). St. Clair Press.

Castro, A. (1994). A liderança no contexto das novas práticas de gestão. *IAPMEI: Pequena e Média Empresa, 13*, 31–34.

Cattell, R. B., & Stice, G. F. (1954). Four formulae for selecting leaders on the basis of personality. *Human Relations, 7*(4), 493–507. doi:10.1177/001872675400700404

Ceitil, M. (1994). O management moderno e as novas lideranças. *IAPMEI: Pequena e Média Empresa, 13*, 31–52.

Correia, P. J. N. (1999). *Mudança organizacional no próximo milénio*. https://repositorio.ipv.pt/bitstream/10400.19/836/1/Mudan%C3%A7a%20organizacional%20no%20pr%C3%B3ximo%20mil%C3%A9nio.pdf

Crédito Agrícola, C. A. (2011). *Balanço Social Consolidado do SICAM 2010*. Crédito Agrícola.

Crédito Agrícola, C. A. (2012). *Balanço Social Consolidado do SICAM 2011*. Crédito Agrícola.

Cunha, M. P. (2002). *As duas faces da mudança organizacional – planeada e emergente*. UNL.

Cunha, M. P., & Rego, A. (2009). *Liderar* (2nd ed.). Publicações Dom Quixote.

Cunha, M. P., Rego, A., Cunha, R. C., & Cabral-Cardoso, C. (2005). Manual de Comportamento Organizacional e Gestão (4.ª ed.). Editora RH.

Cunha, M. P., Rego, A., Cunha, R. C., & Cabral-Cardoso, C. (2007). Manual de Comportamento Organizacional e Gestão (6.ª ed.). Editora RH.

Delaire, G. (1986). *Commander ou motiver?* Les Éditions d'Organisation.

Dorfman, P. W., & House, R. J. (2004). Cultural influences on organizational leadership – literature review, theoretical rationale, and GLOBE project goals. In R. J. House, P. J. Hanges, M., Javidan, P. W. Dorfman & V. Gupta (Eds.), Culture, leadership, and organizations: The GLOBE study of 62 societies (pp. 51-73). Sage Publications.

Fiedler, F. E. (1967). *A theory of leadership effectiveness*. McGraw-Hill.

Fiedler, F. E. (1978). The contingency model and the dynamics of the leadership process. In L. Berkowitz (Ed.), Advances in experimental social psychology (pp. 59–112). Academic Press. doi:10.1016/S0065-2601(08)60005-2

Ghiglione, R., & Matalon, B. (2001). *O Inquérito: Teoria e Prática*. Celta Editora.

Ghiselli, E. E. (1971). *Explorations in Managerial Talent*. Goodyear Publishing.

Gibb, C. A. (Ed.). (1969). *Leadership: Selected Reading*. Penguin Books Ltd.

Gil, A. C. (2008). Métodos e técnicas de pesquisa social (6.ª ed.). Editora Atlas.

Goffee, R., & Jones, G. (2005). Managing authenticity – The paradox of great leadership. *Harvard Business Review*, (December), 87–94. PMID:16334584

Gonçalves, C. A. M. (2012). *Cultura e clima organizacional: Contributos da motivação e das competências de gestão de líderes*. Edições Pedago.

Henderson, D., Liden, R., Glibkowski, B., & Chaudhry, A. (2009). Within-group LMX differentiation: A multilevel review and examination of its construct definition, antecedents and outcomes. *The Leadership Quarterly*, *20*(4), 517–534. doi:10.1016/j.leaqua.2009.04.003

Hersey, P., & Blanchard, K. H. (1993). *Management of organizational behavior – utilizing human resources* (International Editions). Prentice-Hall.

House, R. J. (1971). A path-goal theory of leadership effectiveness. *Administrative Science Quarterly*, *16*(3), 321–339. doi:10.2307/2391905

House, R. J. (1977). A 1976 theory of charismatic leadership. In J. G. Hunt & L. L. Larson (Eds.), *Leadership: The cutting edge* (pp. 189–207). Southern Illinois University Press.

House, R. J., & Mitchell, R. R. (1974). Path-goal theory of leadership. *Journal of Contemporary Business*, *3*, 81–98.

Jago, A. G. (1982). Leadership: Perspectives in theory and research. *Management Science*, *28*(3), 315–336. doi:10.1287/mnsc.28.3.315

Jesuíno, J. C. (2005). Processos de liderança (2.ª ed.). Livros Horizonte.

Klein, K. J., & House, R. J. (1995). On fire: Charismatic leadership and levels of analysis. *The Leadership Quarterly*, *6*(2), 193–198. doi:10.1016/1048-9843(95)90034-9

Lickert, R. (1967). *The Human Organization: Its Management and Value*. McGraw-Hill.

Luthans, F. (2002). *Organizational Behavior*. McGraw-Hill.

Machado, M. S. (1994). Gestor ou líder – o que marca a diferença? *IAPMEI: Pequena e Média Empresa*, *13*, 37–40.

Maillet, L. (1988). *Psychologie et organisations: l'individu dans son milieu de travail*. Vigot.

Mayo, A. J., & Nohria, N. (2005). Zeitgeist Leadership. *Harvard Business Review*, (October), 45–60. PMID:16250624

McGregor, D. (1960). *The Human Side of Enterprise*. McGraw-Hill.

Messeghem, K. & Pierson, F. (2002). Leadership et innovation au sein des groupes de project. *Revue Gestion 2000, 2*, 51-65.

Montserrat, X. (2006). *Como motivar – Dinâmicas para o sucesso*. Edições ASA.

Northcraft, G. B., & Neale, M. A. (1994). *Organizational Behavior – A Management Challenge*. The Dryden Press.

Padilla, A., Hogan, R., & Kaiser, R. B. (2007). The toxic triangle: Destructive leaders, susceptible followers and conductive environments. *The Leadership Quarterly, 18*(3), 176–194. doi:10.1016/j.leaqua.2007.03.001

Palma, P. J. (2012). Liderança. In A. L. Neves & R. F. Costa (Eds.), *Gestão de Recursos Humanos de A a Z*. Editora RH.

Parasuraman, A., Zeithaml, V. A., & Berry, L. L. (1988). SERVQUAL: A multiple item scale for measuring consumer perceptions of service quality. *Journal of Retailing, 64*(1), 12–40.

Pfeffer, J. (1992). *Managing with power: politics and influence in organizations*. Harvard Business School Press.

Pina, L. (2014). A importância da Banca Cooperativa. *Inforbanca, 101*, 4–5.

Preston, P. (2001). Cascading leadership motivates followers. *Healthcare Executive*, (July/August), 62–63. PMID:11434137

Quinn, R. (1984). Applying the Competing Values Approach to Leadership: Toward an Integrative Model. In J. G. Hunt, R. Stewart, C. A. Schriesheim, & D. M. Hosking (Eds.), *Managers and leaders: An International Perspective* (pp. 10–27). Pergamon. doi:10.1016/B978-0-08-030943-9.50010-3

Richardson, R. J., Peres, J. A. S., Correia, L. M., Peres, M. H. M., & Wanderley, J. C. V. (1985). *Pesquisa Social: métodos e técnicas*. Atlas.

Rojot, J. (2000). Les theories et leur portée. *Personnel, 407*, 17–20.

Silva, H. F. (1994). Líder entre iguais. *IAPMEI: Pequena e Média Empresa, 13*, 41.

Stogdill, R. M. (1948). Personal factors associated with leadership: A survey of the literature. *The Journal of Psychology, 26*(1), 35–71. doi:10.1080/00223980.1948.9917362 PMID:18901913

Stogdill, R. M. (1974). *Handbook of leadership: A survey of the literature*. Free Press.

Tannenbaum, R., & Schmidt, W. H. (1973). How to Choose a Leadership Pattern. *Harvard Business Review*, (May-June), 162–175.

Teixeira, S. (1998). *Gestão das Organizações*. McGraw-Hill.

Tichy, N. M., & Cohen, E. (1998). The teaching organization. *Training & Development*, (July), 27–33.

Tomei, P. A., & Fortunato, G. (2008). A solidão no Poder nas Organizações. *Organizações & Sociedade*, *15*(47), 13-31. https://www.redalyc.org/pdf/4006/400638303003.pdf

Urrutikoetxea, K. L. (2003). Técnicas de entrenamiento deportivas para el desarrollo de colaboradores. In M. V. Elorza (Ed.), *Coaching directivo: Desarrollando el liderazgo – fundamentos y práctica del coaching* (pp. 167–202). Ariel.

Yukl, G. (1989). Managerial Leadership: A Review of Theory and Research. *Journal of Management*, *15*(2), 251–289. doi:10.1177/014920638901500207

Yukl, G. (1999). An evaluation essay on current conceptions of effective leadership. *European Journal of Work and Organizational Psychology*, *8*(1), 33–48. doi:10.1080/135943299398429

Zenger, J. H., Musselwhite, E., Hurson, K., & Perrin, C. (1991). Leadership in a Team Environment. *Training & Development*, (October), 47–52.

ENDNOTES

1 Approved by Decree-Law nr. 24/91 of 11 January –Consolidated and republished by Decree-Law nr. 142/2009 of June 16.

2 The supervision of these institutions is directly held by Banco de Portugal (the Portuguese Central Bank).

3 Both individuals and corporate bodies can be associates of CA, it sufficing to subscribe to a minimum of 100 Share Capital Securities with a unit price of €5 (total value of €500), at a branch of the CA in the municipality of residence or where the economic activity is carried out, except for branches of the *Caixa Central*.

4 A further education qualification that is no longer offered, like a degree course but one year shorter and without the final dissertation.

5 The professional groups of the CA are: *Group I* (which comprises employees carrying out activities proper to the institution covered by the CBA: coordinating, commercial, administrative and technical functions); *Group II* (comprises qualified employees in professions that are not specifically banking, namely call operators and employees in the auxiliary services of maintenance and patrimony); *Group III* (comprises employees in professions and functions that provide general support to the activities of the institutions and the non-qualified employees for the professions and functions in Group II, as well as those who perform auxiliary tasks for those professions and functions, with the exception of those included in Group IV, i.e. porters and drivers); *Group*

IV (comprises employees who perform blanket auxiliary functions, namely cleaning).

6 Coordination/management functions: *Senior Management:* Central Director, Director, Deputy Central Director, Deputy Director, Deputy to the Board of Directors, General Manager, Manager, Area Manager, Head of Service / Area, Divisional Head, Assistant to the Board of Directors and Systems Analyst; *Middle Management:* Grade II Technical Expert and Auditing Expert II; *Intermediate Management:* Deputy Head of Service, Under-Manager, Head of Section, Administrative Head of Establishment, Head of Sector, Deputy Head of Section, Deputy Administrative Head of Establishment, Auditing Services Coordinator, Deputy Auditing Services Coordinator, Sata Coordinator, Area Coordinator, Branch Coordinator and IT Analyst.

Compilation of References

A3ES. (n.d.). *Acreditação e Auditoria. Acreditação de Ciclos de Estudos.* Retrieved from http://www.a3es.pt/pt/acreditacao-e-auditoria/resultados-dos-processos-de acreditacao/acreditacao-de-ciclos-de-estudos

Achieng, S., Majuto, C. M., Aseka, P., & Astiaya, E. (2019). Replacing Humans with Machines: Threats and Opportunities. *East African Journal of Business and Economics, 1*(2), 54–68.

Afonso, A. J. (1998). Políticas Educativas e Avaliação Educacional. Para uma Análise Sociológica da Reforma Educativa em Portugal (1985-1995). Braga, University of Minho: Instituto de Educação e Psicologia, Centro de Estudos em Educação e Psicologia.

Afonso, P. L. (2009). *Liderança - Elementos-chave do processo.* Escolar Editora.

Aghajani, M., Keramati, A., & Javadi, B. (2012). Determination of number of kanban in a cellular manufacturing system with considering rework process. *International Journal of Advanced Manufacturing Technology, 63*(9-12), 1177–1189. doi:10.100700170-012-3973-y

Agrawal, A., Gans, J., & Goldfarb, A. (2018). The economics of artificial intelligence. *The McKinsey Quarterly.*

Aguilera, R. V., Rupp, D. E., Williams, C. A., & Ganapathi, J. (2007). Putting the S back in corporate social responsibility: A multilevel theory of social change in organizations. *Academy of Management Review, 32*(3), 836–863. doi:10.5465/amr.2007.25275678

Aguinis, H., & Glavas, A. (2012). What we know and don't know about corporate social responsibility: A review and research agenda. *Journal of Management, 38*(4), 932–968. doi:10.1177/0149206311436079

Ahmad, S., Wong, K. Y., & Rajoo, S. (2019). Sustainability indicators for manufacturing sectors: A literature survey and maturity analysis from the triple-bottom line perspective. *Journal of Manufacturing Technology Management, 30*(2), 312–334. doi:10.1108/JMTM-03-2018-0091

Ahn, J., Lee, S., & Yun, S. (2018). Leaders' core self-evaluation, ethical leadership, and employees' job performance: The moderating role of employees' exchange ideology. *Journal of Business Ethics, 148*(2), 457–470. doi:10.100710551-016-3030-0

Albert Edward, T. (2019). AI in talent acquisition: A review of AI-applications used in recruitment and selection. *Strategic HR Review*, *18*(5), 215–221. doi:10.1108/SHR-04-2019-0024

Albulescu, C. T., Artene, A. E., Luminosu, C. T., & Tămăşilă, M. (2020, September). **Art**ene, A. E., Luminosu, C. T., & Tamaşila, M. (2019). CO 2 emissions, renewable energy, and environmental regulations in the EU countries. *Environmental Science and Pollution Research International*, *27*(27), 33615–33635. Advance online publication. doi:10.100711356-019-06155-1 PMID:31473925

Althor, G., Watson, J. E., & Fuller, R. A. (2016). Global mismatch between greenhouse gas emissions and the burden of climate change. *Scientific Reports*, *6*(1), 20281. doi:10.1038rep20281 PMID:26848052

Amit, R., & Schoemaker, P. J. (1993). Strategic assets and organizational rent. *Strategic Management Journal*, *14*(1), 33–46. doi:10.1002mj.4250140105

Aoyagi-Usui, M., Vinken, H., & Kuribayashi, A. (2003). Pro-environmental attitudes and behaviors: An international comparison. *Human Ecology Review*, *10*(1), 23–31.

Appelbaum, E., Bailey, T., Berg, P., Kalleberg, A. L., & Bailey, T. A. (2000). *Manufacturing advantage: Why high-performance work systems pay off.* Cornell University Press.

Aquino, K. F., & Reed, A. (2002). The self-importance of moral identity. *Journal of Personality and Social Psychology*, *83*(6), 1423–1440. doi:10.1037/0022-3514.83.6.1423 PMID:12500822

Arena, C., Michelon, G., & Trojanowski, G. (2018). Big egos can be green: A study of CEO hubris and environmental innovation. *British Journal of Management*, *29*(2), 316–336. doi:10.1111/1467-8551.12250

Arlbjørn, J. S., Freytag, P. V., & de Haas, H. (2011). Service supply chain management: A survey of lean application in the municipal sector. *International Journal of Physical Distribution & Logistics Management*, *41*(3), 277–295. doi:10.1108/09600031111123796

Armstrong, M., & Taylor, S. (2020). *Armstrong's handbook of human resource management practice*. Kogan Page Publishers.

Arnold, J. E., Graesch, A. P., Ragazzini, E., & Ochs, E. (2012). Life at home in the twenty-first century: 32 families open their doors. Los Angeles, CA: The Cotsen institute of Archaeology Press. doi:10.2307/j.ctvdjrr2c

Arrow, K., Bolin, B., Costanza, R., Dasgupta, P., Folke, C., Holling, C. S., Jansson, B., Levin, S., Mäler, K., Perrings, C., & Pimentel, D. (1995). Economic growth, carrying capacity, and the environment. *Ecological Economics*, *15*(2), 91–95. doi:10.1016/0921-8009(95)00059-3 PMID:17756719

Arulrajah, A. A., & Opatha, H. H. (2016). Analytical and theoretical perspectives on green human resource management: A simplified underpinning. *International Business Research*, *9*(12), 153–164. doi:10.5539/ibr.v9n12p153

Ashley, M. (2019). Forget Darwinian Evolution. Humanity May Soon Evolve Itself Through A.I. *Forbes*. Retrieved 28 July from https://www.forbes.com/sites/cognitiveworld/2019/01/15/forget-darwinian-evolution-humanity-may-soon-evolve-itself-through-a-i/#3790240c466d

Azevedo, J. M. (2005). Avaliação das escolas: Fundamentar modelos e operacionalizar processos. In M. I. Miguéns (Ed.), *Avaliação das escolas. Modelos e processos*. Conselho Nacional de Educação.

Babu, V. R. (2012). *Industrial engineering in apparel production*. Woodhead Publishing Limited. doi:10.1533/9780857095541

Bales, R. F. (1950). *Interaction process analysis: A method for the study of small groups*. University of Chicago Press.

Ball, S. J. (2002). Reformar escolas/reformar professores e os terrores da performatividade. *Revista Portuguesa de Educação*, *15*(2), 3–23.

Bandura, A. (1977). Self-efficacy: Toward a unifying theory of behavioral change. *Psychological Review*, *84*(2), 191–215. doi:10.1037/0033-295X.84.2.191 PMID:847061

Barbuto, J. E. Jr, & Wheeler, D. W. (2006). Scale development and construct clarification of servant leadership. *Group & Organization Management*, *31*(3), 300–326. doi:10.1177/1059601106287091

Barker, V. L. III, & Mueller, G. C. (2002). CEO characteristics and firm R&D spending. *Management Science*, *48*(6), 782–801. doi:10.1287/mnsc.48.6.782.187

Barney, J. B. (1991). Firm resources and sustained competitive advantage. *Journal of Management*, *17*(1), 99–120. doi:10.1177/014920639101700108

Barracho, C. (2012). *Liderança em Contexto Organizacional*. Escolar Editora.

Barracho, C., & Martins, C. (2010). *Liderança e género*. Edições Sílabo.

Barrick, M. R., Mount, M. K., & Li, N. (2013). The theory of purposeful work behavior: The role of personality, higher-order goals, and job characteristics. *Academy of Management Review*, *38*(1), 132–153. doi:10.5465/amr.2010.0479

Barrick, M. R., Stewart, G. L., & Piotrowski, M. (2002). Personality and job performance: Test of the mediating effects of motivation among sales representatives. *The Journal of Applied Psychology*, *87*(1), 43–51. doi:10.1037/0021-9010.87.1.43 PMID:11916215

Barroso, J. (1996). O estudo da autonomia da escola: da autonomia decretada à autonomia construída. In J. Barroso (Ed.), *O estudo da Escola*. Porto Editora.

Bass, B. M. (1985). *Leadership and performance beyond expectations*. Free Press.

Bass, B. M. (1990). From Transactional to Transformational Leadership: Learning to share the vision. *Organizational Dynamics*, *18*(3), 19–31. doi:10.1016/0090-2616(90)90061-S

Bass, B. M. (1990a). *Handbook of Leadership: A survey of theory and research*. Free Press.

Baylão, A., & Rocha, A. (2014). *A Importância do Processo de Recrutamento e Seleção de Pessoal na Organização Empresarial.* XI Simpósio de Excelência em Gestão e Tecnologia.

BBC. (2019). *Robots 'to replace up to 20 million factory jobs' by 2030.* BBC. Retrieved 04 September from https://www.bbc.co.uk/news/business-48760799

Beattie, V., & Smith, S. J. (2013). Value creation and business models: Refocusing the intellectual capital debate. *The British Accounting Review, 45*(4), 243–254. doi:10.1016/j.bar.2013.06.001

Becker, B., & Gerhart, B. (1996). The impact of human resource management on organizational performance: Progress and prospects. *Academy of Management Journal, 39*(4), 779–801.

Becker, B., & Huselid, M. A. (2006). Strategic human resources management: Where do we go from here? *Journal of Management, 32*(6), 898–925. doi:10.1177/0149206306293668

Becker, T. (2020). 2020/06/30). A decomposition heuristic for rotational workforce scheduling. *Journal of Scheduling, 23*(5), 539–554. Advance online publication. doi:10.100710951-020-00659-2

Bernazzani, R., Cantoni, F., & Piva, M. (2019). The Future Role of Machine Learning in HR Development. In F. Cantoni & G. Mangia (Eds.), *Human Resource Management and Digitalization.* Routledge.

Berrone, P., & Gomez-Mejia, L. R. (2009). The pros and cons of rewarding social responsibility at the top. *Human Resource Management, 48*(6), 959–971. doi:10.1002/hrm.20324

Bhanot, N., Rao, P. V., & Deshmukh, S. G. (2017). An integrated approach for analyzing the enablers and barriers of sustainable manufacturing. *Journal of Cleaner Production, 142*(4), 4412–4439. doi:10.1016/j.jclepro.2016.11.123

Bhardwaj, R. (2019). How AI is revolutionizing the human resource functions. *Entrepreneur Asia Pacific, 4.*

Bheda, R., Narag, A. S., & Singla, M. L. (2003). Apparel manufacturing: A strategy for productivity improvement. *Journal of Fashion Marketing and Management, 7*(1), 12–22. doi:10.1108/13612020310464331

Bilge, P., Badurdeen, F., Seliger, G., & Jawahir, I. S. (2016). A novel manufacturing architecture for sustainable value creation. *CIRP Annals, 65*(1), 455–458. doi:10.1016/j.cirp.2016.04.114

Bissola, R., & Imperatori, B. (2020). HRM 4.0: The digital transformation of the HR Department. In F. Cantoni & G. Mangia (Eds.), *Human Resource Management and Digitalization.* Routledge.

Blake, R., & Mouton, J. S. (1964). *The managerial grid.* Gulf Publishing Co.

Blanchard, K. (2000). Leadership by the book. *Executive Excellence, 17*(3), 4–5.

Blau, P. M. (1964). *Exchange and power in social life.* John Wiley and Sons.

Bocken, N. M., Short, S. W., Rana, P., & Evans, S. (2014). A literature and practice review to develop sustainable business model archetypes. *Journal of Cleaner Production, 65*, 42–56. doi:10.1016/j.jclepro.2013.11.039

Bogue, R. (2014). Sustainable manufacturing: A critical discipline for the twenty-first century. *Assembly Automation, 34*(2), 117–122. doi:10.1108/AA-01-2014-012

Boiral, O. (2009). Greening the corporation through organizational citizenship behaviors. *Journal of Business Ethics, 87*(2), 221–236. doi:10.100710551-008-9881-2

Boiral, O., Cayer, M., & Baron, C. M. (2009). The action logics of environmental leadership: A developmental perspective. *Journal of Business Ethics, 85*(4), 479–499. doi:10.100710551-008-9784-2

Bonvoisin, J., Chang, J., Neugebauer, S., Lehmann, A., Scheumann, R., & Finkbeiner, M. (2017). Glossary. In Sustainable Manufacturing. Challenges, Solutions and Implementation Perspectives (pp. 277–283). New York, NY: Springer.

Bos-Nehles, A. C., Van Riemsdijk, M. J., & Kees Looise, J. (2013). Employee perceptions of line management performance: Applying the AMO theory to explain the effectiveness of line managers' HRM implementation. *Human Resource Management, 52*(6), 861–877. doi:10.1002/hrm.21578

Bowman, C., & Ambrosini, V. (2000). Value creation versus value capture: Towards a coherent definition of value in strategy. *British Journal of Management, 11*(1), 1–15. doi:10.1111/1467-8551.00147

Boxall, P., & Purcell, J. (2003). *Strategy and human resource management.* Blackwell.

Boxall, P., & Steeneveld, M. (1999). Human resource strategy and competitive advantage: A longitudinal study of engineering consultancies. *Journal of Management Studies, 36*(4), 443–463. doi:10.1111/1467-6486.00144

Bragg, S. M. (2003). *Essentials of Payroll: Management and Accounting*. John Wiley & Sons.

Buelens, M., Kreitner, R., & Kinicki, A. (2002). *Organizational behaviour*. McGraw Hill.

Caldeira, P., & Correia, M. F. (1996). *Psicologia das Organizações*. APB-ISGB.

Calder, B. J. (1977). An attribution theory of leadership. In B. M. Staw & G. R. Salancik (Eds.), *New directions in organizational behaviour* (pp. 179–204). St. Clair Press.

Camara, P., Guerra, P., & Rodrigues, J. (2016). HUMANATOR XXI - Recursos Humanos e Sucesso Empresarial (7ª ed.). Alfragide: Publicações Dom Quixote.

Camara, P. (2017). *Dicionário de Competências* (1st ed.). Editora RH, Lda.

Canário, R. (2000). Gestão da escola: Como elaborar o plano de Formação. In *Cadernos de Organização e Gestão Curricular*. Instituto de Inovação Educacional.

Cardoso, A. (2010). Recrutamento & Seleção de Pessoal (7ª ed.). Lisboa, Porto: Lidel – edições técnicas, Lda.

Carpenter, M. A., Geletkanycz, M. A., & Sanders, W. G. (2004). Upper echelons research revisited: Antecedents, elements, and consequences of top management team composition. *Journal of Management, 30*(6), 749–778. doi:10.1016/j.jm.2004.06.001

Castro, A. (1994). A liderança no contexto das novas práticas de gestão. *IAPMEI: Pequena e Média Empresa, 13*, 31–34.

Cattell, R. B., & Stice, G. F. (1954). Four formulae for selecting leaders on the basis of personality. *Human Relations, 7*(4), 493–507. doi:10.1177/001872675400700404

Ceitil, M. (1994). O management moderno e as novas lideranças. *IAPMEI: Pequena e Média Empresa, 13*, 31–52.

Ceitil, M. (2016). *Gestão e Desenvolvimento de Competências* (2nd ed.). Edições Sílabo.

Cellan-Jones, R. (2020). *Is Elon Musk over-hyping his brain-hacking Neuralink tech?* BBC. Retrieved 02 September from https://www.bbc.co.uk/news/technology-53987919

Central Intelligence Agency. (2017). *The World Factbook*. Retrieved November 14, 2019 from https://www.cia.gov/library/publications/the-world-factbook/geos/xx.html

Chace, C. (2015). *Surviving AI: The promise and peril of artificial intelligence*. Three Cs.

Chams, N., & García-Blandón, J. (2019). On the importance of sustainable human resource management for the adoption of sustainable development goals. *Resources, Conservation and Recycling, 141*, 109–122. doi:10.1016/j.resconrec.2018.10.006

Chan, F. T. S., Li, N., Chung, S. H., & Saadat, M. (2016). Management of sustainable manufacturing systems-a review on mathematical problems. *International Journal of Production Research, 55*(4), 1210–1225. doi:10.1080/00207543.2016.1229067

Chen, G., & Kanfer, R. (2006). Toward a systems theory of motivated behavior in work teams. *Research in Organizational Behavior, 27*, 223–267. doi:10.1016/S0191-3085(06)27006-0

Chen, Y. S., & Chang, C. H. (2013). The determinants of green product development performance: Green dynamic capabilities, green transformational leadership, and green creativity. *Journal of Business Ethics, 116*(1), 107–119. doi:10.100710551-012-1452-x

Chiarini, A. (2013). *Lean Organization: from the Tools of the Toyota Production System to Lean Office*. Springer. doi:10.1007/978-88-470-2510-3

Chiavenato, I. (1988). *Recursos Humanos*. Editora Atlas.

CIPD. (2020a). *People analytics: Explore the CIPD's point of view on people analytics, including actions for Government and recommendations for employers*. Chartered Institute of Personnel and Development. Retrieved 19 August from https://www.cipd.co.uk/news-views/viewpoint/people-analytics

CIPD. (2020b). *People analytics: Understand what people analytics is, different types of HR data, and the importance of people analytics strategy and process*. Chartered Institute of Personnel and Development. Retrieved 19 August from https://www.cipd.co.uk/knowledge/strategy/analytics/factsheet

Clutterbuck, D. (2020). The challenges of coaching and mentoring in a digitally connected world. *Coaching im digitalen Wandel.*

CNE. (2005). *Avaliação das Escolas: Fundamentar Modelos e Operacionalizar Processos.* Conselho Nacional de Educação.

Colantonio, A. (2007). Social sustainability: an exploratory analysis of its definition, assessment, methods, metrics and tools. In *Measuring social sustainability: Best practice from urban renewal in the EU 2007/01* (EIBURS Working paper series). Oxford, UK: Oxford Institute for Sustainable Development, Oxford Brookes University.

Colbert, B. A. (2004). The complex resource-based view: Implications for theory and practice in strategic human resource management. *Academy of Management Review, 29*(3), 341–358. doi:10.5465/amr.2004.13670987

Combley, R. (Ed.). (2011). *Cambridge business English dictionary*. Cambridge University Press.

Correia, P. J. N. (1999). *Mudança organizacional no próximo milénio*. https://repositorio.ipv.pt/bitstream/10400.19/836/1/Mudan%C3%A7a%20organizacional%20no%20pr%C3%B3ximo%20mil%C3%A9nio.pdf

Costa, N. (2015). *A importância das competências transversais (soft skills) na formação do Engenheiro*. Masters Dissertation, Lorena Engineering School, University of São Paulo, Brazil. Available in: https://sistemas.eel.usp.br/bibliotecas/monografias/2015/MIQ15031.pdf

Couto, V. (2018). *Análise e Reformulação de um Descritivo de Funções*. Masters Project, School of Economics and Management, University of Minho, Portugal. Available in: http://hdl.handle.net/1822/56970

Coutu, D. L. (2002). How resilience works. *Harvard Business Review, 80*(5), 46–56. PMID:12024758

Crédito Agrícola, C. A. (2011). *Balanço Social Consolidado do SICAM 2010*. Crédito Agrícola.

Crédito Agrícola, C. A. (2012). *Balanço Social Consolidado do SICAM 2011*. Crédito Agrícola.

Cunha, I. (2016). *The efficacy of 1-day soft skills training on master students' performance.* Masters Dissertation, ISCTE Business School, University Institute of Lisbon, Portugal. Available in: http://hdl.handle.net/10071/13243

Cunha, M. P., Rego, A., Cunha, R. C., & Cabral-Cardoso, C. (2005). Manual de Comportamento Organizacional e Gestão (4.ª ed.). Editora RH.

Cunha, M. P., Rego, A., Cunha, R. C., & Cabral-Cardoso, C. (2007). Manual de Comportamento Organizacional e Gestão (6.ª ed.). Editora RH.

Cunha, M. P. (2002). *As duas faces da mudança organizacional – planeada e emergente*. UNL.

Cunha, M. P., & Rego, A. (2009). *Liderar* (2nd ed.). Publicações Dom Quixote.

Dansereau, F. Jr, Graen, G. B., & Haga, W. (1975). A vertical dyad linkage approach to leadership in formal organizations. *Organizational Behavior and Human Performance, 13*(1), 46–78. doi:10.1016/0030-5073(75)90005-7

Dastin, J. (2018). *Amazon scraps secret AI recruiting tool that showed bias against women*. Reuters. Retrieved 03 September from https://www.reuters.com/article/us-amazon-com-jobs-automation-insight/amazon-scraps-secret-ai-recruiting-tool-that-showed-bias-against-women-idUSKCN1MK08G

Dattner, B., Chamorro-Premuzic, T., Buchband, R., & Schettler, L. (2019). The legal and ethical implications of using AI in hiring. *Harvard Business Review*, 25.

Davis, S. (2012). *Success Stories in Industrial Energy Efficiency*. Retrieved from https://ccap.org/resource/success-stories-in-industrial-energy-efficiency/

Decreto-Lei 144/2008. Diário da República, 1.ª série, n.º 144 (28-07-2008).

Decreto-Lei n.º 1/98. Diário da República, 1.ª série -A, n.º1 (02-01-1998).

Decreto-Lei n.º 137/2012. Diário da República, 1.ª série, n.º 126 (2-07-2012).

Decreto-Lei n.º 15/2007. Diário da República, 1.ª série, n.º 14 (19-01-2007).

Decreto-Lei n.º 22/2014. Diário da República, 1.ª série, n.º 29 (11-02-2014).

Decreto-Lei n.º 270/2009. Diário da República, 1.ª série, n.º 190 (30-09-2009).

Decreto-Lei n.º 299/2007. Diário da República, 1.ª série, n.º 161 (22-08-2007).

Decreto-Lei n.º 314/97. Diário da República, 1.ª série -A, n.º265 (15-11-1997).

Decreto-Lei n.º 75/2008. Diário da República, 1.ª série, n.º 79 (22-04-2008).

Delaire, G. (1986). *Commander ou motiver?* Les Éditions d'Organisation.

Dennis, P. (2008). *Produção Lean Simplificada - Um Guia para Entender o Sistema de Produção mais Poderoso do Mundo*. Bookman.

Despacho Conjunto n.º 258/97. Diário da República, 2.ª série, n.º 192 (21-08-1997).

Despacho Conjunto n.º 268/97. Diário da República, 2.ª série, n.º 195 (25-08-1997).

Despacho n.º 13096/2008. Diário da República, 2.ª série, n.º 90 (09-05-2008).

Despacho normativo n.º 10-A/2015. Diário da República, 2.ª série, n.º 118 (19-06-2015).

Despacho normativo n.º 7-B/2015. Diário da República, 2.ª série, n.º 88 (7-05-2015).

Deveau, D. (2017). Machine learning helps large companies hire better, potentially cutting turnover. *Financial Post*. Retrieved 23 August from https://financialpost.com/entrepreneur/0508-biz-dd-hr

DGEEC, DSEE, & DEEBS. (2016). *Educação em números Portugal 2015*. Lisboa: DGEEC. Retrieved from http://www.dgeec.mec.pt/np4/96/%7B$clientServletPath%7D/?newsId=145&fileName=Educac_a_o_em_nu_meros_2015.pdf

DGEEC. (2016). *Perfil do docente 2014/2015*. Lisbon: DGEEC. Retrieved from http://www.dgeec.mec.pt/np4/98/%7B$clientServletPath%7D/?newsId=148&fileName=DGEEC_DSEE_DEEBS_DEES_2016_PerfilDocente.pdf

Dienesch, R. M., & Liden, R. C. (1986). Leader-member exchange model of leadership: A critique and further development. *Academy of Management Review*, *11*(3), 618–634. doi:10.5465/amr.1986.4306242

Dorfman, P. W., & House, R. J. (2004). Cultural influences on organizational leadership – literature review, theoretical rationale, and GLOBE project goals. In R. J. House, P. J. Hanges, M., Javidan, P. W. Dorfman & V. Gupta (Eds.), Culture, leadership, and organizations: The GLOBE study of 62 societies (pp. 51-73). Sage Publications.

Druskat, V. U., & Pescosolido, A. T. (2002). The content of effective teamwork mental models in self-managing teams: Ownership, learning, and heedful interrelating. *Human Relations*, *55*(3), 283–314. doi:10.1177/0018726702553001

Du Gay, P. (1996). *Consumption and Identity at Work*. Sage.

Duflou, J.R., Sutherland, J.W., Dornfeld, D., Herrmann, C., Jeswiet, J., Kara, S., Hauschild, M., & Kellens, K. (2012). Towards energy and resource efficient manufacturing: A processes and systems approach. *CIRP Annals - Manufacturing Technology*, *61*(2), 587–609. doi:10.1016/j.cirp.2012.05.002

Dussault, G., & Souza, L. (1999). *Gestão de Recursos Humanos em Saúde*. University of Montreal.

Dyer, L. (1983). Bringing Human resource into the strategy formulation process. *Human Resource Management*, *22*(3), 257–271. doi:10.1002/hrm.3930220306

Ehst, M., Kesrelioglu, S., & Lee, H. (2014). *Building competitive green industries: the climate and technology opportunity for developing countries*. World Bank Group. Retrieved from https://documents.worldbank.org/curated/en/837201468165876160/Building-competitive-green-industries-the-climate-and-clean-technology-opportunity-for-developing-countries

Eiadat, Y., Kelly, A., Roche, F., & Eyadat, H. (2008). Green and competitive? An empirical test of the mediating role of environmental innovation strategy. *Journal of World Business*, *43*(2), 131–145. doi:10.1016/j.jwb.2007.11.012

ElMaraghy, H. A. (2005). Flexible and reconfigurable manufacturing systems paradigms. *International Journal of Flexible Manufacturing Systems, 17*(4), 261–276. doi:10.100710696-006-9028-7

Environmental Protection Agency. (2019). *Learn About Sustainability*. Retrieved from: https://www.epa.gov/sustainability/learn-about-sustainability

Esfahbodi, A., Zhang, Y., Watson, G., & Zhang, T. (2016). Governance pressures and performance outcomes of sustainable supply chain management - an empirical analysis of UK manufacturing industry. *Journal of Cleaner Production, 155*(2), 66–78. doi:10.1016/j.jclepro.2016.07.098

Estevão, C. (1998). Gestão estratégica nas escolas. In *Cadernos de Organização e Gestão Curricular*. Instituto de Inovação Educacional.

Eubanks, B. (2018). *Artificial intelligence for HR: use AI to support and develop a successful workforce*. Kogan Page Publishers.

European Commission. (2009). *Ecodesign requirements for energy-related products directive 2009/125/EC*. Retrieved from: https://eur-lex.europa.eu/eli/dir/2009/125/oj

European Commission. (2011). *Communication from the Commission to the European Parliament, The Council, the European Economic and Social Committee and the Committee of the Regions – Energy Roadmap 2050 (COM (2011) 885/2)*. Retrieved from: https://eur-lex.europa.eu/LexUriServ/LexUriServ.do?uri=COM:2011:0885:FIN:EN:PDF

European Commission. (2020). *IMS2020: Supporting global research for IMS 2020 Vision*. Retrieved from: https://cordis.europa.eu/project/id/233469

Eurydice. (2004). *A avaliação dos estabelecimentos de ensino à lupa*. Retrieved from https://eacea.ec.europa.eu/ressources/eurydice/pdf/0_integral/060PT.pdf

Fantini, P., Palasciano, C., & Taisch, M. (2015). Back to Intuition: Proposal for a Performance Indicators Framework to Facilitate Eco-factories Management and Benchmarking. *Procedia CIRP, 26*, 1–6. doi:10.1016/j.procir.2014.07.099

Fernandes, D. S., & Machado, C. F. (2020). How to Look at Organizations and Human Resource Management in the Economy of the Future? In Circular Economy and Engineering A New Ecologically Efficient Model (pp. 89-117). Springer.

Fernandes, V. (2017). *Recrutamento e Seleção: as práticas de Recrutamento e Seleção nas Startups*. Masters Dissertation, Higher Institute of Accounting and Administration, Polytechnic Institute of Porto, Portugal. Available in: http://hdl.handle.net/10400.22/11283

Fialho, I. (2009). A qualidade de ensino e a avaliação das escolas em Portugal. Contributos para a sua história recente. *Educação Temas e problemas – Avaliação, qualidade e formação, 7*(4), 99-116.

Fiedler, F. E. (1967). *A theory of leadership effectiveness*. McGraw-Hill.

Fiedler, F. E. (1978). The contingency model and the dynamics of the leadership process. In L. Berkowitz (Ed.), Advances in experimental social psychology (pp. 59–112). Academic Press. doi:10.1016/S0065-2601(08)60005-2

Fish, S., & Hardy, M. (2015). Complex issues, complex solutions: Applying complexity theory in social work practice. *Nordic Social Work Research, 5*(sup1), 98-114.

Florea, L., Cheung, Y. H., & Herndon, N. C. (2013). For all good reasons: Role of values in organizational sustainability. *Journal of Business Ethics, 114*(3), 393–408. doi:10.100710551-012-1355-x

Formosinho, J., Ferreira, F. I., & Machado, J. (2000). *Políticas Educativas e Autonomia das Escolas*. Edições ASA.

Fox, A. A. (2018). *Discrimination concerns in job ads: Important social media changes you need to know*. Recruitics. Retrieved 03 September from https://info.recruitics.com/blog/discrimination-concerns-in-job-ads-important-social-media-changes-you-need-to-know

Freixo, M. (2011). *Metodologia Científica: Fundamentos Métodos e Práticas* (3rd ed.). Piaget Institute.

Funder, D. C. (2006). Towards a resolution of the personality triad: Persons, situations, and behaviors. *Journal of Research in Personality, 40*(1), 21–34. doi:10.1016/j.jrp.2005.08.003

Gardner, W. L., Avolio, B. J., Luthans, F., May, D. R., & Walumbwa, F. (2005). "Can you see the real me?" A self-based model of authentic leader and follower development. *The Leadership Quarterly, 16*(3), 343–372. doi:10.1016/j.leaqua.2005.03.003

Garetti, M., Rosa, P., & Terzi, S. (2012). Life cycle simulation for the design of product–service systems. *Computers in Industry, 63*(4), 361–369. doi:10.1016/j.compind.2012.02.007

Garetti, M., & Taisch, M. (2012). Sustainable manufacturing: Trends and research challenges. *Production Planning & Control: The Management of Operations, 23*(2-3), 83–104. doi:10.1080/09537287.2011.591619

Gbededo, M. A., Liyanage, K., & Garza-Reyes, J. A. (2018). Towards a Life Cycle Sustainability Analysis: A systematic review of approaches to sustainable manufacturing. *Journal of Cleaner Production, 184*, 1002–1015. doi:10.1016/j.jclepro.2018.02.310

GDPR. (2016). *What is GDPR, the EU's new data protection law?* Retrieved 03 September from https://gdpr.eu/what-is-gdpr/

Gerhart, B. (2005). Human resource and business performance: Findings, unanswered questions and an alternative approach. *Management Review, 16*(2), 174–185.

Gerhart, B. (2005). Human resources and business performance: Findings, unanswered questions, and an alternative approach. *Management Review, 16*, 174–185.

Gerstner, C. R., & Day, D. V. (1997). Meta-Analytic review of leader–member exchange theory: Correlates and construct issues. *The Journal of Applied Psychology*, *82*(6), 827–844. doi:10.1037/0021-9010.82.6.827

Ghiglione, R., & Matalon, B. (2001). *O Inquérito: Teoria e Prática*. Celta Editora.

Ghiselli, E. E. (1971). *Explorations in Managerial Talent*. Goodyear Publishing.

Gibb, C. A. (Ed.). (1969). *Leadership: Selected Reading*. Penguin Books Ltd.

Gil, A. C. (2008). Métodos e técnicas de pesquisa social (6.ª ed.). Editora Atlas.

Goffee, R., & Jones, G. (2005). Managing authenticity – The paradox of great leadership. *Harvard Business Review*, (December), 87–94. PMID:16334584

Gonçalves, C. A. M. (2012). *Cultura e clima organizacional: Contributos da motivação e das competências de gestão de líderes*. Edições Pedago.

Goodman, E. (2017). Biometrics won't solve our data-security crisis. *Harvard Business Review*. Retrieved 1 March from https://hbr.org/2017/12/biometrics-wontsolve-our-data-security-crisis

Gouldner, A. W. (1960). The norm of reciprocity: A preliminary statement. *American Sociological Review*, *25*(2), 161–178. doi:10.2307/2092623

Graban, M. (2016). *Lean Hospitals Improving Quality, Patient Safety, and Employee Engagement*. Productivity Press.

Graen, G. B., & Uhl-Bien, M. (1995). Development of leader-member exchange (LMX) theory of leadership over 25 years: Applying a multi-level multi-domain perspective. *The Leadership Quarterly*, *6*(2), 219–247. doi:10.1016/1048-9843(95)90036-5

Grasso, A. (2015). Will machines replace us or work with us? *Wired*. Retrieved 03 September from https://www.wired.com/insights/2015/01/will-machines-replace-us-or-work-with-us/

Graves, L. M., Sarkis, J., & Gold, N. (2019). Employee proenvironmental behavior in Russia: The roles of top management commitment, managerial leadership, and employee motives. *Resources, Conservation and Recycling*, *140*, 54–64. doi:10.1016/j.resconrec.2018.09.007

Grint, K. (2008). Wicked problems and clumsy solutions: The role of leadership. *Clinical Leader*, *1*(2), 11–25.

Guest, D. E. (2011). Human resource management and performance: Still searching for some answers. *Human Resource Management Journal*, *21*(1), 3–13. doi:10.1111/j.1748-8583.2010.00164.x

Guimarães, M., & Arieira, J. (2005). O Processo de Recrutamento e Seleção como uma ferramenta de Gestão, *Rev. Ciências Empresariais da UNIPAR*, *6*(2), 203–214.

Gumusluoglu, L., & Ilsev, A. (2009). Transformational leadership, creativity, and organizational innovation. *Journal of Business Research*, *62*(4), 461–473. doi:10.1016/j.jbusres.2007.07.032

Habraken, M., & Bondarouk, T. (2017). Smart industry research in the field of HRM: Resetting job design as an example of upcoming challenges. In T. Bondarouk, H. Ruël, & E. Parry (Eds.), *Electronic HRM in the smart era* (pp. 221–259). Emerald Publishing Limited. doi:10.1108/978-1-78714-315-920161009

Hackett, R. D., Wang, A. C., Chen, Z., Cheng, B. S., & Farh, J. L. (2018). Transformational Leadership and Organisational Citizenship Behaviour: A Moderated Mediation Model of Leader-Member-Exchange and Subordinates' Gender. *Applied Psychology*, *67*(4), 617–644. doi:10.1111/apps.12146

Haleblian, J., & Finkelstein, S. (1993). Top management team size, CEO dominance, and firm performance: The moderating roles of environmental turbulence and discretion. *Academy of Management Journal*, *36*(4), 844–863.

Hämäläinen, P., Takala, J., & Tan, B. K. (2017). Estimates of occupational accidents and work-related illnesses 2017. In *Proceedings of XXI World Congress on Safety and Health at Work*. Singapore: Workplace Safety and Health Institute.

Hambrick, D. C. (2007). Upper echelons theory: An update. *Academy of Management Review*, *32*(2), 334–343. doi:10.5465/amr.2007.24345254

Hao, K. (2019). A robot hand taught itself to solve a Rubik's Cube after creating its own training regime. *MIT Technology Review*. Retrieved 28 July from https://www.technologyreview.com/2019/10/15/75292/a-robot-hand-taught-itself-to-solve-a-rubiks-cube-after-creating-its-own-training-regime/

Harris, S. M. (2009). *Smooth and cool or warm and soft, investigating the properties of cloth in prehistory*. Oxbow Books.

Harteis, C. (2018). Machines, Change and Work: An Educational View on the Digitalization of Work. In C. Harteis (Ed.), *The Impact of Digitalization in the Workplace* (pp. 1–10). Springer International Publishing. doi:10.1007/978-3-319-63257-5_1

Harter, J. K., Schmidt, F. L., & Hayes, T. L. (2002). Business-unit-level relationship between employee satisfaction, employee engagement, and business outcomes: A meta-analysis. *The Journal of Applied Psychology*, *87*(2), 268–279. doi:10.1037/0021-9010.87.2.268 PMID:12002955

Hart, S. L., & Dowell, G. (2011). Invited editorial: a natural-resource-based view of the firm: fifteen years after. *Journal of Management*, *37*(5), 1464–1479. doi:10.1177/0149206310390219

Hayes, R. H., & Pisano, G. P. (1994). Beyond world-class: The new manufacturing strategy. *Harvard Business Review*, *72*(1), 77–86.

Hayward, M. L., & Hambrick, D. C. (1997). Explaining the premiums paid for large acquisitions: Evidence of CEO hubris. *Administrative Science Quarterly*, *42*(1), 103–127. doi:10.2307/2393810

Hayward, M. L., Shepherd, D. A., & Griffin, D. (2006). A hubris theory of entrepreneurship. *Management Science*, *52*(2), 160–172. doi:10.1287/mnsc.1050.0483

Henderson, D., Liden, R., Glibkowski, B., & Chaudhry, A. (2009). Within-group LMX differentiation: A multilevel review and examination of its construct definition, antecedents and outcomes. *The Leadership Quarterly*, *20*(4), 517–534. doi:10.1016/j.leaqua.2009.04.003

Herrmann, C., Schmidt, C., Kurle, D., Blume, S., & Thiede, S. (2014). Sustainability in manufacturing and factories of the future. *International Journal of Precision Engineering and Manufacturing-Green Technology*, *1*(4), 283–292. doi:10.100740684-014-0034-z

Herrmann, P., & Datta, D. K. (2005). Relationships between top management team characteristics and international diversification: An empirical investigation. *British Journal of Management*, *16*(1), 69–78. doi:10.1111/j.1467-8551.2005.00429.x

Hersey, P., & Blanchard, K. H. (1993). *Management of organizational behavior – utilizing human resources* (International Editions). Prentice-Hall.

Hines, P., Holweg, M., & Rich, N. (2004). Learning to evolve: A review of contemporary lean thinking. *International Journal of Operations & Production Management*, *24*(10), 994–1011. doi:10.1108/01443570410558049

Hirshleifer, D., Low, A., & Teoh, S. H. (2012). Are overconfident CEOs better innovators? *The Journal of Finance*, *67*(4), 1457–1498. doi:10.1111/j.1540-6261.2012.01753.x

Hoffman, B. J., Bynum, B. H., Piccolo, R. F., & Sutton, A. W. (2011). Person-organization value congruence: How transformational leaders influence work group effectiveness. *Academy of Management Journal*, *54*(4), 779–796. doi:10.5465/amj.2011.64870139

Holbeche, L. (2009). *Aligning human resources and business strategy*. Routledge. doi:10.4324/9780080941424

House, R. J. (1971). A path-goal theory of leadership effectiveness. *Administrative Science Quarterly*, *16*(3), 321–339. doi:10.2307/2391905

House, R. J. (1977). A 1976 theory of charismatic leadership. In J. G. Hunt & L. L. Larson (Eds.), *Leadership: The cutting edge* (pp. 189–207). Southern Illinois University Press.

House, R. J. (1996). Path-goal theory of leadership: Lessons, legacy, and a reformulated theory. *The Leadership Quarterly*, *7*(3), 323–352. doi:10.1016/S1048-9843(96)90024-7

House, R. J., & Mitchell, R. R. (1974). Path-goal theory of leadership. *Journal of Contemporary Business*, *3*, 81–98.

Hribar, P., & Yang, H. (2016). CEO overconfidence and management forecasting. *Contemporary Accounting Research*, *33*(1), 204–227. doi:10.1111/1911-3846.12144

Hu, H., Huang, T., Cheng, Y., & Lu, H. (2019). The evolution of sustainable business model innovation: Evidence from a sharing economy platform in China. *Sustainability*, *11*(15), 4207. doi:10.3390u11154207

Hülsheger, U. R., Anderson, N., & Salgado, J. F. (2009). Team-level predictors of innovation at work: A comprehensive meta-analysis spanning three decades of research. *The Journal of Applied Psychology, 94*(5), 1128–1145. doi:10.1037/a0015978 PMID:19702361

IBM. (2019). *Prepare for re-entry with a resilient talent management strategy*. Retrieved 02 September from https://www.ibm.com/talent-management

ICO. (2017). *Big data, artificial intelligence, machine learning and data protection*. Information Commissioner's Office. Retrieved 20 August from https://ico.org.uk/media/for-organisations/documents/2013559/big-data-ai-ml-and-data-protection.pdf

Ilies, R., Fulmer, I. S., Spitzmuller, M., & Johnson, M. D. (2009). Personality and citizenship behavior: The mediating role of job satisfaction. *The Journal of Applied Psychology, 94*(4), 945–959. doi:10.1037/a0013329 PMID:19594236

Intergovernmental Panel on Climate Change. (2014). *Climate Change 2014: Synthesis Report. Contribution of Working Groups I, II and III to the Fifth Assessment Report of the Intergovernmental Panel on Climate Change* (R. K. Pachauri & L. A. Meyer, Eds.). IPCC.

International Labour Organization. (2017). *Global Estimates of Child Labour, Results and Trends, 2012-2016*. Retrieved from: https://www.ilo.org/global/publications/books/WCMS_575499/lang--en/index.htm

Isa, K., & Tsuru, T. (2002). Cell production and workplace innovation in Japan: Toward a new model for Japanese manufacturing? *Industrial Relations, 41*(4), 548–578. doi:10.1111/1468-232X.00264

Iwamuro, H. (2004). *An Easy Book About Seru Production*. Nikkan Kogyo Shimbun. (in Japanese)

Jackson, T. (2007). The process of trend development leading to a fashion season. In *Fashion Marketing* (pp. 192–211). Routledge.

Jackson, T. (2009). *Prosperity without Growth: Economics for a Finite Planet*. Earthscan Publications. doi:10.4324/9781849774338

Jago, A. G. (1982). Leadership: Perspectives in theory and research. *Management Science, 28*(3), 315–336. doi:10.1287/mnsc.28.3.315

Jani, N. (2019). Sentiment Analysis and Employee Engagement: How Companies can Leverage AI? *Medium*. Retrieved 1 September from https://becominghuman.ai/sentiment-analysis-and-employee-engagement-how-companies-can-leverage-ai-3aa7d2e0dd4b

Jesuíno, J. C. (2005). Processos de liderança (2.ª ed.). Livros Horizonte.

Jia, Q., Guo, Y., Li, R., Li, Y., & Chen, Y. (2018). A conceptual artificial intelligence application framework in human resource management. *Proceedings of the International Conference on Electronic Business*.

Jovane, F., Yoshikawa, H., Alting, L., Boer, C. R., Westkamper, E., Williams, D., Tseng, M., Seliger, G., & Paci, A. M. (n.d.). The incoming global technological and industrial revolution towards competitive sustainable manufacturing. *CIRP Annals - Manufacturing Technology*, *57*(2), 641–659. doi:10.1016/j.cirp.2008.09.010

Jung, D. I., & Avolio, B. J. (2000). Opening the black box: An experimental investigation of the mediating effects of trust and value congruence on transformational and transactional leadership. *Journal of Organizational Behavior*, *21*(8), 949–964. doi:10.1002/1099-1379(200012)21:8<949::AID-JOB64>3.0.CO;2-F

Kark, R., Shamir, B., & Chen, G. (2003). The two faces of transformational leadership: Empowerment and dependency. *The Journal of Applied Psychology*, *88*(2), 246–255. doi:10.1037/0021-9010.88.2.246 PMID:12731708

Kaswan, M. S., Rathi, R., & Singh, M. (2019). Just in time elements extraction and prioritization for health care unit using decision making approach. *International Journal of Quality & Reliability Management*, *36*(7), 1243–1263. doi:10.1108/IJQRM-08-2018-0208

Kepplinger, D., Templ, M., & Upadhyaya, S. (2013). Analysis of energy intensity in manufacturing industry using mixed-effects models. *Energy*, *59*(15), 754–763. doi:10.1016/j.energy.2013.07.003

Khan, M. A., Jianguo, D., Ali, M., Saleem, S., & Usman, M. (2019). Interrelations Between Ethical Leadership, Green Psychological Climate, and Organizational Environmental Citizenship Behavior: A Moderated Mediation Model. *Frontiers in Psychology*, *10*, 1–12. doi:10.3389/fpsyg.2019.01977 PMID:31555173

Khan, N., & Millner, D. (2020). *Introduction to People Analytics: A Practical Guide to Data-driven HR*. Kogan Page Publishers.

Kidwell, R. E. Jr, Mossholder, K. W., & Bennett, N. (1997). Cohesiveness and organizational citizenship behavior: A multilevel analysis using work groups and individuals. *Journal of Management*, *23*(6), 775–793. doi:10.1177/014920639702300605

Kim, A., Kim, Y., Han, K., Jackson, S. E., & Ployhart, R. E. (2017). Multilevel influences on voluntary workplace green behavior: Individual differences, leader behavior, and coworker advocacy. *Journal of Management*, *43*(5), 1335–1358. doi:10.1177/0149206314547386

Kiritsis, D., Nguyen, V. K., & Stark, J. (2008). How closed-loop PLM improves Knowledge Management over the complete product lifecycle and enables the factory of the future. *International Journal of Product Lifecycle Management*, *3*(1), 54–77. doi:10.1504/IJPLM.2008.019970

Klein, K. J., & House, R. J. (1995). On fire: Charismatic leadership and levels of analysis. *The Leadership Quarterly*, *6*(2), 193–198. doi:10.1016/1048-9843(95)90034-9

Kono, H. (2006). Fundamental Principles and Viewpoints of Manufacturing Revolution Activities,«. *IEMS*, *5*, 48–56.

Kuipers, B. S., & Giurge, L. M. (2017). Does alignment matter? The performance implications of HR roles connected to organizational strategy. *International Journal of Human Resource Management, 28*(22), 3179–3201. doi:10.1080/09585192.2016.1155162

Kulik, C. T., & Ambrose, M. L. (1992). Personal and situational determinants of referent choice. *Academy of Management Review, 17*(2), 212–237. doi:10.5465/amr.1992.4279534

Kuo, T., Huang, S. H., & Zhang, H. C. (2001). Design for manufacture and design for 'X': Concepts, applications, and perspectives. *Computers & Industrial Engineering, 41*(3), 241–260. doi:10.1016/S0360-8352(01)00045-6

Lamm, E., Tosti-Kharas, J., & King, C. E. (2015). Empowering employee sustainability: Perceived organizational support toward the environment. *Journal of Business Ethics, 128*(1), 207–220. doi:10.100710551-014-2093-z

Law, K. S., Wang, H., & Hui, C. (2010). Currencies of exchange and global LMX: How they affect employee task performance and extra-role performance. *Asia Pacific Journal of Management, 27*(4), 625–646. doi:10.100710490-009-9141-8

Lawler, E. III, & Boudreau, J. (2009). What makes HR a strategic partner? *People and Strategy, 14*(1), 14–22.

Lee, L. H., Abernathy, F. H., & Ho, Y. C. (2000). Production scheduling for apparel manufacturing systems. *Production Planning and Control, 11*(3), 281–290. doi:10.1080/095372800232243

Legnani, E., Cavalieri, S., & Ierace, S. (2009). A framework for the configuration of after-sales service processes. *Production Planning & Control: The Management of Operations, 20*(2), 113–124. doi:10.1080/09537280802707522

Lengnick-Hall, C., & Lengnick-Hall, M. (1988). Strategic Human Resources Management: A Review of the Literature and a Proposed Typology. *Academy of Management Review, 13*(3), 454–470. doi:10.5465/amr.1988.4306978

Levin, B. (1998). An Epidemic of Education Policy: What can we learn for each other? *Comparative Education, 34*(2), 131–142. doi:10.1080/03050069828234

Lickert, R. (1967). *The Human Organization: Its Management and Value.* McGraw-Hill.

Liden, R. C., & Graen, G. (1980). Generalizability of the vertical dyad linkage model of leadership. *Academy of Management Journal, 23*(3), 451–465.

Liden, R. C., Sparrowe, R. T., & Wayne, S. J. (1997). Leader-member exchange theory: The past and potential for the future. In J. Ferris (Ed.), *Research in personnel and human resource management* (pp. 47–119). JAI Press.

Liker, J. (2003). *The Toyota Way: 14 Management principles from the world's greatest manufacturer.* McGraw Hill Professional.

Lima, L. (1992). *A Escola Como Organização e a Participação na Organização Escolar. Um Estudo à Escola Secundária em Portugal (1974-1988)*. I.E.P.

Lim, B. C., & Ployhart, R. E. (2004). Transformational leadership: Relations to the five-factor model and team performance in typical and maximum contexts. *The Journal of Applied Psychology*, *89*(4), 610–621. doi:10.1037/0021-9010.89.4.610 PMID:15327348

Lin, S. H., Kincade, D. H., & Warfield, C. (1994). Productivity and production in the apparel industry. *International Journal of Clothing Science and Technology*, *6*(1), 20–27. doi:10.1108/09556229410054477

Litecky, C. R., Arnett, K. P., & Prabhakar, B. (2004). The paradox of soft skills versus technical skills in IS hiring. *Journal of Computer Information Systems*, *45*(1), 69–76.

Littig, B., & Griessler, E. (2005). Social sustainability: A catchword between political pragmatism and social theory. *International Journal of Sustainable Development*, *8*(1), 65–79. doi:10.1504/IJSD.2005.007375

Liu, C., Lian, J., Yin, Y., & Li, W. (2010). Seru Seisan-an innovation of the production management Mode in Japan. *Asian Journal of Technology Innovation*, *18*(2), 89–113. doi:10.1080/19761597.2010.9668694

Liu, C., Stecke, K. E., Lian, J., & Yin, Y. (2014). An implementation framework for seru production. *International Transactions in Operational Research*, *21*(1), 1–19. doi:10.1111/itor.12014

Liu, C., Yang, N., Li, W., Lian, J., Evans, S., & Yin, Y. (2013). Training and assignment of multi-skilled workers for implementing seru production systems. *International Journal of Advanced Manufacturing Technology*, *69*(5-8), 937–959. doi:10.100700170-013-5027-5

Lockett, A., Thompson, S., & Morgenstern, U. (2009). The development of the resource-based view of the firm: A critical appraisal. *International Journal of Management Reviews*, *11*(1), 9–28. doi:10.1111/j.1468-2370.2008.00252.x

López-Cotarelo, J. (2009). *HR discretion: understanding line managers' role in Human Resource Management*. Industrial Relations Unit, Warwick Business.

Luthans, F. (2002). *Organizational Behavior*. McGraw-Hill.

Luthans, F. (2002). The need for and meaning of positive organizational behavior. Journal of Organizational Behavior: *The International Journal of Industrial. Occupational and Organizational Psychology and Behavior*, *23*(6), 695–706. doi:10.1002/job.165

Luthans, F., Luthans, K. W., & Luthans, B. C. (2004). Positive psychological capital: Beyond human and social capital. *Business Horizons*, *47*(1), 45–50. doi:10.1016/j.bushor.2003.11.007

Luthans, F., Youssef, C. M., & Avolio, B. J. (2007). Psychological capital: Investing and developing positive organizational behavior. *Positive Organizational Behavior*, *1*(2), 9–24. doi:10.4135/9781446212752.n2

Machado, M. S. (1994). Gestor ou líder – o que marca a diferença? *IAPMEI: Pequena e Média Empresa*, *13*, 37–40.

Magalhães, J. (2010). *Da Cadeira ao Banco - Escola e modernização (séculos XVIII-XX).* EducaUnidade de I&D de Ciências da Educação.

Mahmoud, A. B., Fuxman, L., Mohr, I., Reisel, W. D., & Grigoriou, N. (2020). "We aren't your reincarnation!" workplace motivation across X, Y and Z generations. *International Journal of Manpower*, *42*(1), 193–209. doi:10.1108/IJM-09-2019-0448

Mahmoud, A. B., Grigoriou, N., Fuxman, L., Reisel, W. D., Hack-Polay, D., & Mohr, I. (2020). A generational study of employees' customer orientation: A motivational viewpoint in pandemic time. *Journal of Strategic Marketing*, •••, 1–18. doi:10.1080/0965254x.2020.1844785

Mahmoud, A. B., Reisel, W. D., Fuxman, L., & Mohr, I. (2021). A motivational standpoint of job insecurity effects on organizational citizenship behaviors: A generational study. *Scandinavian Journal of Psychology*, *62*(2), 267–275. doi:10.1111jop.12689

Mahmoud, A. B., Reisel, W. D., Grigoriou, N., Fuxman, L., & Mohr, I. (2020). The reincarnation of work motivation: Millennials vs older generations. *International Sociology*, *35*(4), 393–414. doi:10.1177/0268580920912970

Mahmoud, A. B., Sitladin, M., Tehseen, S., Fuxman, L., Reisel, W. D., Purkayastha, H., & Grigoriou, N. (2020). *Job satisfaction as a mediator between performance appraisal effectiveness and organisational outcomes in the Caribbean. British Academy of Management.* Alliance Manchester Business School, University of Manchester.

Mahmoud, A. B., Tehseen, S., & Fuxman, L. (2020). The Dark Side of Artificial Intelligence in Retail Innovation. In E. Pantano (Ed.), *Retail Futures* (1st ed., pp. 165–180). Emerald Publishing Limited. doi:10.1108/978-1-83867-663-620201019

Maillet, L. (1988). *Psychologie et organisations: l'individu dans son milieu de travail.* Vigot.

Malek, J., & Desai, T. N. (2020). A systematic literature review to map literature focus of sustainable manufacturing. *Journal of Cleaner Production*, *256*, 120345. doi:10.1016/j.jclepro.2020.120345

Malmendier, U., & Tate, G. (2005). Does overconfidence affect corporate investment? CEO overconfidence measures revisited. *European Financial Management*, *11*(5), 649–659. doi:10.1111/j.1354-7798.2005.00302.x

Mani, V., Gunasekaran, A., & Delgado, C. (2018). Enhancing supply chain performance through supplier social sustainability: An emerging economy perspective. *International Journal of Production Economics*, *195*, 259–272. doi:10.1016/j.ijpe.2017.10.025

Marafa, L. M. (2002). Socio-ecological impact and risk assessments in the urban environment: A multidisciplinary concept from Hong Kong. *The Environmentalist*, *22*(4), 377–285. doi:10.1023/A:1020723132203

Marcolin, L., Miroudot, S., & Squicciarini, M. (2016). *Routine jobs, employment and technological innovation in global value chains*. OECD Publishing. Retrieved 20 August from https://www.oecd.org/sti/ind/GVC-Jobs-Routine-Content-Occupations.pdf

Maurer, R. (2019). *Employee Sentiment Analysis Shows HR All the Feels*. SHRM. Retrieved 1 September from https://www.shrm.org/resourcesandtools/hr-topics/technology/pages/employee-sentiment-analysis-shows-hr-all-the-feels.aspx

May, G., Stahl, B., & Taisch, M. (2016). Energy management in manufacturing: Towards eco-factories of the future - A focus group study. *Applied Energy*, *164*, 628–638. doi:10.1016/j.apenergy.2015.11.044

Mayo, A. J., & Nohria, N. (2005). Zeitgeist Leadership. *Harvard Business Review*, (October), 45–60. PMID:16250624

McGregor, D. (1960). *The Human Side of Enterprise*. McGraw-Hill.

Meadows, D. H., Meadows, D. L., Jorgen, R., & Behren, W. H. (1972). *The Limits to Growth; A Report for the Club of Rome's Project on the Predicament of Mankind*. Universe Books., doi:10.1349/ddlp.1

Meadows, D. H., Randers, J., & Meadows, D. L. (2004). *Limits to growth: The 30-year update*. Chelsea Green Publishing.

Meier, H., Roy, R., & Selinger, G. (2010). Industrial Product-Service Systems – IPS^2. *CIRP Annals - Manufacturing Technology*, *59*(2), 607–627. doi:10.1016/j.cirp.2010.05.004

Meister, J. (2019). Ten HR Trends In The Age Of Artificial Intelligence. *Forbes*. Retrieved 20 August from https://www.forbes.com/sites/jeannemeister/2019/01/08/ten-hr-trends-in-the-age-of-artificial-intelligence/

Mendes, I. M. (2012). *Gestão estratégica de recursos humanos na Administração Pública: mito ou realidade? Perceção dos dirigentes do I.G.F.S.S. I.P. sobre a gestão estratégica de recursos humanos e os possíveis impactos na performance*. Master Thesis in Strategic Human Resources Management, Escola Superior de Ciências Empresariais, Instituto Politécnico de Setúbal, Setúbal.

Messeghem, K. & Pierson, F. (2002). Leadership et innovation au sein des groupes de project. *Revue Gestion 2000, 2*, 51-65.

Mitchell, G., Skinner, L., & White, B. (2010). Essential Soft Skills for success in the twenty-first century workforce as perceived by business educators. *Delta Pi Epsilon Journal*, *52*, 43–53.

Mittal, S., & Dhar, R. L. (2016). Effect of green transformational leadership on green creativity: A study of tourist hotels. *Tourism Management*, *57*, 118–127. doi:10.1016/j.tourman.2016.05.007

Miyake, D. I. (2006). The shift from belt conveyor line to work-cell based assembly systems to cope with increasing demand variation in Japanese industries. *International Journal of Automotive Technology and Management*, *6*(4), 419–439. doi:10.1504/IJATM.2006.012234

Moldavska, A., & Welo, T. (2017). The concept of sustainable manufacturing and its definitions: A content-analysis based literature review. *Journal of Cleaner Production, 166*, 744–755. doi:10.1016/j.jclepro.2017.08.006

Montserrat, X. (2006). *Como motivar – Dinâmicas para o sucesso*. Edições ASA.

Moore, C., Mayer, D. M., Chiang, F. F., Crossley, C., Karlesky, M. J., & Birtch, T. A. (2019). Leaders matter morally: The role of ethical leadership in shaping employee moral cognition and misconduct. *The Journal of Applied Psychology, 104*(1), 123–145. doi:10.1037/apl0000341 PMID:30221953

Morgan McKinley. (2018). *The risks of Artificial Intelligence recruitment tools*. Retrieved 03 September from https://www.morganmckinley.com/uk/article/risks-artificial-intelligence-recruitment-tools

Nejati, M., Rabiei, S., & Jabbour, C. J. (2017). Envisioning the invisible: Understanding the synergy between green human resource management and green supply chain management in manufacturing firms in Iran in light of the moderating effect of employees resistance to change. *Journal of Cleaner Production, 168*, 163–172. doi:10.1016/j.jclepro.2017.08.213

Newman, A., Schwarz, G., Cooper, B., & Sendjaya, S. (2017). How servant leadership influences organizational citizenship behavior: The roles of LMX, empowerment, and proactive personality. *Journal of Business Ethics, 145*(1), 49–62. doi:10.100710551-015-2827-6

Niwa, M. (2002). The importance of clothing science and prospects for the future. *International Journal of Clothing Science and Technology, 14*(3/4), 238–246. doi:10.1108/09556220210437202

Northcraft, G. B., & Neale, M. A. (1994). *Organizational Behavior – A Management Challenge*. The Dryden Press.

Norton, T. A., Parker, S. L., Zacher, H., & Ashkanasy, N. M. (2015). Employee green behavior: A theoretical framework, multilevel review, and future research agenda. *Organization & Environment, 28*(1), 103–125. doi:10.1177/1086026615575773

Nóvoa, A. (1992). Para uma análise das instituições escolares. *As organizações escolares em análise, 3*, 13-43.

Ocampo, L. A., & Clark, E. E. (2015). A sustainable manufacturing strategy framework: The convergence of two fields. *Asian Academy of Management Journal, 20*(2), 29–57. doi:10.1108/SCM-01-2014-0007

OECD. (2014 a). *TALIS 2013: Results An International Perspective on Teaching and Learning. Teaching and Learning International Survey*. OECD Publishing. doi:10.1787/9789264196261-

OECD. (2014 b). *PISA 2012 Results in Focus What 15-year-olds know and what they can do with what they know*. Paris: OECD Publishing. Retrieved from https://www.oecd.org/pisa/keyfindings/pisa-2012-results-overview.pdf

OECD. (2016). *PISA 2015 PISA results in focus*. Paris: OECD Publishing. Retrieved from https://www.oecd.org/pisa/pisa-2015-results-in-focus.pdf

Ohno, T. (1998). *Toyota production system: beyond large-scale production.* CRC Press.

Organization for Economic Cooperation and Development. (2013). *Green entrepreneurship, eco-innovation and SMEs.* Retrieved from https://oecdinfo.oecd.org/info.aspx?app=OLIScoteEN&Ref=CFE/SME(2011)9/FINAL

Organization for Economic Cooperation and Development. (2018). *Environmental Policy Toolkit for SME Greening in EU Eastern Partnership Countries.* OECD Publishing.

Ortiz-Avram, D., Domnanovich, J., Kronenberg, C., & Scholz, M. (2018). Exploring the integration of corporate social responsibility into the strategies of small-and medium-sized enterprises: A systematic literature review. *Journal of Cleaner Production, 201*, 254–271. doi:10.1016/j.jclepro.2018.08.011

Osterwalder, A., Pigneur, Y., & Tucci, C. (2005). Clarifying Business Models: Origins, Present, and Future of the Concept. *Communications of the Association for Information Systems, 16.* Advance online publication. doi:10.17705/1CAIS.01601

Padilla, A., Hogan, R., & Kaiser, R. B. (2007). The toxic triangle: Destructive leaders, susceptible followers and conductive environments. *The Leadership Quarterly, 18*(3), 176–194. doi:10.1016/j.leaqua.2007.03.001

Paiva, C. (2013). *As redes sociais online – A perspetiva do potencial candidato.* Masters Dissertation, ISCTE Business School, University Institute of Lisbon, Portugal. Available in: http://hdl.handle.net/10071/6658

Palma, P. J. (2012). Liderança. In A. L. Neves & R. F. Costa (Eds.), *Gestão de Recursos Humanos de A a Z.* Editora RH.

Pantelic, D., Sakal, M., & Zehetner, A. (2016). Marketing and sustainability from the perspective of future decision makers. *South African Journal of Business Management, 47*(1), 37–47. doi:10.4102ajbm.v47i1.51

Papadakis, V. M., & Barwise, P. (2002). How much do CEOs and top managers matter in strategic decision-making? *British Journal of Management, 13*(1), 83–95. doi:10.1111/1467-8551.00224

Parasuraman, A., Zeithaml, V. A., & Berry, L. L. (1988). SERVQUAL: A multiple item scale for measuring consumer perceptions of service quality. *Journal of Retailing, 64*(1), 12–40.

Patacsil, F. F., & Tablatin, C. L. S. (2017). Exploring the importance of soft and hard skills as perceived by IT internship students and industry: A gap analysis. *Journal of Technology and Science Education, 7*(3), 347–368. doi:10.3926/jotse.271

Peñuelas, J., Poulter, B., Sardans, J., Ciais, P., Van Der Velde, M., Bopp, L., Boucher, O., Godderis, Y., Hissinger, P., Llusia, J., Nardin, E., Vicca, S., Obersteiner, M., & Hanssens, I. (2013). Human-induced nitrogen–phosphorus imbalances alter natural and managed ecosystems across the globe. *Nature Communications*, *4*(2934), 1–10. doi:10.1038/ncomms3934 PMID:24343268

Peretti, J.-M. (1998). Recursos Humanos (2nd ed.). Lisbon: Sílabo.

Peteraf, M. A. (1993). The cornerstones of competitive advantage: A resource-based view. *Strategic Management Journal*, *14*(3), 179–191. doi:10.1002mj.4250140303

Peterson, S. J., & Luthans, F. (2003). The positive impact and development of hopeful leaders. *Leadership and Organization Development Journal*, *24*(1), 26–31. doi:10.1108/01437730310457302

Pfeffer, J. (1992). *Managing with power: politics and influence in organizations*. Harvard Business School Press.

Pfeffer, J. (1994). *Competitve Advantage through people*. Harvard Business School Press.

Piaskowska, D., & Trojanowski, G. (2014). Twice as smart? The importance of managers' formative-years' international experience for their international orientation and foreign acquisition decisions. *British Journal of Management*, *25*(1), 40–57. doi:10.1111/j.1467-8551.2012.00831.x

Pina, L. (2014). A importância da Banca Cooperativa. *Inforbanca*, *101*, 4–5.

Plageras, A. P., Psannis, K. E., Stergiou, C., Wang, H., & Gupta, B. B. (2018). Efficient IoT-based sensor BIG Data collection–processing and analysis in smart buildings. *Future Generation Computer Systems*, *82*, 349–357. doi:10.1016/j.future.2017.09.082

Pordata. (2016). *Portugal. População. Nascimentos e Fecundidade. Nados-vivos de mães residentes em Portugal: total e fora do casamento – Portugal Atualização 2016-04-28.* Retrieved from https://www.pordata.pt/Portugal/Nados+vivos+de+m%C3%A3es+residentes+em+Portugal+total+e+fora+do+casamento-14

Preston, P. (2001). Cascading leadership motivates followers. *Healthcare Executive*, (July/August), 62–63. PMID:11434137

Prince, A. E. R., & Schwarcz, D. (2019). Proxy discrimination in the age of artificial intelligence and big data. *Iowa Law Review*, *105*, 1257.

Quinn, R. (1984). Applying the Competing Values Approach to Leadership: Toward an Integrative Model. In J. G. Hunt, R. Stewart, C. A. Schriesheim, & D. M. Hosking (Eds.), *Managers and leaders: An International Perspective* (pp. 10–27). Pergamon. doi:10.1016/B978-0-08-030943-9.50010-3

Quintela, D. (2016). *Recrutamento e Seleção de Pessoas: A perceção dos avaliadores*. Masters Dissertation, Higher School of Industrial and Management Studies, Polytechnic Institute of Porto, Portugal. Available in: http://hdl.handle.net/10400.22/7946

Quivy, R.; Campenhoudt, L. (1995). *Manual de investigação em ciências sociais*. Lisboa: Gradiva – Publicações, S. A.

Ramus, C. A., & Steger, U. (2000). The roles of supervisory support behaviors and environmental policy in employee "Ecoinitiatives" at leading-edge European companies. *Academy of Management Journal*, *43*(4), 605–626.

Randers, J. (2012). *2052: A Global Forecast for the Next Forty Years*. Chelsea Green Publishing.

Rasmunssen, B. (2007). *Business Models and the Theory of the Firm* (Working Paper). Melbourne, Australia: Victoria University of Technology.

Rasmussen, T., & Ulrich, D. (2015). Learning from practice: How HR analytics avoids being a management fad. *Organizational Dynamics*, *44*(3), 236–242. doi:10.1016/j.orgdyn.2015.05.008

Reilly, P. (2018). *The impact of artificial intelligence on the HR function*. Institute for Employment Studies. Retrieved 04 September from https://www.employment-studies.co.uk/system/files/resources/files/mp142_The_impact_of_Artificial_Intelligence_on_the_HR_function-Peter_Reilly.pdf

Reinhardt, I. C., Oliveira, J. C., & Ring, D. T. (2020). Current perspectives on the development of Industry 4.0 in the pharmaceutical sector. *Journal of Industrial Information Integration*, *18*, 100131. doi:10.1016/j.jii.2020.100131

Renwick, D. W., Redman, T., & Maguire, S. (2013). Green human resource management: A review and research agenda. *International Journal of Management Reviews*, *15*(1), 1–14. doi:10.1111/j.1468-2370.2011.00328.x

Resolução do Conselho de Ministros 95/2003. Diário da República, 1.ª série-B, n.º 174 (30-07-2003).

Reynolds, S. J. (2008). Moral attentiveness: Who pays attention to the moral aspects of life? *The Journal of Applied Psychology*, *93*(5), 1027–1041. doi:10.1037/0021-9010.93.5.1027 PMID:18808223

Ribeiro, E. (2017). *Soft Skills no mundo laboral atual. A criação de uma nova empresa*. Masters Project, Department of Languages and Cultures, Aveiro University, Portugal. Available in: http://hdl.handle.net/10773/18347

Richardson, R. J., Peres, J. A. S., Correia, L. M., Peres, M. H. M., & Wanderley, J. C. V. (1985). *Pesquisa Social: métodos e técnicas*. Atlas.

Robertson, J. L., & Barling, J. (2013). Greening organizations through leaders' influence on employees' pro-environmental behaviors. *Journal of Organizational Behavior*, *34*(2), 176–194. doi:10.1002/job.1820

Robles, M. (2012). Executive Perceptions of the Top 10 Soft Skills Needed in Today's Workplace. *Business Communication Quarterly*, *75*(4), 453–465. doi:10.1177/1080569912460400

Rojot, J. (2000). Les theories et leur portée. *Personnel*, *407*, 17–20.

Rosa, E. (1999). *Recrutamento e Seleção de Pessoal. Quais as vantagens deste processo para uma organização?* Recuperado em 3 setembro, 2018, available in: https://www.oswaldocruz.br/download/artigos/social22.pdf

Ross Peter, K., Ressia, S., & Sander Elizabeth, J. (2017). Prelims. In *Work in the 21st Century* (pp. i–xxvi). Emerald Publishing Limited., doi:10.1108/978-1-78714-577-120171009

Rother, M., & Shook, J. (2013). *Learning to see: value stream mapping to add value and eliminate muda*. Lean Enterprise Institute.

Sakazume, Y. (2005). Is Japanese Cell Manufacturing a New System?: A Comparative Study between Japanese Cell Manufacturing and Cellular Manufacturing (< Special English Issue> Production and Logistics). *Journal of Japan Industrial Management Association*, *55*(6), 341–349.

Sakazume, Y. (2006). Conditions for successful implementation of assembly cells. *Industrial Engineering and Management Systems*, *5*(2), 67–73.

Salkin, C., Oner, M., Ustundag, A., & Cevikcan, E. (2018). A conceptual framework for Industry 4.0. In Industry 4.0: Managing the Digital Transformation (pp. 3-23). Springer.

Samuel, A. L. (1959). Some Studies in Machine Learning Using the Game of Checkers. *IBM Journal of Research and Development*, *3*(3), 210–229. doi:10.1147/rd.33.0210

Santos, M. E. B. (1997). Qualidade das escolas. *Inovação*, *10*(2-3), 161–170.

Schwab, K., & Davis, N. (2018). *Shaping the future of the fourth industrial revolution*. Currency.

Sekhri, A., & Cheema, D. J. (2019). The new era of HRM: AI reinventing HRM functions. *International Journal of Scientific Research and Review*, *7*(3).

Seligman, M. (2002). *Authentic happiness: Using the new positive psychology to realize your potential for lasting fulfillment*. Free Press.

Semuels, A. (2020). Millions of Americans Have Lost Jobs in the Pandemic—And Robots and AI Are Replacing Them Faster Than Ever. *Time*. Retrieved 04 September from https://time.com/5876604/machines-jobs-coronavirus/

Senanayake, M. M. (2004). *Mixed Mass Production and Mass Customization: Best Practices for Apparel* (PhD Thesis). NCSU.

Settoon, R. P., Bennett, N., & Liden, R. C. (1996). Social exchange in organizations: Perceived organizational support, leader–member exchange, and employee reciprocity. *The Journal of Applied Psychology*, *81*(3), 219–227. doi:10.1037/0021-9010.81.3.219

Shamir, B., Arthur, M. B., & House, R. J. (1994). The rhetoric of charismatic leadership: A theoretical extension, a case study, and implications for research. *The Leadership Quarterly*, *5*(1), 25–42. doi:10.1016/1048-9843(94)90004-3

Shao, L., Zhang, Z., & Yin, Y. (2017). Production System Performance Improvement by Assembly Line-seru Conversion. In *Proceedings of the Tenth International Conference on Management Science and Engineering Management* (pp. 1165-1180). Springer. 10.1007/978-981-10-1837-4_95

Shih, H. A., & Chiang, Y. H. (2005). Strategy alignment between HRM, KM, and corporate development. *International Journal of Manpower*, *26*(6), 582–603. doi:10.1108/01437720510625476

Sidekick. (2020). *Organizations thrive with remote-ready managers.* Retrieved 2 September from https://sidekick-hq.com/

Silva, H. F. (1994). Líder entre iguais. *IAPMEI: Pequena e Média Empresa*, *13*, 41.

Singh, S. (2017, July). A study on seru production system. In *3rd International Conference on Emerging Technologies in Engineering, Biomedical, Management and Science [3rd ETEBMS-2017]* (*Vol. 9*). Academic Press.

Smalley, A. (2004). *Creating level pull: a lean production-system improvement guide for production-control, operations, and engineering professionals.* Lean Enterprise Institute.

Smith, C., Organ, D. W., & Near, J. P. (1983). Organizational citizenship behavior: Its nature and antecedents. *The Journal of Applied Psychology*, *68*(4), 653–663. doi:10.1037/0021-9010.68.4.653

Smith, W. K., Besharov, M. L., Wessels, A. K., & Chertok, M. (2012). A paradoxical leadership model for social entrepreneurs: Challenges, leadership skills, and pedagogical tools for managing social and commercial demands. *Academy of Management Learning & Education*, *11*(3), 463–478. doi:10.5465/amle.2011.0021

Snyder, C. R., Lopez, S. J., Shorey, H. S., Rand, K. L., & Feldman, D. B. (2003). Hope theory, measurements, and applications to school psychology. *School Psychology Quarterly*, *18*(2), 122–139. doi:10.1521cpq.18.2.122.21854

Sondhi, P. (2018). *5 Important Guidelines to Increase HR Values In Your Organization*. Entrepreneur. Retrieved 02 August from https://www.entrepreneur.com/article/312696

Song, Z., & Moon, Y. (2017). Assessing sustainability benefits of cybermanufacturing systems. *International Journal of Advanced Manufacturing Technology*, *90*(5-8), 1365–1382. doi:10.100700170-016-9428-0

Sotarauta, M. (2016). Place leadership, governance and power. *Administration*, *64*(3-4), 45–58. doi:10.1515/admin-2016-0024

Spearman, M., Woodruff, D., & Hopp, W. (1990). CONWIP: A pull alternative to kanban. *International Journal of Production Research*, *28*(5), 879–894. doi:10.1080/00207549008942761

Spieth, P., Schneckenberg, D., & Ricart, J. E. (2014). Business model innovation–state of the art and future challenges for the field. *R & D Management*, *44*(3), 237–247. doi:10.1111/radm.12071

Spreitzer, G. M. (1995). Psychological empowerment in the workplace: Dimensions, measurement, and validation. *Academy of Management Journal*, *38*(5), 1442–1465.

Spreitzer, G. M. (2008). Taking stock: A review of more than twenty years of research on empowerment at work. *Handbook on Organizational Behaviour*, *1*, 54–72. doi:10.4135/9781849200448.n4

Stajkovic, A. D., & Luthans, F. (2003). Behavioral management and task performance in organizations: Conceptual background, meta-analysis, and test of alternative models. *Personnel Psychology*, *56*(1), 155–194. doi:10.1111/j.1744-6570.2003.tb00147.x

Stecke, K. E., Yin, Y., & Kaku, I. (2014). Seru production: an extension of just-in-time approach for volatile business environments. In Analytical Approaches to Strategic Decision-Making: Interdisciplinary Considerations (pp. 45-58). IGI Global. doi:10.4018/978-1-4666-5958-2.ch003

Stecke, K. E., Yin, Y., Kaku, I., & Murase, Y. (2012). Seru: The organizational extension of JIT for a super-talent factory. *International Journal of Strategic Decision Sciences*, *3*(1), 106–119. doi:10.4018/jsds.2012010104

Steckel, J. C., Edenhofer, O., & Jakob, M. (2015). Drivers for the renaissance of coal. In M.G. Morgan (Ed.), *Proceedings of the National Academy of Sciences (PNAS)* (vol. 112, pp. E3775–E378). Washington DC: PNAS. 10.1073/pnas.1422722112

Stewart, J., & Ranson, S. (1994). Management in the public domain. In D. McKevitt & A. Lawton (Eds.), *Public Sector Management. Theory, Critique & Practice*. Sage.

Stogdill, R. M. (1948). Personal factors associated with leadership: A survey of the literature. *The Journal of Psychology*, *26*(1), 35–71. doi:10.1080/00223980.1948.9917362 PMID:18901913

Stogdill, R. M. (1974). *Handbook of leadership: A survey of the literature*. Free Press.

Stone, A. G., Russell, R. F., & Patterson, K. (2004). Transformational versus servant leadership: A difference in leader focus. *Leadership and Organization Development Journal*, *25*(4), 349–361. doi:10.1108/01437730410538671

Stubbs, W., & Cocklin, C. (2008). Conceptualizing a "Sustainability business model.". *Organization & Environment*, *21*(2), 103–127. doi:10.1177/1086026608318042

Suedfeld, P. (1992). Cognitive managers and their critics. *Political Psychology*, *13*(3), 435–453. doi:10.2307/3791607

Suzuki, Y. (2004). Structure of the Japanese production system: Elusiveness and reality. *Asian Business & Management*, *3*(2), 201–219. doi:10.1057/palgrave.abm.9200088

Swiatkiewicz, O. (2014). Competências transversais, técnicas ou morais: Um estudo exploratório sobre as competências dos trabalhadores que as organizações em Portugal mais valorizam. *Cadernos EBAPE.BR*, *12*(3), 663–687. doi:10.1590/1679-395112337

Szilárd, S., Benedek, A., & Ionel-Cioca, L. (2018). Soft Skills Development Needs and Methods in Micro-Companies of ICT Sector. *Procedia: Social and Behavioral Sciences*, *238*, 94–103. doi:10.1016/j.sbspro.2018.03.012

Tadvi, S., Rangari, S., & Rohe, A. (2020). *HR Based Interactive Chat bot (PowerBot).* In *2020 International Conference on Computer Science, Engineering and Applications (ICCSEA)*, Gunupur, India. 10.1109/ICCSEA49143.2020.9132917

Tajfel, H. (1979). Individuals and groups in social psychology. *The British Journal of Social and Clinical Psychology*, *18*(2), 183–190. doi:10.1111/j.2044-8260.1979.tb00324.x

Takeuchi, N. (2006). *Seru Seisan* [Seru Production System]. JMA Management Center.

Takeuchi, R., Lepak, D. P., Wang, H., & Takeuchi, K. (2007). An empirical examination of the mechanisms mediating between high-performance work systems and the performance of Japanese organizations. *The Journal of Applied Psychology*, *92*(4), 1069–1083. doi:10.1037/0021-9010.92.4.1069 PMID:17638466

Tambe, P., Cappelli, P., & Yakubovich, V. (2019). Artificial intelligence in human resources management: Challenges and a path forward. *California Management Review*, *61*(4), 15–42. doi:10.1177/0008125619867910

Tang, G., Cai, Z., Liu, Z., Zhu, H., Yang, X., & Li, J. (2015). The importance of ethical leadership in employees' value congruence and turnover. *Cornell Hospitality Quarterly*, *56*(4), 397–410. doi:10.1177/1938965514563159

Tannenbaum, R., & Schmidt, W. H. (1973). How to Choose a Leadership Pattern. *Harvard Business Review*, (May-June), 162–175.

Tata. (2018). *Cognitive Diversity: AI & The Future of Work*. Tata Consultancy Services. Retrieved 20 August from https://www.tatacommunications.com/wp-content/uploads/2018/09/Report_Cognitive-Diversity_AI-and-The-Future-of-Work.pdf

Tavares, S., & Caetano, A. (1999). A emergência da Gestão de Recursos Humanos Estratégica. In A. Caetano (Ed.), *Coord), Mudança Organizacional e Gestão de Recursos Humanos*. Observatório do Emprego e Formação Profissional.

Teixeira, S. (1998). *Gestão das Organizações*. McGraw-Hill.

Terkaj, W., & Tolio, T. (2019). The Italian Flagship Project: Factories of the Future. In T. Tolio, G. Copani, & W. Terkaj (Eds.), *Factories of the Future* (pp. 3–35). Springer. doi:10.1007/978-3-319-94358-9_1

Thies, C., Kieckhäfer, K., Spengler, T. S., & Sodhi, M. S. (2019). Operations research for sustainability assessment of products: A review. *European Journal of Operational Research*, *274*(1), 1–21. doi:10.1016/j.ejor.2018.04.039

Thompson, G. M., & Pullman, M. E. (2007). Scheduling workforce relief breaks in advance versus in real-time. *European Journal of Operational Research*, *181*(1), 139–155. doi:10.1016/j.ejor.2006.05.018

Tichy, N. M., & Cohen, E. (1998). The teaching organization. *Training & Development*, (July), 27–33.

Tomastik, R. N., Luh, P. B., & Liu, G. (1995, December). Optimization-based scheduling of flexible manufacturing systems for apparel production. In *Proceedings of 1995 34th IEEE Conference on Decision and Control* (Vol. 3, pp. 3152-3157). IEEE. 10.1109/CDC.1995.478632

Tomei, P. A., & Fortunato, G. (2008). A solidão no Poder nas Organizações. *Organizações & Sociedade, 15*(47), 13-31. https://www.redalyc.org/pdf/4006/400638303003.pdf

Tuan, L. T. (2019). Catalyzing Employee OCBE in Tour Companies: Charismatic Leadership, Organizational Justice, and Pro-Environmental Behaviors. *Journal of Hospitality & Tourism Research (Washington, D.C.), 43*(5), 682–711. doi:10.1177/1096348018817582

U.S. Department of Labor. (2020). *Employee Polygraph Protection Act*. U.S. Department of Labor. Retrieved 03 September from https://www.dol.gov/agencies/whd/polygraph

Ulrich, D. (1998). A New mandate for Human Resources. *Harvard Business Review*, (January-February), 124–135. PMID:10176915

United Nations. (1987). *Report of the World Commission on Environment and Development: Our Common Future*. Oxford University Press.

Upadhyay Ashwani, K., & Khandelwal, K. (2019). Artificial intelligence-based training learning from application. *Development and Learning in Organizations, 33*(2), 20–23. doi:10.1108/DLO-05-2018-0058

Urrutikoetxea, K. L. (2003). Técnicas de entrenamiento deportivas para el desarrollo de colaboradores. In M. V. Elorza (Ed.), *Coaching directivo: Desarrollando el liderazgo – fundamentos y práctica del coaching* (pp. 167–202). Ariel.

Usman, M., Hameed, A. A., & Manzoor, S. (2018). Exploring the links between Ethical Leadership and Organizational Unlearning: A Case Study of a European Multinational Company. *Business and Economic Review, 10*(2), 29–54. doi:10.22547/BER/10.2.2

Van der Werff, E., Steg, L., & Keizer, K. (2014). I am what I am, by looking past the present: The influence of biospheric values and past behavior on environmental self-identity. *Environment and Behavior, 46*(5), 626–657. doi:10.1177/0013916512475209

Van Dyne, L., & Pierce, J. L. (2004). Psychological ownership and feelings of possession: Three field studies predicting employee attitudes and organizational citizenship behavior. *Journal of Organizational Behavior, 25*(4), 439–459. doi:10.1002/job.249

Van Esch, P., Black, J. S., & Ferolie, J. (2019). Marketing AI recruitment: The next phase in job application and selection. *Computers in Human Behavior, 90*, 215–222. doi:10.1016/j.chb.2018.09.009

Vandewalle, D., Van Dyne, L., & Kostova, T. (1995). Psychological ownership: An empirical examination of its consequences. *Group & Organization Management, 20*(2), 210–226. doi:10.1177/1059601195202008

Vehkamäki, S. (2005). The concept of sustainability in modern times. In *Sustainable use of renewable natural resources: From principles to practices* (pp. 23–35). University of Helsinki – Department of Forest Sciences.

Villa, A., & Taurino, T. (2013). From JIT to Seru, for a production as lean as possible. *Procedia Engineering, 63*, 956–965. doi:10.1016/j.proeng.2013.08.172

Vinzant, J. C., & Vinzant, D. H. (1996). Strategic Management and total Quality Management: Challenges and choices. *Public Administration Quarterly, 20*(2), 201–219.

Waddell, G. (2013). *How fashion works: Couture, ready-to-wear and mass production.* John Wiley & Sons.

Waldman, D. A., & Siegel, D. (2008). Defining the socially responsible leader. *The Leadership Quarterly, 19*(1), 117–131. doi:10.1016/j.leaqua.2007.12.008

Walger, C., Roglio, K. D. D., & Abib, G. (2016). HR managers' decision-making processes: A "reflective practice" analysis. *Management Research Review, 39*(6), 655–671. doi:10.1108/MRR-11-2014-0250

Wall, F. M. (2019). 2019 Law Journal: Navigating the risks of processing employee-generated data. *Business North Carolina Magazine*. Retrieved 04 September from https://businessnc.com/2019-law-journal-navigating-the-risks-of-processing-employee-generated-data/

Wang, J., Liu, H., Qu, P., & Yin, Y. (2013). Design and operations of seru manufacturing: case study. In *The 6th PSU-UNS International Conference on Engineering and Technology (ICET-2013)* (pp. 15-17). Academic Press.

Wang, J., Schroer, B. J., & Ziemke, M. C. (1991). Understanding modular manufacturing in the apparel industry using simulation. In *1991 Winter Simulation Conference Proceedings*. (pp. 441-447). IEEE. 10.1109/WSC.1991.185645

Wang, S., Wan, J., Zhang, D., Li, D., & Zhang, C. (2016). Towards smart factory for industry 4.0: A self-organized multi-agent system with big data based feedback and coordination. *Computer Networks, 101*, 158–168. doi:10.1016/j.comnet.2015.12.017

Wayne, S. J., & Green, S. A. (1993). The effects of leader-member exchange on employee citizenship and impression management behavior. *Human Relations, 46*(12), 1431–1440. doi:10.1177/001872679304601204

Wayne, S. J., Shore, L. M., Bommer, W. H., & Tetrick, L. E. (2002). The role of fair treatment and rewards in perceptions of organizational support and leader-member exchange. *The Journal of Applied Psychology, 87*(3), 590–598. doi:10.1037/0021-9010.87.3.590 PMID:12090617

Weller, S. A. (2004). *Fashion's influence on garment mass production: Knowledge, commodities and the capture of value* (Doctoral dissertation). Victoria University.

Wernerfelt, B. (1984). A resource-based view of the firm. *Strategic Management Journal, 5*(2), 171–180. doi:10.1002mj.4250050207

Werner, J. M., & DeSimone, R. L. (2012). *Human resource development*. Cengage Learning.

West-Burnham, J. (1994). Strategy, policy and planning. In T. Bush & J. West-Burnham (Eds.), *The Principles of Educational Management*. Longman.

Williams, B. A., Brooks, C. F., & Shmargad, Y. (2018). How Algorithms Discriminate Based on Data They Lack: Challenges, Solutions, and Policy Implications. *Journal of Information Policy*, *8*, 78–115. doi:10.5325/jinfopoli.8.2018.0078

Womack, J. P., & Jones, D. T. (2007). Lean thinking: banish waste and create wealth in your corporation. New York: Simon and Schuster.

Wong, W. K., Chan, C. K., & Ip, W. H. (2001, July). Application of fuzzy concept in the apparel industry. In *Proceedings Joint 9th IFSA World Congress and 20th NAFIPS International Conference (Cat. No. 01TH8569)* (pp. 2547-2550). IEEE. 10.1109/NAFIPS.2001.943623

World Health Organization. (2018). *Preventing disease through a healthier and safer workplace.* Retrieved from https://www.who.int/quantifying_ehimpacts/publications/healthier-workplace/en/

Wright, P. M., Gardner, T. M., & Moynihan, L. M. (2003). The impact of HR practices on the performance of business units. *Human Resource Management Journal*, *13*(3), 21–36. doi:10.1111/j.1748-8583.2003.tb00096.x

Wright, P. M., & Snell, S. A. (1998). Toward a unifying framework for exploring fit and flexibility in strategic human resource management. *Academy of Management Review*, *23*(4), 756–772. doi:10.5465/amr.1998.1255637

Wright, P., & McMahan, G. (1992). Theoretical perspectives for strategic human resource Management. *Journal of Management*, *18*(2), 295–320. doi:10.1177/014920639201800205

Wright, P., & McMahan, G. (2011). Exploring Human Capital: Putting Human back into Strategic Human Resource Management. *Human Resource Management Journal*, *21*(2), 93–104. doi:10.1111/j.1748-8583.2010.00165.x

Xia, X., Govindan, K., & Zhu, Q. (2015). Analyzing internal barriers for automotive parts remanufacturers in China using grey-DEMATEL approach. *Journal of Cleaner Production*, *87*, 811–825. doi:10.1016/j.jclepro.2014.09.044

Yabanci, O. (2020, 2020/04/20). From human resource management to intelligent human resource management: a conceptual perspective. *Human-Intelligent Systems Integration*. doi:10.100742454-020-00007-x

Yagyuu, S. (2003). *Synchronized seru seisan mode: beginning from one*. Nikkan Kogyo Shimbun.

Yao, M., Zhou, A., & Jia, M. (2018). *Applied Artificial Intelligence: A Handbook for Business Leaders*. Topbots Inc.

Yin, Y., Kaku, I., & Stecke, K. E. (2008). *The evolution of seru production systems throughout Canon*. Neilson Journals Publishing. doi:10.4135/9781526462060

Yin, Y., Stecke, K. E., & Li, D. (2018). The evolution of production systems from Industry 2.0 through Industry 4.0. *International Journal of Production Research, 56*(1-2), 848–861. doi:10.1080/00207543.2017.1403664

Yin, Y., Stecke, K. E., Swink, M., & Kaku, I. (2017). Lessons from seru production on manufacturing competitively in a high cost environment. *Journal of Operations Management, 49*(1), 67–76. doi:10.1016/j.jom.2017.01.003

Youssef, C. M., & Luthans, F. (2005). Resiliency development of organizations, leaders and employees: Multi-level theory building for sustained performance. *Authentic leadership theory and practice: Origins, effects and development, 3*(1), 303-343.

Yukl, G. (1989). Managerial Leadership: A Review of Theory and Research. *Journal of Management, 15*(2), 251–289. doi:10.1177/014920638901500207

Yukl, G. (1999). An evaluation essay on current conceptions of effective leadership. *European Journal of Work and Organizational Psychology, 8*(1), 33–48. doi:10.1080/135943299398429

Zenger, J. H., Musselwhite, E., Hurson, K., & Perrin, C. (1991). Leadership in a Team Environment. *Training & Development*, (October), 47–52.

Zhang, X., Liu, C., Li, W., Evans, S., & Yin, Y. (2017). Effects of key enabling technologies for seru production on sustainable performance. *Omega, 66*, 290–307. doi:10.1016/j.omega.2016.01.013

Zhu, J., Liao, Z., Yam, K. C., & Johnson, R. E. (2018). Shared leadership: A state-of-the-art review and future research agenda. *Journal of Organizational Behavior, 39*(7), 834–852. doi:10.1002/job.2296

Zientara, P., & Zamojska, A. (2018). Green organizational climates and employee pro-environmental behaviour in the hotel industry. *Journal of Sustainable Tourism, 26*(7), 1142–1159. doi:10.1080/09669582.2016.1206554

Zott, C., Amit, R., & Massa, L. (2011). The business model: Recent developments and future research. *Journal of Management, 37*(4), 1019–1042. doi:10.1177/0149206311406265

About the Contributors

Carolina Machado received her PhD degree in Management Sciences (Organizational and Policies Management area / Human Resource Management) from the University of Minho in 1999, Master degree in Management (Strategic HRM) from Technical University of Lisbon in 1994, and Degree in Business Administration from University of Minho in 1989. Teaching in the HRM subjects since 1989 at University of Minho, she is since 2004 Associated Professor, with experience and research interest areas in the field of HRM, International HRM, HRM in SMEs, Training and Development, Emotional Intelligence, Management Change, Knowledge Management and Management/HRM in the Digital Age. She is Head of the Human Resources Management Work Group at the School of Economics and Management of University of Minho, Coordinator of Advanced Training Courses at the Interdisciplinary Centre of Social Sciences, Member of the Interdisciplinary Centre of Social Sciences (CICS.NOVA.UMinho), University of Minho, as well as Chief Editor of the International Journal of Applied Management Sciences and Engineering (IJAMSE), Guest Editor of journals, books Editor and book Series Editor, as well as reviewer in different international prestigious journals. In addition, she has also published both as editor/co-editor and as author/co-author several books, book chapters and articles in journals and conferences.

J. Paulo Davim is a Full Professor at the University of Aveiro, Portugal. He is also distinguished as honorary professor in several universities/colleges/institutes in China, India and Spain. He received his Ph.D. degree in Mechanical Engineering in 1997, M.Sc. degree in Mechanical Engineering (materials and manufacturing processes) in 1991, Mechanical Engineering degree (5 years) in 1986, from the University of Porto (FEUP), the Aggregate title (Full Habilitation) from the University of Coimbra in 2005 and the D.Sc. (Higher Doctorate) from London Metropolitan University in 2013. He is Senior Chartered Engineer by the Portuguese Institution of Engineers with an MBA and Specialist titles in Engineering and Industrial Management as well as in Metrology. He is also Eur Ing by FEANI-Brussels and Fellow (FIET) of IET-London. He has more than 30 years of teaching and research experience in Manufacturing, Materials, Mechanical and Industrial Engineering, with special emphasis in Machining & Tribology. He has also interest in Management, Engineering Education and Higher Education for Sustainability. He has guided large numbers of postdoc, Ph.D. and master's students as well as has coordinated and participated in several financed research projects. He has received several scientific awards and honors. He has worked as evaluator of projects for ERC-European Research Council and other international research agencies as well as examiner of Ph.D. thesis for many universities in different countries. He is the Editor in Chief of several international journals, Guest Editor of journals, books Editor, book Series Editor and Scientific Advisory for many international journals and conferences. Presently, he is an Editorial Board member of 30 international journals and acts as reviewer for more than 100 prestigious Web of Science journals. In addition, he has also published as editor (and co-editor) more than 200 books and as author (and co-author) more than 15 books, 100 book chapters and 500 articles in journals and conferences (more than 280 articles in journals indexed in Web of Science core collection/h-index 59+/11000+ citations, SCOPUS/h-index 63+/14000+ citations, Google Scholar/h-index 81+/23000+ citations). He has listed in World´s Top 2% Scientists by Stanford University study.

* * *

Emre Bilgin Sarı is a research assistant in the Faculty of Economics and Administrative Sciences at the Dokuz Eylul University, Turkey. She is working on operations management. Dr. Sarı has undergraduate education on Business Administration with double majority on Logistics Management. In 2010 she started to Production Management and Industrial Relations master program and wrote her thesis on reverse logistics model and focused on low cost raw materials on manufacturing inputs. Dr. Sarı completed her Ph.D on World Class Manufacturing System at 2016. Her doctoral research investigated on the operational performance of manufacturing firms and managing the improvements of production facilities. Her interest areas are on production systems, the use and management of production technologies on these systems. In addition, she implements applications with the decision models used in the establishment of these systems. She works with the performance indicators of production systems management.

Robisom Calado is an Adjunct Professor in the Production Engineering course at the Institute of Science and Technology at the Universidade Federal Fluminense (UFF), Rio das Ostras-Brazil. He is a Production Engineer and PhD in Mechanical Engineering and has over 25 years of experience in Industrial Engineering in the automotive industry (Randon), Whirlpool (Brastemp), Bosch und Siemens Hausgeräte GmbH (BSH), where he served as an engineer, leader of Lean Manufacturing program and head of Industrial Engineering. He is currently a research supervisor at the Laboratory of Design Thinking, Management and Industrial Engineering, where he leads a Lean Healthcare Implementation Project in 50 emergency departments of the Ministry of Health of Brazil, with a team of 30 researchers directly involved with the topic. He has published some books on Lean Six Sigma (amazon.com) and his research can be found in high-impact journals, such as the International Journal of Lean Six Sigma and the International Journal of Productivity and Performance Management.

Gema Calleja Sanz holds a PhD and MSc. in Engineering Management from Universitat Politècnica de Catalunya (UPC). She is currently working as a Lecturer at UPC, within the framework of the Serra Hunter program. Previously, she worked as a Management Engineer, in the construction and automotive industries. Her research interests focus on the application of optimization methods to engineering and management problems in the field of renewable energies.

Milena dos Reis is an Associate Professor in the Production Engineering Department at the Federal University of Rio de Janeiro (UFRJ / Campus Macaé), holds a PhD in Production Engineering from the Federal University of Rio de Janeiro, a Msc. in Production Engineering from the Federal University of Minas Gerais (UFMG), Bachelor's degree in Production Engineering from the Federal University of Ouro Preto (UFOP) and a Black Belt Certification - Lean Six Sigma. She teaches Production Management, Operations Research and Product Engineering and is a researcher at the DGE Laboratory (Design Thinking, Management and Industrial Engineering). She is a researcher in the area of Production Management on topics such as Production Planning and Control, Production Management, Lean healthcare, Lean Six Sigma and in the areas of Operational Research and Product Engineering, Active learning methodologies in Production Engineering education.

Sabri Erdem is full time professor at Dokuz Eylül University, Faculty of Business, Chair of Quantitative Methods Division and member of Faculty Board. He is Industrial Engineer since 1996 and holds two MSc degrees both in Computer Engineering and Business Administration. He received PhD in Computer Engineering in 2007. He who received Associated Professor title by April 2011, published many articles in national and international journals and has many proceedings at national and international symposium and congress and partner of three patents about automated medical dispensing machines. He is also member of many associations such as Chamber of Turkey Mechanical Engineers (Industrial Engineers Branch), Information Association of Turkey, Turkish Operations Research Association, Head of DEU Entrepreneurship, Business and Economics Applied Research Center, advisory board member of Dokuz Eylül University Information Technology Research Center. Main interest areas are optimization algorithms, nature inspired algorithms, intelligent systems, data mining, healthcare information technologies, healthcare materials management, data envelopment analysis and structural equation modeling. He is also owner of an online survey and data management platform named veritopla (https://veritopla.biz.tr). Specialties: Intelligent Systems, Optimization, Heuristics, Artificial Intelligence, RD Projects, Statistical Researches, Structural Equation Modeling, Data Mining, Busines Intelligence, Computer Crimes, Law of Information Tech. Orcid ID: 0000-0001-6766-3202 Researcher Id: N-6239-2014.

Diana Fernandes received her Master degree in International Business in 2017, and her Degree in International Relations in 2015, both from University of Minho. She currently develops professional activity in Adidas GBS Porto (Porto, Portugal), in the HR Services Department. In the meanwhile, she dedicates to academic research, being her main research interest areas in the field of International Business, Human Resources Management, Green Economy and Change Management.

Ali B. Mahmoud, PhD (Marketing), PhD (HRM), MRes (Management), MSc (Finance), BSc (Economics), FCIM, MABP, FHEA, researches in the area of business psychology from an interdisciplinary angle (like digital consumer behaviour and people analytics) and has published over 50 journal articles, book chapters, and conference papers. His work has appeared in outlets like the Journal of Family Studies, BMC Public Health, Journal of Strategic Marketing, International Journal of Manpower, International Sociology, Scandinavian Journal of Psychology, Journal of Research in Interactive Marketing, Nonprofit Management and Leadership, Media, War and Conflict, Higher Education Quarterly, Journal of Promotion Management and many others. Dr Mahmoud serves as an associate editor at the International Journal of Public Sociology and Sociotherapy, academic editor at PLOS ONE and a member of the editorial advisory board at Quality Assurance in Education.

Maria Heliodora Vieira Geraldes Matos received her PhD degree in Business Sciences from the University of Minho in 2013, Master's degree in Accounting and Administration from University of Minho in 2007, Post-graduation degree in Accounting and Administration from University of Minho in 2002, Degree in Banking Management from the Portuguese School of Bank Management in 1998, and Qualifying Training in Financial Analysis from the Portuguese Bank Training Institute in 1996. She held administrative, technical, and managerial roles at the Portuguese Central Bank (Banco de Portugal) from 1982 to 2018. She was tutor at the Portuguese Bank Training Institute from 1998 to 2005. She was Assistant

Professor, member of the Credentialing of Academic Competences Jury of the Management and HR Management bachelors, and member of the Scientific Council of the Faculty of Economics, Social and Enterprise Science, at Lusofona University of Porto from 2014 to 2020. Since 2019 she is Invited Assistant Professor at the University of Minho. She is Certified Trainer from the Institute of Employment and Professional Training since 1999, and member of the Portuguese Association of Financial Analysts.

Pedro Novo Melo is a professor of business sciences at Polytechnic Institute of Cávado and Ave, Portugal. He received his PhD in business sciences from University of Minho in 2015. His main research areas are human resource management, SMEs, people analytics, HR and technology and organizational studies.

Jordi Olivella Nadal graduated in Economy from the Universitat de Barcelona (UB). Insurance Actuary. PhD in Business Administration from the Universitat Politècnica de Catalunya (UPC). He specialises in Technology management and innovation management methods, Operations Management, in particular cross-training and process improvement and Project Management. His research activity includes more than 150 contributions.

Eduardo Guilherme Satolo is Associate Professor in School of Science and Engineering at São Paulo State University (UNESP), Tupã Campus, Brazil lecturing on Business Administration course and Graduate Program in Agribusiness and Development. Post-Doctorate in Production Engineering from UFF (2018), doctor in Production Engineering from the Universidade Metodista de Piracicaba (UNIMEP). Has a master's in Production Engineering from UNIMEP is a graduate in Production Engineering from UNIMEP.

Index

A

AI-Powered Human Resource Management 20
artificial intelligence 1-5, 9-10, 12-19, 70-71

B

big data 1, 3-4, 10, 15, 17, 19
Business Ethics 37-40, 43-44
business models 55, 59-61, 70-71, 73, 77-78

C

case study 45-46, 137, 193-194, 206
CONWIP 80, 100-103, 112
corporate consciousness 23, 31, 34, 36, 49
COVID-19 11

D

digitalisation 3-4, 20
digitisation 4, 20
Divisional Seru 119, 132, 138

E

Eco-factory 51
Employee Sentiment Analysis 8, 16, 20
energy efficiency 54, 64-66, 74
Environmental Management 21-22, 34, 47-48

F

Flexible Manufacturing 74, 115, 137-138
flexible manufacturing systems 74, 137

G

green transformational leadership 21-23, 26, 29, 31, 33-34, 36, 39, 43, 49
Group Technology 113, 118, 130-131, 138

H

HR analytics 6, 17, 20
human resource management 1-2, 4-5, 12-13, 15-16, 19-20, 37-40, 43-44, 47-48, 140, 166-167, 169, 185, 210

I

Industry 4.0 1, 3-4, 17, 19, 138
information and communication technologies 64, 161, 185, 189, 193-194
innovation 4, 12, 16, 20, 27, 36-37, 39-40, 42, 51, 55, 57, 60, 62, 67, 69-71, 75, 78, 118-119, 135-136, 142, 186, 189, 197, 228, 253
Institutional leadership 209
internet of things 3
IT Professionals 185

J

Just in time 89, 102, 110, 112

K

kanban 80, 90-91, 93-96, 98-103, 109-112

L

leadership 21-31, 33-49, 147, 153, 155, 162, 185, 189, 195-196, 206, 209-210, 214-231, 233-234, 237, 240, 243, 246, 248-255
lean production 80-81, 103, 111, 113, 117, 130, 134, 138
life cycle management 62-63

M

machine learning 1-2, 8, 13-15, 17, 20
manufacturing 3-4, 11, 37, 44, 51-78, 80-83, 85-86, 88-89, 91-93, 103-104, 106-107, 110-111, 113-115, 118-119, 122, 124, 126-127, 130-131, 133, 135-138
manufacturing paradigms 52, 57, 61, 67, 69, 71-72
manufacturing processes 61-62, 64
mass customization 115-118, 121, 134, 136, 138
mass personalization 113, 116-117, 131, 133, 138
motivation 16, 22-23, 26-27, 29-30, 34, 36, 38, 121, 128, 154, 185, 206, 209-210, 223-225, 231, 233, 240-241, 244, 248-250
muda 80, 86-87, 111-112

O

organizational citizenship behaviors 16, 24, 32, 34, 38, 49
Organizational Psychological Contract 21-22, 29, 33-34, 36, 49

P

people 1-3, 5-7, 9-10, 14-15, 20, 26-27, 30, 32, 36, 49, 56, 63, 68, 70, 116, 118, 121, 123-124, 141, 144-145, 149-151, 167-168, 182, 187-188, 193-196, 199, 210-211, 214-216, 218-219, 222, 224, 229-230, 232-233, 240, 242-243, 246-247, 250
personal development 32, 209
personality traits 21-23, 25-26, 30, 34-35, 48-49, 216-219, 224, 230
Positive Organizational Behavior 33, 43, 49
product design 62, 116, 127
production planning and control 107, 135
psychological capital 21-22, 25-26, 28, 31-34, 36, 43, 49
psychological empowerment 32-33, 36, 45
Psychological Ownership 32, 36, 46-47, 49
public education 139-140, 146-149, 153, 156-161, 164, 183
pull production systems 80, 84, 89

R

recruitment and selection 6, 13, 186-189, 191-192, 194-195, 198-199, 206
Rotating Seru 119, 132, 138

S

satisfaction 7, 16, 24, 32, 41-42, 89, 104, 116, 142, 159, 164, 209-210, 222, 224-225, 231, 233, 240
Seru 113, 117-122, 128-129, 131-138
Seru Seisan 118-119, 121-122, 137-138
SMEs 60-61, 66-67, 77
soft skills 185-186, 189-190, 196, 198-199, 207-208
strategy 5, 14-16, 18, 25, 38-39, 60, 63, 67, 77, 81, 102, 119, 135, 139, 143-145, 167, 169, 188, 229, 244, 247, 250
superhuman cognition 12, 20
supply chain 4, 44, 51, 58, 65-67, 70, 74, 76, 83, 89, 91-92, 95, 99-100, 111
sustainability 21-22, 30-31, 33-36, 40, 43, 48-49, 51-64, 66-70, 72, 74-78, 141, 147, 161, 165, 169, 191, 213

V

value congruence 21-22, 30-31, 36, 41-42, 46, 50

www.ingramcontent.com/pod-product-compliance
Lightning Source LLC
LaVergne TN
LVHW061541290926
849162LV00004B/871